Power Systems

Electrical power has been the technological foundation of industrial societies for many years. Although the systems designed to provide and apply electrical energy have reached a high degree of maturity, unforeseen problems are constantly encountered, necessitating the design of more efficient and reliable systems based on novel technologies. The book series Power Systems is aimed at providing detailed, accurate and sound technical information about these new developments in electrical power engineering. It includes topics on power generation, storage and transmission as well as electrical machines. The monographs and advanced textbooks in this series address researchers, lecturers, industrial engineers and senior students in electrical engineering.

Power Systems is indexed in Scopus

Tariq Masood • Jamil Abdo
Arshad Ali • Atif Iqbal

Protection of Electrical Power Transmission Systems

Smart Grid – Microgrid, AI, and Cybersecurity

Springer

Tariq Masood
College of Business, Engineering,
Computing & Analytics (BECA)
Frostburg State University
Frostburg, MD, USA

Jamil Abdo
College of Business, Engineering,
Computing & Analytics (BECA)
Frostburg State University
Frostburg, MD, USA

Arshad Ali
Software Productivity Strategists, Inc.
Rockville, MD, USA

Atif Iqbal
College of Engineering
Qatar University
Doha, Qatar

ISSN 1612-1287 ISSN 1860-4676 (electronic)
Power Systems
ISBN 978-3-031-94311-9 ISBN 978-3-031-94312-6 (eBook)
https://doi.org/10.1007/978-3-031-94312-6

This Springer imprint is published by the registered company Springer Nature Switzerland AG
The registered company address is: Gewerbestrasse 11, 6330 Cham, Switzerland

Dedication:
This book is dedicated to our cherished parents and family members, whose unwavering support and sacrifice of time have been instrumental in bringing this project to fruition.

Additionally, it honors all engineers whose innovative contributions have played a vital role in advancing Smart Grids, Microgrids, and FACTS technology.

A heartfelt dedication goes to the legendary ***Professor Dr. Abdel-Aty Edris*** *(late), a pioneer in FACTS technology. As my PhD mentor and external supervisor, he has been an inspiring role model and a former Director of* ***EPRI*** *(Electrical Power Research Institute), leaving an indelible mark on the field.*

Foreword

In an era defined by the rapid transformation of the electrical power industry, the protection and control of power transmission systems have become increasingly complex and critical. The integration of renewable energy sources, the emergence of innovative grid technologies, and heightened concerns about cybersecurity demand creative solutions and a profound understanding of advanced protection mechanisms. This book, *Protection of Electrical Power Transmission Systems: Smart Grid—Microgrid, AI, and Cybersecurity*, is a comprehensive guide that decisively addresses the pivotal challenges in the electrical power industry, providing essential knowledge for professionals and students alike.

This book is not just a theoretical guide. It offers a thorough exploration of both fundamental and cutting-edge aspects of power system protection, equipping readers with the practical knowledge they need to succeed in the field. Readers will understand how to set up the basic equivalent circuits of voltage transformers (VT) and current transformers (CT) for real-world protection relay measurements and learn to define steady-state and transient errors (including saturation effects) and understand the accuracy requirements of measuring transformers in practical applications.

This book rigorously outlines the differences between various overcurrent relays and their characteristics. More importantly, it emphasizes the critical need for precise coordination of relay characteristics. With this knowledge, readers can methodically describe the processes for determining relay settings and utilize overcurrent relays for adequate backup protection. The discussion extends to the essential role of computer applications in achieving sophisticated and intelligent coordination.

Readers will thoroughly understand distance protection's main characteristics and advantages and disadvantages. The text provides detailed insights into the measurements used for distance relay comparators, the operation of central electrotechnical distance relays, and the mechanics of block average comparators. It addresses the application of three-zone stepped distance protection and covers the derivation of the residual compensation factor and different polarization methods and their responses to close-up faults. Additionally, the book elucidates the function of standard communication-enhanced distance protection schemes and how directional earth fault protection schemes can bolster the performance of distance protection systems. It also examines the behavior of distance relays under power swings and the necessity of inhibiting their operation during such events.

Readers will master the fundamental principles of current differential protection, including the critical concept of bias. They will recognize the necessity of bias and be adept at deriving biases and operating quantities, underscoring their expertise in this area. The book evaluates the advantages and disadvantages of various current differential protection schemes and emphasizes the importance of a reliable communication system in differential protection. The book explicitly discusses the leading causes of faults in transmission systems, equipping readers with practical knowledge to assess their implications. They will learn to articulate the likelihood of faults becoming permanent and the conditions under which an arc may re-strike. The text clarifies the role of delay settings in auto-reclose relays, highlighting their significance in preventing system failures in real-world contexts.

In addition, the book provides essential insights into how a microprocessor communicates with its peripherals, identifies the primary components within the hardware of a numeric relay, and explains the conversion of analog busbar signals into numeric values. Readers will understand the causes of signal aliasing, how digital filters function, and the rationale behind using analog and digital filters in numeric relays. Furthermore, they will master converting values into phasor form using the discrete Fourier transform, culminating in a robust understanding of numeric distance relays and current differential relays.

Finally, the book delves into electric grid performance, covering conventional and innovative transmission technologies and the operational and control mechanisms of state-of-the-art devices such as FACTS and HVDC. It also outlines the prospects for smart device operations. Readers will adopt a strategic approach to calculate operating conditions for overcurrent protection systems safeguarding radial feeders, determine settings for directional overcurrent relays in ring distribution systems, and assess conditions in distance protection systems utilizing stepped characteristics. Moreover, this book will provide valuable insights into the current differential protection network used for unit protection and the critical concept of microgrids, which is highly relevant in electrical engineering.

Manuel Perez Donsion

European Association for the Development of Renewable Energies, Environment (EA4EPQ/AEDERMACP)
European Association for the Development of Electrical Engineering (EADEE)
Spanish Association for the Development of Electrical Engineering (AEDIE)
Vice-President of the "Portuguese Association for the Development of Electrical Engineering (APDEE)"
School of Industrial Engineering Department of Eléctrical Engineering
University of Vigo, Campus of Lagoas-Marcosende Vigo, Spain
http://www.donsion.org

Preface

This book does not discuss the details of distribution system design and system analysis, which have already been the subject of several good, published books.

Chapter 1: This book describes the main types of protection in transmission systems and includes sections on transducers, overcurrent relays, distance protection, differential protection, and auto-reclose relaying. The final section introduces the fast-growing field of numeric protection, where relaying decisions are made through calculations performed by microprocessor-based relay hardware. The fast-operating speed and high dependability of numeric relays make them especially suited to the transmission environment.

Chapter 2: By the end of this chapter, you will be able to apply the principles of setting up the basic equivalent circuits of a voltage transformer (VT) and a current transformer (CT) in real-world protection relay measurements. This knowledge will enable you to clearly distinguish between the two transformers' primary differences in design criteria and set up an equivalent circuit of a capacitor voltage transformer (CVT) as used in protection relays. You will also be equipped to define the steady-state and transient errors (including the effect of saturation) and the accuracy requirements of the measuring transformer in practical scenarios.

Chapter 3: Upon completing this chapter, you will be equipped to list the differences between the different types of overcurrent relays and their characteristics. More importantly, you will understand the crucial need for relay coordination of relay characteristics. This understanding will enable you to describe the procedure required to determine relay settings and the use of an overcurrent relay as backup protection. You will also appreciate the need to apply computers for complex and intelligent coordination.

Chapter 4: When you have finished this chapter, you should be able to: List the main types of distance protection characteristics and their advantages and disadvantages. List the measurements used for distance relay comparators. Describe the central electrotechnical distance relays. Explain the operation of the block average comparator. Describe the use of three-zone stepped distance protection. Give the derivation of the residual compensation factor. Describe the operation of different polarization methods and their response to close-up faults. Describe the operation of the standard communication to enhance distance protection schemes. Show how the operation of a directional earth fault protection scheme is used to strengthen a distance protection scheme's performance. Describe the performance of a distance relay to power swings and the need to inhibit its operation.

Chapter 5: When you have completed this chapter, you should be able to define the basic principle of current differential protection. This includes a clear understanding of the concept of bias, which is a crucial principle in current differential protection. Identifying the need for bias and being able to derive both biases and operate quantities is a key skill that will make you feel more skilled and proficient. You will also be able to assess the advantages and disadvantages of different types of current differential protection schemes and understand the requirement for a sound, reliable communications system in differential protection.

Chapter 6: Upon completing this chapter, you will be able to list the leading causes of faults in transmission systems and understand the practical implications of this knowledge. This understanding will enable you to describe the likelihood of faults remaining permanently and the mechanism by which an arc may be re-struck. You will also be able to explain the delay settings on auto-reclose relays and their role in preventing system failures in real-world scenarios.

Chapter 7: When you have completed this chapter, you will be able to: Describe how a microprocessor communicates with its peripherals. List the primary components found within the hardware of a numeric relay. Explain how analog busbar signals are converted into numeric values. Explain the cause of the aliasing of sampled signals. Describe how a digital filter works and why analog and digital filters are used in a numeric relay. Describe how values may be converted into phasor form using the discrete Fourier transform. In addition, you will have a comprehensive understanding of the operation of a numeric distance relay and a numeric current differential relay, which will make you feel confident and competent in dealing with these critical components in protection systems.

Chapter 8: When you have completed this chapter, you will be able to understand electric grid performance—conventional transmission technologies, smart transmission technologies, state-of-the-art (FACTS and HVDC) device operations and control, and the future outlook of smart device operations. You will be able to make smart and rigorous approach to calculate the operating conditions for an overcurrent protection system used for the protection of radial feeders and calculation for the settings of directional overcurrent relays in a ring distribution system. You will be able to use simple and smart calculate conditions in distance protection system using stepped characteristics and calculate conditions in a current differential protection network used for unit protection.

Chapter 9: At the end of this chapter, you will learn about the Microgrid, a topic directly relevant to the field of study in electrical engineering. This includes independent active and reactive power control, correction of voltage sag and system imbalances, and meeting the grid's load dynamics requirements. This chapter also covers the classification of microgrids, the merits and demerits of AC and DC microgrid functionality, and the functionality of hybrid microgrids. Finally, this chapter will discuss the performance comparison of DC and AC microgrids, providing insights directly applicable to the work and studies in electrical engineering. Cyber and information programs are also discussed in this chapter.

Chapter 10: When you have completed this chapter, you will be able to: Determine the capacitance and inductance of lines and cables in simple regular arrangement. Calculate the voltage induced between parallel arrangements of conductors. Setup a line representation using four-terminal network to calculate the conditions on lines. Use characteristic impedance to calculate changes in conditions on a transmission line. Use ***π and T*** models of transmission lines to make simple calculation of voltage, current, load flow and power factor conditions on a line. Construct a simple chart to show how the operating condition of a line may be changed by changes in terminal voltages. Use the simple chart, or mathematical analysis, to determine changes in power and var flow caused by changes in reactive loads.

Chapter 11: At the end of this chapter, you will have been equipped with the knowledge of the general principles of AI components used for industrial applications. This includes probabilistic thinking, AI machine power and disruption, designing reliable systems, AI system reliable technology, AI machine prediction rules, machine learning in microgrid, **tutorial on simulating probability AI models**, and the **challenges and future direction of AI in smart and microgrid operations and control.** This knowledge will empower you to apply these principles in your professional endeavors.

Chapter 12: By the conclusion of this chapter, you will have a solid grasp of effective control and mitigation strategies essential for cybersecurity in industrial applications, particularly in the energy sector. You will gain insights into core cybersecurity principles and the operational challenges that arise within Electrical Power Systems, along with their implications and practical solutions. More importantly, you will have the opportunity to validate your understanding through cybersecurity process mapping and data collection techniques, boosting your confidence significantly. Finally, you will learn how to construct two distinct levels of project scope: first, for IT penetration testing and assessment; and second, specifically tailored to Electrical Power Systems and distributed electrical power generation. Embrace this knowledge to enhance your cybersecurity proficiency and secure our energy infrastructure effectively!

Frostburg, MD, USA — Tariq Masood
Frostburg, MD, USA — Jamil Abdo
Rockville, MD, USA — Arshad Ali
Doha, Qatar — Atif Iqbal

Competing Interests The authors have no competing interests to declare that are relevant to the content of this manuscript.

Acknowledgments

As a principal author, I express my deep gratitude to the **Academic Senior Leadership** for their unprecedented support and encouragement in making this project a grand success, as listed **in their short profile**. Their pivotal role in arranging my workshops and lectures on **Electrical Power Systems (smart grid, microgrid, and renewable energy resources)** from time to time since 2004 has been a cornerstone of my professional journey. The sheer number of these events, totaling over 200 workshops/lectures, underscores the breadth of their support and the depth of my dedication. I have contributed as an **Adjunct Professor, Visiting Professor, and full-time Professor and Invited Speaker at** different Universities Internationally.

I have delivered these workshops and lectures and taught different courses part-time and full-time with great enthusiasm and commitment, and I am deeply thankful for their support.

I also sincerely appreciate the **Academic Senior Leadership's** encouragement and supporting role in making this project a grand success: **Mark H. Weichold,** former Dean, Executive Vice President, and CEO at Texas A&M University; **Mazen O. Hasan,** Chairman of the National Committee for qualifications and academic accreditation in the State of Qatar, former **professor**, Dean college of Engineering, and the Vice President and CEO, **Qatar University**, Doha, Qatar; **Ken L. Fincher,** Vice President and Executive Director Board secretary at Oregon Tech**, Manuel Pérez-Donsión**, President of the European Association for the Development of Renewable Energies, Environment and Power Quality (EA4EPQ) at University of Vigo Spain; **Abdul Jabbar**, Managing Director of the National Transmission Dispatch Company at Lahore, Pakistan; **Akhtar Kalam,** Professor and Head of the **Smart Energy Research Unit** at the Institute for Sustainable Industries and Liveable Cities at Victoria University, Australia; **Afzal Haque,** Vice-Chancellor Sir Syed University, former VC, NED UET, Karachi, Pakistan; Adel **El Banna,** former Head of asset integrity and optimization, Qatar Energy**; Hassan El-Kishky**, Professor and Chair, Department of Electrical Engineering at the University of Texas, Tyler; **Muhammad Kamran**, Vice-Chancellor of Muhammad Nawaz Sharif University of Engineering & Technology at Multan, Pakistan; **Saeed Minhas is** the vice chancellor of **DHA/SUFF** University in Karachi, Pakistan; **Suhail A Qureshi,** former Professor and Dean of UET at Lahore, Pakistan; **Tahir Imran Qureshi,**

Professor and the head **of** the environmental sciences and engineering Department at NFC UET Multan, Pakistan; **D.P. Kothari,** former vice chancellor of IIT, India. **Arshad Ali,** former Professor and Dean at NUST Pakistan**; Bhawani Shankar Chowdhry,** former Professor, and Dean at MUET Pakistan; **Ahad Ali,** Professor and Director of Industrial Engineering at LTU; **Amir Shahirinia**, Professor and Director of the Center of Excellence for Renewable Energy (CERE) at the University of the District of Columbia; **Amjad Ali,** Professor at Hamad bin Khalifa University, Qatar. **Esther T. Ososanya**, Professor and Chair, Department of Electrical and Computer Engineering, University of the District of Columbia; Florin **POPENTIU**, Professor of the Department of Control Engineering and Industrial Informatics at University "**Politehnica**" of Bucharest; **Michael Giesselmann**, Professor and Chair of the Department of electrical and electronics engineering at the Texas Tech University; **Navarun Gupta,** Professor and Chair, Department of electrical engineering at the University of Bridgeport; **Aziz Rahman,** Professor at Texas A&M University—Qatar (TAMUQ); **Raziq Yaqub**, a Professor at Alabama A&M University; **Robert Melendy**, Professor of electrical engineering at George Fox University Newberg, Oregon; **Farhan Malik,** Professor at ADPoly at Abu Dhabi; **Abdul Sattar Malik,** Professor and Chair of the Department of electrical and electronics engineering at Bahauddin Zakariya University Pakistan; **Hardeep Singh,** Professor at Amritsar Group of colleges, India; **Adnan Kavak,** Professor of the Department of electrical and electronics engineering at Kocaeli University, Turkey; **Murtaza Hashami is** an equipment specialist at ARAMCO, Saudi Arabia; **Muhammad Zubair Ikram** at Bose corporation, meta-reality labs research, and Texas Instruments.

Their time, support, and willingness to provide their short profile with full conviction have been invaluable as listed below. Their endorsement of this project as a reference book for their institutions' **undergraduate and graduate students and a copy in the central library** is not just a testament to its value but also a significant influence on its success. I am genuinely grateful for their important and impactful role in this project.

I also express my deep gratitude to Professor **Dr. Saifur Rahman,** the **IEEE President and CEO in 2023** and the president of the **IEEE Power** and Energy Society (PES) for 2018 and 2019, for being part of my book acknowledgment page. As an **Executive Secretary/Treasurer**, I contributed to the **IEEE Qatar Chapter from 2004 to 2014** and received numerous outstanding performance awards honored by the **IEEE President.** These serve as a testament to my value and integral role in dynamically growing the **IEEE Qatar Chapter**. I also led and organized the **IEEE Regional** meeting in Qatar**,** a source of pride and honor for all of us and a testament to our value and integral role in the IEEE community.

I am also profoundly grateful to the **Senior leadership at FSU**, including the **Jamil Abdo**, Professor and Chair, Department of Engineering **Sudhir Singh, Dean**, College of Business, Engineering, Computational & Mathematical Sciences, **Tarki L. Tyler** Senior Advisor to the President**, former Provost and Vice President** for Academic Affairs, and **Robert Al Delia** Acting president at **Frostburg State University**. Their recognition of my contribution as a faculty member at FSU and their inclusion on my acknowledgment page is not just a token of appreciation, but a reflection of their significant role in our research. Their influence and guidance have been invaluable.

Last, but not least, I am very much indebted to and appreciate the Honorable **Nasser Jeham Abdel-Aziz Al Kuwari**. I have had the privilege of working under his leadership at **Qatar Energy**[1] for over 12 years. We led high-level technical committees and accomplished many projects, particularly the GCC Production and Maintenance Committee. We also shared and implemented many Productions Best Practices initiatives across the GCC *(Qatar-QE, Oman-PDO, Bahrain-BP, Kuwait-KOC, United Arab* Emirates-ADNOC, *Saudi-ARAMCO)* Oil and Gas Industries in the Energy Sector. Over this extended period, his guidance and mentorship played a pivotal role in my career progression.

[1] Former Name **Qatar Petroleum**, upstream production facilities, at Dukhan Operations, this is the main field of QP, encompassing almost 1500 oil and gas, production wells, gas and water injection wells, 70 miles from Doha Metropolitan City.

Contents

1 Introduction 1

2 Transducers 5
- 2.1 Introduction 6
- 2.2 Voltage Transformers (VTs) 6
- 2.3 The Capacitor Voltage Transformer (CVT) 8
- 2.4 Current Transformer (CT) 12
- Bibliography 19

3 Overcurrent Relays 23
- 3.1 Introduction 24
- 3.2 Inverse Definite Minimum Time Overcurrent Relays 25
  - 3.2.1 Operating Characteristics 25
  - 3.2.2 Relay Types 27
  - 3.2.3 Overcurrent Relay Setting 31
- 3.3 Instantaneous Overcurrent Relays 41
- 3.4 Conclusion 41
- 3.5 Self-Assessment Questions 41
- Bibliography 42

4 Distance Protection 45
- 4.1 Introduction 46
- 4.2 Impedance Measurement 47
- 4.3 Three-Phase Distance Protection 63
  - 4.3.1 Relay Input Measurands 64
  - 4.3.2 Load Encroachment and Power Swings 72
- 4.4 Three-Phase Distance Protection Schemes 78
  - 4.4.1 Three-Zone Stepped Distance 78
  - 4.4.2 Full and Switched Distance Protection Scheme 83
  - 4.4.3 Carrier-Assisted Distance Protection 85
- 4.5 Directional Earth Fault Protection Schemes 90
  - 4.5.1 The Basic Directional Earth Fault Protection Scheme 91
  - 4.5.2 Directional Earth Fault Protection Schemes Integrated with Distance Relaying 94
- Self-Assessment Questions 97
- Bibliography 98

5 Current Differential Protection 101
5.1 Introduction 102
5.2 Fundamental Principles 103
5.3 Summation Arrangements 108
5.4 Phase Comparison Carrier Protection 110
5.5 Communication Media in Differential Protection 113
5.5.1 Pilot Wires 113
5.5.2 Power Line Carrier Circuits 114
5.5.3 Microwave Radio Links 115
5.5.4 Fiber Optical Links 115
5.6 Differential Protection of Teed Circuits 115
Self-Assessment Questions 118
Bibliography 119

6 Auto-Reclose for Transmission Systems 123
6.1 Introduction 124
6.2 The Main Causes of Faults on Transmission Lines 124
6.3 Transient and Permanent Faults 125
6.4 Fault Clearance 125
6.5 Circuit Breaker Operation 126
6.6 Auto-Reclosing Relays 127
6.6.1 The Auto-Reclosing Relay Sequence 127
6.6.2 Multiple End-Fed Transmission Lines 128
6.6.3 Delaying Auto-Reclosing Relay Schemes 128
6.6.4 High-Speed Auto-Reclosing Relay Schemes 128
6.6.5 Check Synchronism Relays 130
6.6.6 Live Line/Dead Bus, Live Bus/Deadline Check Relays 131
Self-Assessment Questions 131
Bibliography 132

7 Numeric Protection 135
7.1 Introduction 136
7.2 Microprocessors 137
7.2.1 General Description 137
7.2.2 Communication between a Microprocessor and its Peripherals 139
7.2.3 Number Representation 141
7.3 Relay Hardware 142
7.3.1 Introduction 142
7.3.2 Data Acquisition 142
7.3.3 Galvanic Isolation Transformers 145
7.3.4 Memory 146
7.3.5 Timer 147
7.3.6 Microprocessor 147

7.3.7 Serial Communications 148
7.3.8 Digital I/O 148
7.3.9 Keypad and Display 149
7.3.10 Electromagnetic Compatibility 149
7.4 Digital Signal Processing 149
7.4.1 Sampling 149
7.4.2 Sampling Theorem 149
7.4.3 Preventing Aliasing in a Protection Relay 151
7.4.4 Digital Filtering 152
7.4.5 Fourier Transforms 155
7.4.6 The Fast Fourier Transforms 156
7.5 Distance Protection 157
7.5.1 Introduction 157
7.5.2 Filtering 157
7.5.3 Impedance Calculation 158
7.5.4 Fault Evaluation 159
7.5.5 Relay Elements 161
7.6 Differential Protection 162
7.6.1 Introduction 162
7.6.2 Hardware 162
7.6.3 Current Vector Calculation 163
7.6.4 Source of Error in a Digital Differential Relay 163
7.6.5 Propagation Delay Compensation 164
7.6.6 Differential Characteristics 166
Self-Assessment Questions 167
Bibliography 170

8 Smart Grid Operation and Control 173
8.1 Introduction 174
8.2 The Smart Grid Goal 175
8.2.1 Smart Grid (Evolutionary Development) 176
8.3 Smart Grid Driving Factors and Their Influences 177
8.3.1 Fundamental Characteristics of Smart-Grid Operations 177
8.3.2 STATCOM Operations 178
8.3.3 STATCOM: DC Capacitor Control 179
8.3.4 STATCOM: DC Capacitor Control 180
8.4 Understanding Electric Grid Performance 180
8.4.1 Development of Transmission Grid Infrastructure 180
8.4.2 Power System Complexity 181
8.4.3 Power Flow Limits on AC Transmission System 182
8.4.4 Consideration of Reactive Power to Maintain Seamless AC Power Transmission Operations 183
8.4.5 Reactive Power Balance at Different Load Levels with Respect to "SIL" 184
8.4.6 Complexity of Managing Reactive Power 186

8.5 Technology Option for a Smart Transmission Grid . . . 189
8.5.1 Transmission Technologies . . . 189
8.5.2 FACTS Controller for Power Transmission Technology . . . 191
8.5.3 HVDC Power Transmission Technology . . . 198
8.6 State-of-the-Art Smart Technology . . . 201
8.6.1 Case Study . . . 201
8.6.2 Cost Analysis of HVDC and FACTS Technology . . . 202
8.7 Types of Transmission Protection . . . 203
8.7.1 Smart Protection Settings for a Ring Distribution System . . . 208
8.7.2 Smart Distance Protection . . . 209
8.7.3 Unit Protection . . . 213
8.8 Self-Assessment Questions . . . 216
Bibliography . . . 218

9 Microgrid Operation and Control . . . 221
9.1 Introduction . . . 222
9.1.1 Classification of Microgrid . . . 223
9.2 AC Microgrid Operations and Control . . . 225
9.3 DC Microgrid Operations and Control . . . 226
9.4 Hybrid Microgrid Operations and Control . . . 229
9.5 Microgrid Control . . . 230
9.5.1 Off-Grid and Islanded Microgrid . . . 231
9.5.2 Different Types of Islanding . . . 233
9.5.3 Grid Following Inverters . . . 234
9.5.4 Grid-Forming Inverters . . . 236
9.6 Short-Circuit Study . . . 237
9.6.1 Advantages of Short Circuit Ratio Study . . . 237
9.6.2 Traditional Short Circuit Ratio (SCR) Method . . . 237
9.7 Strong and Weak Microgrid . . . 237
Bibliography . . . 238

10 Power Transmission Line Characteristics . . . 241
10.1 Introduction . . . 242
10.2 The Series Inductance of a Two-Conductor Line . . . 243
10.3 Mutual Inductance and Electromagnetic Interference . . . 245
10.4 The Self-Inductance of a Coaxial Cable . . . 247
10.5 The Smart Shunt Capacitance of a Two-Conductor Line . . . 248
10.6 The Concentric Cable . . . 250
10.7 The Simple Line Method . . . 252
10.8 The Line Performance Chart . . . 257
Self-Assessment Questions . . . 262
Bibliography . . . 264

11 Artificial Intelligence Industrial Application . . . 267
11.1 Thinking Probabilistic . . . 268
11.2 AI Machine Power and Disruption . . . 270

11.3 Designing Reliable Systems 270
11.4 Embracing Uncertainty 273
11.5 AI System Prediction Technology 273
11.6 AI Prediction Machine Rules 274
11.7 AI Model Architecture 275
11.8 AI Models Tutorials 275
11.8.1 AI Industrial Assessment Models 275
11.8.2 Introduction to Machine Learning in Microgrid Operations 280
11.8.3 Data Analytics in Microgrid Control 282
11.8.4 Machine Learning Applications in Microgrid Control 283
11.8.5 Case Studies and Practical Implementations 284
11.9 Challenges and Future Directions 286
11.10 Conclusion 288
Self-Assessment Questions 291
Bibliography 292

12 Cybersecurity and Information Program's Power System 293
12.1 Introduction 294
12.2 Cybersecurity Operations 295
12.3 Cybersecurity Impact and Solution 296
12.3.1 Validate Process Mapping 298
12.3.2 Data Collection Framework 298
12.3.3 Data Collection: Introducing System Mapping 299
12.4 How Can You Help? 300
12.5 Cybersecurity Tailored Models 303
12.5.1 Project Plan Work for (IT Penetration Testing, Vulnerability & Risk Assessment Framework) 303
Self-Assessment Questions 309
Bibliography 309

Self-Assessment Question's Answers: Chapter Wise 311

Assignment A 357

Assignment B 367

About the Authors

Tariq Masood is a distinguished professional with a wealth of academic qualifications in electrical power system engineering. His credentials include an **M.Eng, M.Phil,** and a **Ph.D. (Smartgrid/Microgrid Renewable Energy Distributed Power Generation)** from the esteemed University of Bath, Bath, UK. He also obtained **Scientific Teaching Fellowship-2022**, Accredited/Sponsored by the **NSF**—National Science Foundation and **HHMI**—Howard Hughes Medical Institute in the **USA**. These qualifications, coupled with his status as a chartered engineer registered with the Royal Engineering Council in the UK, a senior member of IEEE-USA, a senior member of AIChE-USA, and a member of IET-UK, reassure his extensive expertise. His professional affiliations also include being a certified CET International USA member and a registered ASCET (American Society of Certified Engineers and Technicians) USA member.

Dr. Masood is an Assistant Professor (Tenured-Track) at Frostburg State University in Frostburg, Maryland. His leadership roles at the university, including serving as a Senator ***of the Faculty Senate,*** *a Member of the Academic Affairs Committee, and a Member of the Academic Standard Committee, demonstrate his commitment to fostering academic excellence.*

Dr. Masood is a versatile, collaborative, accomplished, and knowledgeable professional. He has a successful track record as a **Professional Engineer/ Academician. His collaborative approach makes everyone feel included and part of a team. He has also c**ontributed as a Visiting and Adjunct Professor and **Invited speaker at OIT, UTT, Texas A&M Qatar,**

UTA, TTU, UND, U.B., and UDC in the USA. He is also certified by the Lean Six Sigma Green Belt. In the past, he also served as a **Vice President of Technical Operations** at Energy and Utilities Service in New Jersey, USA.

He joined Qatar Petroleum in 1997 as a Technical Coordinator Operations/Advisor. He is a member of the BPTC (Best Practices Technical Committee) and other production integrity committees under the patronage of **H.E. Dr. Mohd Saleh Al-Sada**, Minister of Energy and Industry Qatar. His contributions to the field are evident in his more than 68 technical research papers published in IEEE Conference, journal, papers, and books **ISBN-978-3-659-40,995-0**, **ISBN: 978–3–659-40,995-0**, and **ISBN:978–3–659-40,995-0.** He has received several awards for his outstanding performance and dedication to improving Qatar Petroleum's production operations and control, including two Mubarak awards from the H.E. Minister of Energy and Industry, and three Al-Hasba Awards from the Vice President of Operations. He was the secretary for the GCC oil-producing companies (**QP-Qatar, PDO-Oman ARAMCO-Saudi Arabia, KOC-Kuwait, TATWEER-Bahrain, and ADNOC-United Arab Emirates**) Production and Maintenance Technical Committee in 2008 and 2011. He is also serving as an **Associate Editor of *IEEE Access* (Impact Factor 3.44)**.

You may reach by email: t.masood.dr@ieee.org.

Jamil Abdo, PhD, is Professor of Mechanical Engineering and Chair of the Department of Engineering & Sustainable Technologies at Frostburg State University. He is the Director of the Center for Product Design and Advanced Manufacturing (CPDAM), which advances prototyping, additive manufacturing, and industry partnerships that provide hands-on experience for students and services for the community and regional industry. Dr. Abdo earned his PhD in Mechanical Design from Southern Illinois University–Carbondale. His expertise includes machine element design, failure analysis, product development, and renewable energy systems. He has authored extensive peer-reviewed publications and technical reports and secured competitive grants to enhance research capacity and student innovation. As department chair, he leads ABET-driven continuous improvement, strengthens industry collaboration, and promotes workforce-ready, experiential engineering education.

Arshad Ali is an accomplished academician with extensive experience in academic management, strategy planning, and industry engagement spreading over 40 years. He possesses a deep understanding of skill development and stays abreast of emerging technology trends. Dr. Ali's contributions to the field have been recognized with prestigious awards. He has been awarded Gold Medal by the Pakistan Academy of Sciences and COMSTECH for his outstanding IT research. Additionally, he has received the President's Gold Medal and the "Distinguished Scientist of the Year 2006" award from the Pakistan Academy of Sciences. The Government of Pakistan has honored him with the President's Pride of Performance and the Sitara-i-Imtiaz (Star of Distinction) awards. Throughout his career, Dr. Ali has actively collaborated on numerous research and technology development projects with renowned institutions such as CERN, Stanford, and Caltech Universities. Dr. Ali emphasis that automation should be applied to efficient business processes to amplify productivity. More importantly, applying automation to inefficient processes can lead to amplification of inefficiencies and serious business losses. With his wealth of experience and expertise, Dr. Arshad Ali has made significant contributions to his field and continues to inspire others with his insights and innovative thinking.

Atif Iqbal (Fellow IEEE (USA), Fellow IET (UK), Fellow IE (India), Fellow AIAA (Asia Pacific)) received the B.Sc. and M.Sc. degrees in Electrical Engineering from the Aligarh Muslim University (AMU), Aligarh, India, in 1991 and 1996, respectively, and Ph.D. degree from Liverpool John Moores University, Liverpool, UK, in 2006. He received D.Sc. degree (Habilitation) in Control, Informatics, and Electrical Engineering from the Gdansk University of Technology, Gdansk, Poland in 2019. He is a Full Professor with the Department of Electrical Engineering, Qatar University, Doha, Qatar, and a former Full Professor of the Department of Electrical Engineering, AMU, Aligarh, India. Dr. Iqbal has been listed in top 2% highly cited scientists of the world (Stanford University, USA) since 2019. He has published widely in international journals and conferences on his research findings related to power electronics, variable speed drives, e-mobility, smart grid, complex energy transition, micro- and nanogrids, and renewable energy sources. He has authored or coauthored more than 580 research articles, 8 patents, and four books and several chapters in edited books. He has supervised several large research and development projects worth several million USD. He has supervised 21 PhD. He was also a recipient of the Outstanding Faculty Merit Award for year 2014–2015 and the Research Excellence Awards at Qatar University, in 2015, and 2022. He has received Research Excellence Award from the College of Engineering, Qatar University in 2019. He has received several best research papers awards in top International Conferences. He is serving as the Vice-Chair of the IEEE Qatar Chapter. He is an Associate Editor of the *IEEE Transaction on Industrial Electronics* and Senior Editor of *IEEE Access*.

Abbreviations

EHV	Extra high voltage
RPA	Relay protection and automation
MEPS	Modern Electric Power System
IDMT	Inverse definite minimum time delay overcurrent relay
TMA	Time multiplier setting
PSM	Plug scale (or setting) multiplier
CT	Current transformer
VT	Voltage transformer
HRC	High rupturing capacity
NUG	Nonutility generators
DSGU	Distributed storage and generation units
FM	Frequency modulation
LAN	Local/area network
WAN	Wide area network (WAN)
OLTC	On-load tap changers
DER	Distributed energy resources
AGC	Automatic generation control
SCADA	Supervisory Control and Data Acquisition
DMS	Distribution management system
OMS	Outage management system
FDIR	Fault detection, isolation and restoration
VVO	Volt-var optimization
CVR	Conservation voltage reduction
DOE	Department of Energy
STS	Static transfer switch
PCC	Concept of standard coupling
VSC	Voltage source converter
LVRT	Low-voltage ride-through
UPS	Uninterruptible power supply
SCR	Short circuit ratio
TCC	Time current curve
LSI	Long time, short time, and instantaneous
LTPU	Long time pick up
LTD	Long time delay

STPU	Short time pick up
STD	Short time delay
IPU	Instantaneous pickup
LVPCB	Low voltage power circuit breaker
MCCB	Molded case circuit breaker
EMS	Energy management system
DCS	Distribute control system
QE	Qatar Energy, former name Qatar Petroleum
CPM	Controller performance monitoring
ML	Machine learning

1 Introduction

Abstract

This book describes the main types of protection in transmission systems and includes sections on transducers, overcurrent relays, distance protection, differential protection, and auto-reclose relaying. The final section introduces the fast-growing field of numeric protection, where relaying decisions are made through calculations performed by microprocessor-based relay hardware. The fast-operating speed and high dependability of numeric relays make them especially suited to the transmission environment.

This book describes the main types of protection found in transmission systems and includes chapters on transducers, overcurrent relays, distance protection, differential protection, and auto-reclose relaying. The final chapter is an introduction to the fast-growing field of *numeric protection,* where relaying decisions are made through calculations performed by microprocessor-based relay hardware. The fast-operating speed and high dependability of numeric relays make them especially suited to the transmission environment.

This book is concerned with transmission system protection and builds upon the introduction to protection provided in Book Part 1 (Protection of Power Distribution Systems). You are therefore assumed to be familiar with the basic principles of protection and the difference between unit and non-unit protection systems; these concepts may be used without further explanation in this so, if you are not fully confident

Supplementary Information The online version contains supplementary material available at https://doi.org/10.1007/978-3-031-94312-6_1.

T. Masood et al., *Protection of Electrical Power Transmission Systems*, Power Systems, https://doi.org/10.1007/978-3-031-94312-6_1

on these issues, review the protection Chap. 8 on these issues. You will recall that protection is provided predominantly for three reasons.

- First, to isolate items of the plant to minimize damage through, for example, overheating caused excessive current flow.
- Second, to remove faults from the power systems to prevent instability of the system as a whole, since faults, particularly three-phase faults, control the flow of power and thus upset the real power balance.
- Third, to avoid danger to life and limb. In a transmission system, the latter instability criterion is generally the more important determinant of protection system design since the transmission system will frequently become unstable under faulted conditions before thermal damage to the plant occurs. Additionally, in a transmission system, the minimum fault current can be less than the maximum load current.

Both stability considerations and the high capital cost of transmission plant, together with the high outage costs associated with transmission, dictate that transmission protection needs to clear faults in the shortest possible time. Although the most significant influence on fault clearance is often the circuit breaker operating time, there is nonetheless a requirement for fast operation of the protection system, with the desired operating time commonly being less than one cycle, referred to as *Smart High-Speed Protection*. In contrast, typical transmission circuit breakers can operate significantly faster than this, e.g., within one cycle.

Other essential factors to consider for protection relays include *reliability, selectivity, and economics*. Relay reliability has two aspects:

- *Dependability* is a measure of the protection system's ability to trip correctly infinitely.
- *Security* is a measure of the ability of the protection system not to operate for external faults.

In general, dependability and security are often at odds with each other. Increasing the sensitivity of a relay may enhance dependability by ensuring that it always trips for faults within its zone. However, this heightened sensitivity can also compromise security by occasionally causing the relay to trip for external faults. The goal is to find the optimal balance between dependability and security, ensuring that only the faulted component is disconnected to minimize disruption following a fault. In this respect, it is essential to remember the difference between "*Primary and Backup protection*," where backup protection is used to protect against protection system failure and will necessarily be less selective than primary protection. Finally, economics will ultimately dictate the protection system used for any given application. Taking the view that a protection system, which can be considered a form of insurance, can be valued as a fixed percentage of capital plant assets, it follows that transmission systems, where the plant is expensive compared to distribution systems, can afford high-value protection systems.

This book describes the main types of protection found in transmission systems and includes chapters on transducers, overcurrent relays, distance protection, differential protection, and auto-reclose relaying. The final chapter is an introduction to the fast-growing field of *numeric protection*, where relaying decisions are made through calculations performed by microprocessor-based relay hardware. The fast-operating speed and high dependability of numeric relays make them especially suited to the transmission environment.

Transducers

2

Contents

2.1 Introduction.......... 6
2.2 Voltage Transformers (VTs).......... 6
2.3 The Capacitor Voltage Transformer (CVT).......... 8
2.4 Current Transformer (CT).......... 12
Bibliography.......... 19

Abstract

By the end of this chapter, you will be able to apply the principles of setting up the basic equivalent circuits of a voltage transformer (VT) and a current transformer (CT) in real-world protection relay measurements. This knowledge will enable you to clearly distinguish between the two transformers' primary differences in design criteria and set up an equivalent circuit of a capacitor voltage transformer (CVT) as used in protection relays. You will also be equipped to define the steady-state and transient errors (including the effect of saturation) and the accuracy requirements of the measuring transformer in practical scenarios.

Supplementary Information The online version contains supplementary material available at https://doi.org/10.1007/978-3-031-94312-6_2.

T. Masood et al., *Protection of Electrical Power Transmission Systems*, Power Systems, https://doi.org/10.1007/978-3-031-94312-6_2

Objective

When you have completed this chapter, you should be able to:

- Set up the basic equivalent circuits of a voltage transformer (VT) and a current transformer (CT) used in protection relay measurements.
- Distinguish clearly between the basic differences in design criteria of the two transformers.
- Set up an equivalent circuit of a capacitor voltage transformer (CVT) as used in protection relaying.
- Define the steady-state and transient errors (including the effect of saturation) and the accuracy requirements of the measuring transformer.

2.1 Introduction

A relay receives information from the power system, from which the health of the system can be assessed. Whenever the values of voltage or current in a power circuit are too high to permit convenient direct connection of protection relays, coupling is made through CTs and VTs, which are designed to withstand high voltage and current levels.

Protection has to operate under disturbed system conditions following a fault, so a good transient response is required from the transducers in addition to good steady-state accuracy. CTs and VTs are therefore designed to reproduce an accurate scaled-down replica of the input quantity. Many protection relays are required to measure during the transient period, which invariably follows a fault, so that both the accuracy and transient response of the transducers are important. Errors in CTs and CTs may delay the operation of the protection or cause unnecessary operations.

This looks in detail at both the steady-state and transient performance (for the purposes of ascertaining accuracy) of voltage and current transducers used in protection relaying. The basic transformer models are used for these purposes. Particular emphasis is placed on analyzing the performance of the more commonly used capacitor voltage transformer. Apart from the above, the transformer performance, in particular of the CT, is assessed under saturation conditions [1–3].

2.2 Voltage Transformers (VTs)

VTs are much like small power transformers, differing only in design details, which control ratio accuracy over a specified range of output. The equivalent circuit of a wound electromagnetic transformer is the well-known one shown in Fig. 2.1.

The system voltage V_p is connected across the primary, and the voltage V_s developed across the burden is required to be an accurate replica of the input V_p over a specified range of the output. One way of achieving this is to make the winding

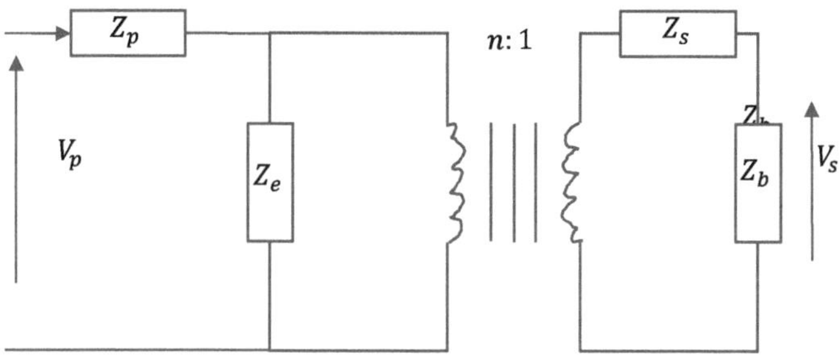

Fig. 2.1 (The equivalent circuit of a wound electromagnetic VT) Diagram of a transformer circuit with primary and secondary sides. The equivalent circuit of a wound electromagnetic VT. The transformer has a turns ratio of \(n:1\)

impedance Z_p and Z_s as low as possible. The Nominal core flux density is then arranged to be well below the saturation level.

Exercise 2.1
What do you understand by the term of burden? In practice, why is it essential that the secondary output of the transformer be independent of the burden?

Burden, also referred to as load impedance, is the total impedance of the measurement equipment, such as protection relays, connected to the secondary of the VT. If the output of VT is affected by the burden over a specific range of outputs, then this can cause serious and unacceptable errors in the input to the protection relays [4–6].

Errors
Two errors are important. First, the ratio error, which is defined as:

$$\%\boldsymbol{error} = n \times V_s - V_p / V_p \times 100 \quad (2.1)$$

It is apparent from the above that if the error is positive, then the secondary voltage exceeds the nominal value. All VTs are required to comply with one of the classes in Table 2.1.

Table 2.1 (Time and rated voltage) Limits of error for measuring voltage transformers

0.8–1.2 times rated voltage 0.25–1.0 times rated burden to 0.8 PF		
Accuracy class	Voltage ratio error (%)	Phase displacement
0.1	±0.1	±5
0.2	±0.2	±10
0.5	±0.5	±20
1.0	±1.0	±40
3.0	±3.0	Not specified

Transient Performance of a VT

Transient errors cause few difficulties in the use of VTs, although some do occur. If a voltage is suddenly applied, *an inrush transient* will occur as with ordinary transformers. The effect is, however, less severe than the power transformers because of the lower flux density for which the VT is designed. Also, when the supply to a VT is interrupted (i.e., the primary is open-circuited), the core flux will not readily collapse, and a more or less exponential current flows in the burden. The condition must not be confused with the collapse of the primary voltage due, for example, to a close-up system fault when the secondary voltage also collapses [7, 8].

Exercise 2.2

What do you understand by the term of burden? In practice, why is it essential that the secondary output of the transformer be independent of the burden?

The phenomenon of magnetizing inrush is a transient condition that occurs primarily when a transformer is energized. The problem is near-zero voltage, as the latter can result in a magnetic flux of nearly twice the nominal peak flux developing in the core. This can cause severe saturation of the transformer and therefore severely distort the output. The effect is less severe in an instrument VT because its core is designed to work at a lower flux density than an ordinary voltage transformer.

2.3 The Capacitor Voltage Transformer (CVT)

VTs for high voltages (above 132 kV) can be expensive owing to the insulation needed to withstand the high voltage ($cost \propto V^2$). The more economical capacitor-type VTs (CVTs) are often used instead. In the CVT, a capacitive potential divider produces an intermediate voltage (typically 12 kV), which is transformed by a small electromagnetic VT to around 110 V (line voltage) to supply the relays. The arrangement is shown in Fig. 2.2. The **Thevenin equivalent circuit** is shown in Fig. 2.3, and

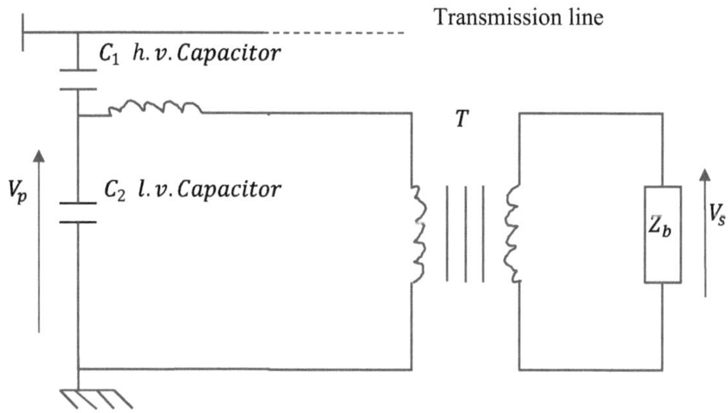

Fig. 2.2 (A CVT arrangement) Diagram of an electrical circuit featuring a transmission line. On the left, a high-voltage capacitor labeled "C1 (h.v. Capacitor)" is connected in series with an inductor and a low-voltage capacitor labeled "C2 (l.v. Capacitor)." The input voltage is marked as "Vp. (A CVT arrangement)"

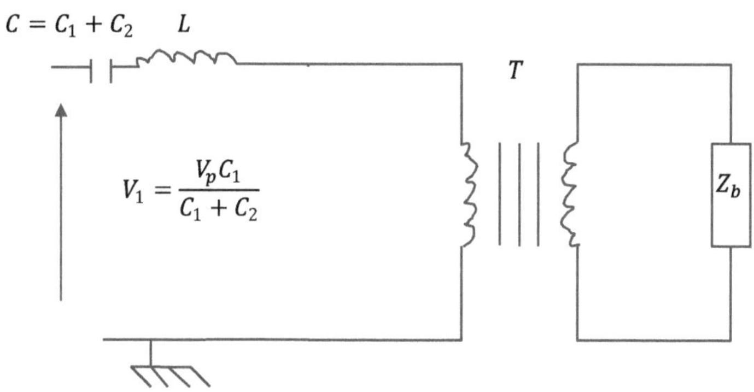

Fig. 2.3 (The Thevenin equivalent circuit CVT) A schematic diagram of an electrical circuit featuring a capacitor, inductor, and transformer (the Thevenin equivalent circuit CVT)

it can be seen that at the nominal system frequency, when C and L are in resonance and therefore cancel, the circuit behaves similarly to a conventional VT [9, 10].

At other frequencies, however, a reactive component exists, which modifies the errors. Providing that the reactive voltages across *C* and *L* are not too large in relation to *Vi*, the change of error with frequency is not excessive. For a typical design, in which *C*1 has a value of 2000 pF and *Vi* = 12 kV, the change in phase error (with 150 VA unity power factor (PF) output) per Hz of frequency change is about 15′ for a 400 kV CVT [11].

Exercise 2.3

The intermediate voltage vi of the CVT arrangement shown in Fig. 2.2 is to be 10 kV for a 400 kV system. If the capacitance of the high-voltage capacitor C1 is 2000 pF, calculate the C1:C2.

After

$$V_i = V_p \frac{C_1}{C_1 + C_2}$$

$$\therefore 10 = \frac{400 \times 2000}{\sqrt{3}\left(2000 + C_1\right)}$$

This gives:

$$C_2 = \frac{400 \times 2000 - 17.32 \times 2000}{17.32} = 44189\, pF$$

$$\therefore ratio\, C_1 : C_2 = 1 : 22.1$$

Transient Behavior of CVTs

The transient response of a CVT is much inferior to that of a conventional VT, and the errors are influenced by many factors. A CVT, as can be seen by the exact equivalent circuit of Fig. 2.4, is a multi-resonant circuit.

For example, under conditions of sudden voltage change (such as collapse due to a fault) on the primary, oscillation caused by the interaction of the equivalent capacitance C and the inductance L_m of the intermediate transformer (typically 10 kHz) will often occur. Oscillations at around 300 Hz will also occur under similar conditions due to the interaction of the series tuning reactance L and the capacitance C_m of the intermediate transformer [12, 13].

Exercise 2.4

If in Fig. 2.4 $C = 0.1\ \mu F$ $L = 100\ H$, evaluate the transformer inductance L_m and capacitance C_m for oscillation to occur under faults at 10 kHz and 300 Hz, respectively. In practice, is it possible for oscillation to occur at other frequencies?

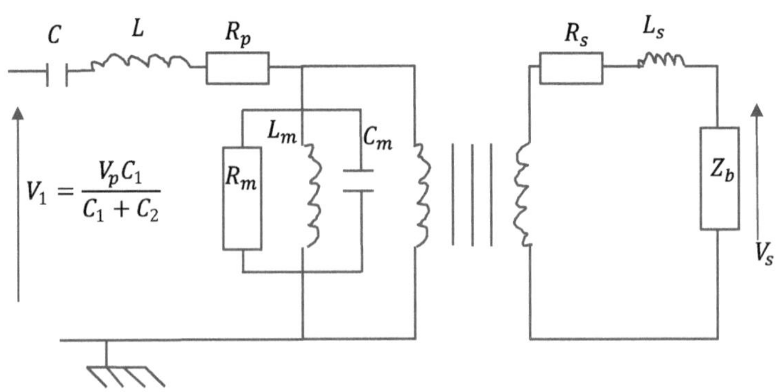

Fig. 2.4 (The exact equivalent circuit of a CVT) Diagram of a transformer electrical circuit, primary and secondary sections, a CVT, an exact equivalent, and a multi-resonant circuit

At any resonant frequency ω

$$\omega L = \frac{1}{\omega C}$$

(i) Thus, for $\omega = (2\pi \times 10 \times 10^3)$ rad,oscillation involving C and L_m, we have

$$L_m = \frac{1}{\omega^2 C} = \frac{1}{\left(2\pi \times 10 \times 10^3\right)^2 \times 0.1 \times 10^{-6}} = 2.5\,mH$$

(ii) For $\omega = (2\pi \times 300)$ rad, oscillation involving C_m and L, we have

$$C_m = \frac{1}{\omega^2 L} = \frac{1}{\left(2\pi \times 300\right)^2 \times 1000} F = 2800\,pF$$

In practice, the transformer inductance L_m is a function of frequency, i.e., its value is different at different frequencies, and since a voltage collapsed signal on the primary side comprises many frequency components, it is therefore possible for L_m to resonate with C at other frequencies.

It should be noted that the power factor of the burden greatly influences the transient responses. It can dominate the CVT behavior, allowing the same CVT with two different burdens to exhibit different transients. In general, an increase in resistive burden results in large amplitude transients but faster decay. On the other hand, a reactive burden constitutes a further tuned circuit and introduces new modes of oscillation, which can persist for several cycles of the system frequency [14–16].

Exercise 2.5

In your opinion, what types of burden should be avoided in CVT? Which types will give less error?

Avoid reactive burdens (both inductive and capacitive) as far as possible, as they can introduce new modes of oscillations in the CVT output, lasting for several cycles. A resistive burden causes any new oscillation introduced to decay very quickly, thus minimizing error.

Ferro-Resonance

It is possible when energizing CVTs to induce sustained subharmonic ferro-resonance, which takes place between the inductance of the exciting impedance and the capacitance of the potential divider (these are L_m and C in Fig. 2.4). Because of its linear nature, L_m in this case takes a value different from that under a fault, as considered in the exercise on pages 2–7. One quite common condition occurs in which a stable one-third subharmonic persists. The result of this is a rise in output voltage of 25–50% above the nominal voltage. Figure 2.5 shows the output waveform. A resistive burden tends to reduce the likelihood of such oscillation, and so it is common to build in tuned circuits. Some compromise is, however, necessary because the latter tend to impair transient response [17, 18].

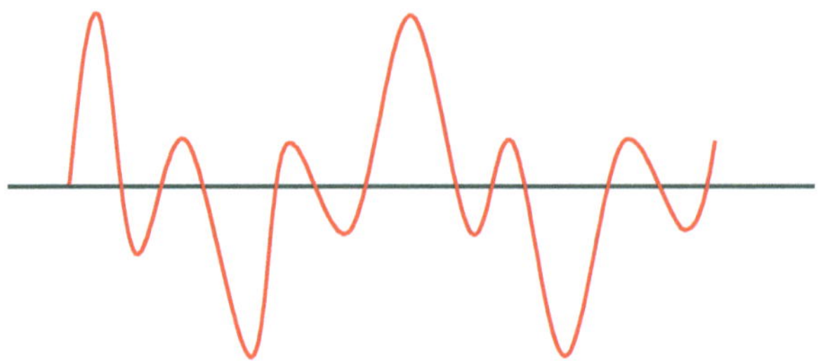

Fig. 2.5 (CVT output due to ferro-resonance) A red waveform with varying amplitude and frequency is displayed against a horizontal axis. The waveform shows peaks and troughs of different heights, indicating changes in signal strength over time

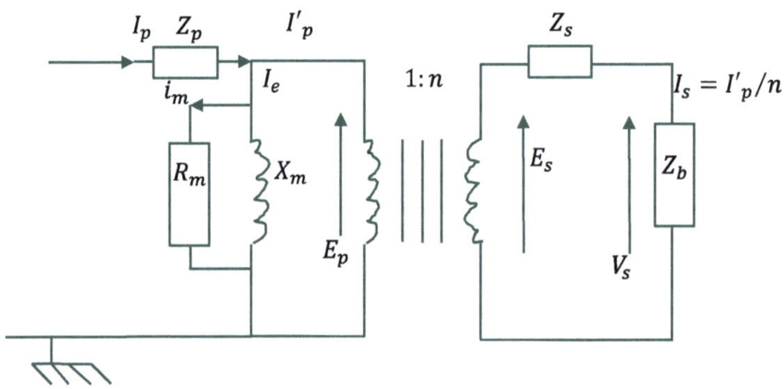

Fig. 2.6 (Equivalent circuit of CT) The image depicts an equivalent circuit diagram of a transformer. On the left, the primary side includes components labeled. The right side represents the secondary circuit with components labeled. The transformer ratio is indicated as 1:n. The diagram illustrates the flow of current and the relationship between primary and secondary components

2.4 Current Transformer (CT)

The usual type of current sensor fitted in the power system is the electromagnetic CT, the equivalent circuit of which is shown in Fig. 2.6. The primary winding of a CT is the same as for any other transformer. The objective is to ensure that:

(a) Minimum interference with the power circuit, i.e., Z_p and the combined value of Z_s and Z_b referred to the primary, is made very small by design.
(b) I_s is a faithful replica of I_p.

The CTs are therefore designed to give as high a value as possible for the magnetizing impedance. This involves high-quality core material. The more closely the value $I_e \rightarrow 0$, the greater is accuracy, but it is not possible to eliminate both ratio and phase errors. The phasor diagram shown in Fig. 2.7 illustrates this point. It can be clearly seen that the exciting current I_e is responsible for both the ratio and phase

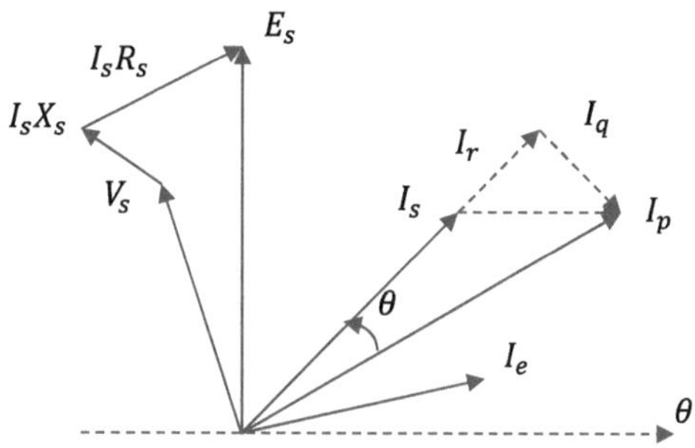

Fig. 2.7 (Phasor representation of the CT circuit (referred to primary)) Diagram illustrating a phasor representation in an electrical circuit. The vectors include, on the left, forming a closed loop. On the right, vectors form a triangle, with angles labeled \(\theta\). The diagram shows relationships between voltage, current, and impedance in the circuit

Table 2.2 Limits of error for measuring current transformers

	± Percentage current (ratio) error at percentage of rated current shown below			Phase displacement at percentage of rated current shown below (')		
Accuracy class	10 up to but not including 20	20 up to but not including 100	100 up to 120	10 up to but not including 20	20 up to but not including 100	100 up to 120
0.1	0.25	0.20	0.1	10	8	5
0.2	0.5	0.35	0.2	20	15	10
0.5	1	0.75	0.5	60	45	30
1	2	1.5	1	120	90	60

error, the component I_r of I_e, which is cophasal with I_s, causing the ratio error, and the component I_q, which is in quadrature with I_s, causing the phase error θ. The problem is that I_e is also dependent on the magnitude of the voltage E_q, which, in turn, is dependent on the magnitude of the power factor (PF) of the burden. It will be seen that, with a moderately inductive burden, with I_s and I_e approximately in phase, there will be little phase error, and the exciting component will result almost entirely in ratio error. It is possible to compensate partially for the ratio error by reducing the secondary winding by one or two turns. The accuracy class of measuring CTs used in British practice is shown in Table 2.2 [19, 20].

Exercise 2.6

In a VT, in order to have a reduced voltage output from the secondary, we need to have fewer turns on the secondary winding than the primary winding. Show that in the case of a CT, the opposite is true, i.e., in order to reduce the magnitude of the output current, more turns are needed on the secondary than on the primary.

In any transformer, irrespective of the primary to secondary turns ratio, the volt-ampere (VA) rating remains the same on each side of the transformer, i.e.,

$$V_P I_P = V_S I_S$$

Thus

$$\frac{I_s}{I_P} = \frac{V_P}{V_S} = \frac{N_P}{N_S}$$

$$\therefore I_s = I_P \frac{N_P}{N_S}$$

It is apparent from the above that in order to have $I_s < I_{P'}\, N_{s'}$ the number of turns on the secondary must be greater $I_{P'}$ than the number of turns on the primary.

Transient Response

CTs generally have a good transient response, with the bandwidth being typically ≈10 kHz.

However, a CT can produce a distorted output if magnetic saturation of the core occurs under heavy current fault conditions. For protective purposes, CT specifications are normally laid out in terms of the knee-point e.m.f. The knee point of the excitation curve is defined as that point at which a further increase of 10% of secondary e.m.f. would require an increment of exciting current of 50% as shown in Fig. 2.8 [21–23].

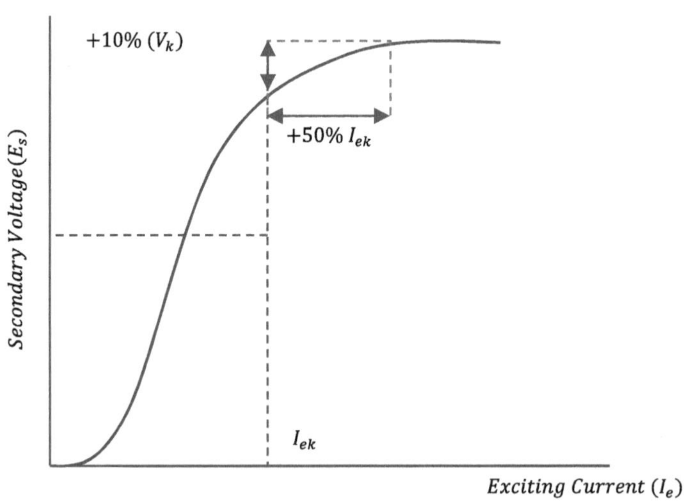

Fig. 2.8 (Definition of knee point of excitation curve) Graph showing the relationship between secondary voltage (E_s) and exciting current (I_e). The curve rises steeply initially and then levels off. A vertical dashed line marks the point I_{ek} on the x-axis, with a horizontal arrow indicating a 50% increase in I_{ek}. A vertical arrow shows a 10% increase in voltage (V_k) at the top of the curve

At voltages above the knee point, saturation quickly occurs. It is important to note that if saturation is reached, a negligible core flux change can occur, and the secondary e.m.f. collapse as does the voltage developed across the burden. Under these conditions, the device no longer behaves as CT. The problem is therefore to ensure that the core flux and therefore the secondary voltage are kept below the knee-point value [24, 25].

To understand the effect of the knee-point voltage on the secondary current, we need to consider the worst-case condition where the input current is fully offset with a long time constant. This may be seen by considering a CT with unity PF of resistance Rb (referred to the primary); this is if the fault current takes a fully offset form:

$$i_P = I_P\left[-\cos\omega_0\, t + e^{-Rt/L}\right] \tag{2.2}$$

The effect of the exciting and winding impedances is neglected, and the primary voltage developed will be

$$i_P = I_P R_b\left[\cos\omega_0\, t + e^{-Rt/L}\right] \tag{2.3}$$

(This is the voltage drop across the burden referred to the primary.) It can be shown that, under such conditions, the maximum flux developed in the CT is

$$\Phi_A\left(1+\frac{\omega_0\, L}{R}\right) \tag{2.4}$$

where Φ_A is the steady-state flux.

Exercise 2.7
Using the relationship given in Eq. 2.3, show that the flux in the CT is given by Eq. 2.4.

$$V_P = I_P R_b\left[\cos\omega_0\, t + e^{-Rt/L}\right] \tag{2.5}$$

Now

$$v = N_P\frac{d\Phi}{dt} or \frac{d\Phi}{dt} = \frac{v}{N_P}$$

Thus, flux in the CT core is

$$\Phi = \frac{1}{N_P}\int_{t1}^{t2} V_p\, dt$$

Integrating each term in Eq. 2.5 in turn gives

$$\Phi_A \text{ the steady state flux} = \frac{I_P R_b}{N_p} \int_{\frac{\pi}{\omega_0}}^{\frac{3\pi}{2\omega_0}} \cos \omega_0 \, t \, dt = \frac{I_P R_b}{N_p \omega_0}$$

(where, to illustrate the worst case, integration has been performed over a quarter cycle at power frequency)

$$\Phi_B \text{ the transient flux} = \frac{I_P R_b}{N_p} \int_0^\infty e^{-Rt/L} dt = \frac{I_P R_b L}{R N_p}$$

(assuming an infinitely long decaying time)
Thus, the ratio

$$\frac{\Phi_B}{\Phi_A} = \frac{\omega_o L}{R} = \omega_o \tau.$$

The CT core has to support both the fluxes so that:

$$\Phi_T = \Phi_A \left(1 + \frac{\omega_o L}{R} \right).$$

It is clear that the flux swing associated with the full offset input current condition is ($\omega_o t$) times the steady-state flux swing caused by a pure sine wave of current. The value of $\omega_o t$ is easily recognized as the (X:R) ratio of the primary system from which the current was derived (this can be as high as ≈ 15 *for* 400 *kV line*). From this, it can be seen that the X:R ratio of the power system is an important feature in the study of protective behavior. The importance of designing on the basis of knee-point voltage thus becomes obvious since the core must have a cross-sectional area sufficient to support a flux swing which is many times larger than that which would be required under normal steady-state conditions [26–29]. Figure 2.9 typifies the problem encountered under CT saturation, resulting in the protection relay being possibly affected.

Exercise 2.8
In what way can CT saturation cause a protection relay to malfunction?

As you can see from Fig. 2.9, CT saturation causes a total collapse of the current developed across the secondary burden. Thus, protection relay principles, based either on a comparison of currents between line ends or measurement of line impedance based on voltage and current, will give erroneous results. The problem is particularly acute for external faults where a relay may indicate an internal fault following CT saturation [30–32].

Following a heavy current, the core has a high remanent flux as shown in Fig. 2.10, thus making saturation on a subsequent fault even worse. Some CTs are

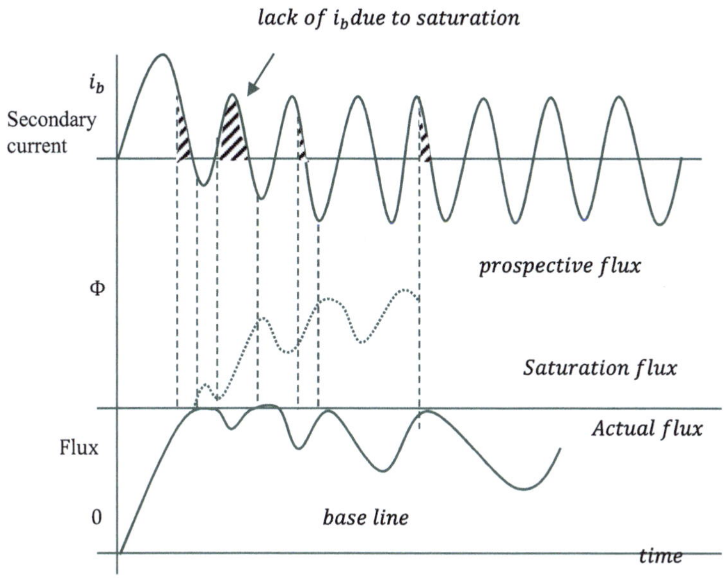

Fig. 2.9 (Effect of CT saturation) The image is a graph showing the relationship between secondary current and flux over time

Fig. 2.10 (Effect of remanence)

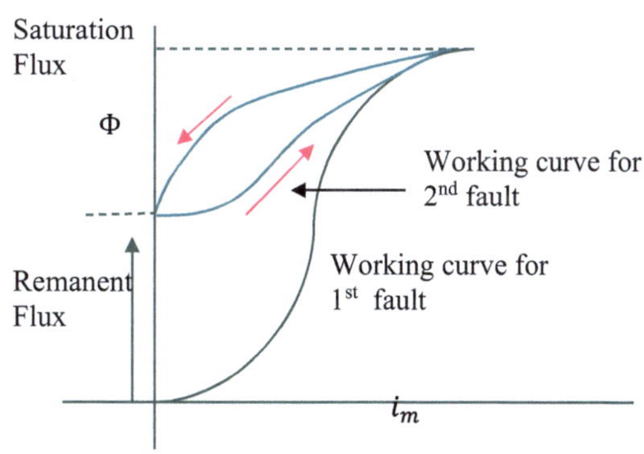

designed with a small air gap to reduce the effects of saturation. Others have a large air gap, which totally eliminates the problem, but this leads to a much larger ratio and phase errors [33].

Exercise 2.9
What do you understand by the term remanent flux? In practice, how can the effect of saturation be reduced? Why is an air-cored CT (as opposed to an iron-cored CT) unsuitable?

Reduced from a high fault current to a low no-fault current, the core flux does not change proportionality (due to the magnetic properties of the iron core) but settles at a higher level. This is called the remanent flux. In practice, the effect of saturation can be reduced by having a small air gap in the core, thereby increasing the threshold level at which saturation occurs. Although an air-cooled CT can support a very large flux (as compared to an iron-cored CT, the leakage flux is also proportionately higher, leading to a much larger ratio and phase error [34, 35].

Self-Assessment Questions

? Self-assessment question 2.1

2.1 a

Discuss the transient response of a capacitor voltage transformer (CVT).

2.1 b

A CVT has a high voltage capacitance of 1000 pF and is connected to a 400-kV system, the interchange voltage being 10 kV. Deduce.

- The low-voltage capacitors.
- The tuning inductance.
- Given that the magnetizing inductance of the interstage transformer is 4 kH, the frequency at which this might resonate with the capacitance of the potential divider. Under what condition, is this likely to occur, and how can such an abnormality be avoided.

? Self-assessment question 2.2

2.2. a

Outline the important factors that must be considered when designing a current transformer (CT) as opposed to a voltage transformer. With the aid of suitable diagram, illustrate how the exciting current mainly responsible for the errors in a CT and suggest methods by which these errors may be partially compensated.

2.2. b

A protective CT is subjected to a primary current of the form:

$$i_P = I_P\left[\sin\left(\omega_o t - \pi/2\right) + e^{-Rt/L}\right]$$

Show that the flux developed across the core of the CT many cycles after the fault is approximately given by:

$$\Phi_T = \Phi_A\left(1 + X/R\right)$$

where Φ_A is the steady-state peak flux; $X = \omega_o L$; and L and R are the primary system parameters.

2.2. c

A CT is designed to transform satisfactorily a sinusoidal load current I. The saturation flux density of the CT is 100 times the flux associated with the full-load current. What is the maximum primary system X:R ratio at which satisfactory transformation for fault current ten times full load current occur?

Bibliography

1. Bin Li; Chao Li; Fengrui Guo; Ying Xin "Overcurrent Protection Coordination in a Power Distribution Network With the Active Superconductive Fault Current Limiter" Published in IEEE Transactions on Applied Superconductivity
2. Bruno Fazio; Sanja Bajic; Serge St-Antoine "Relay Protection Coordination Analysis using Fault Current Measurements:" published in 2022 IEEE Power & Energy Society General Meeting (PESGM).
3. Dong Chen; Bing Hu; Mingjun Xue; Yinian Qi; Chang Song "Research and Application of the System of Relay Protection and Self-Healing for Distribution Network with Mixed Use of Circuit Breaker and Load Switch" published in 2023 International Conference on Power System Technology (PowerCon)
4. Feng Deng; Xiang Ji; Guo-Fang Huang "A centralized distribution protection system" published in 2012 China International Conference on Electricity Distribution.
5. Hana A. Halim; Melaty Amirruddin; Nur S. Noorpi; Nurhakimah M. Mukhtar "An Improved Protection Scheme for Smart Distribution Grid published in 2013 1st International Conference on Artificial Intelligence, Modelling and Simulation.
6. Hang Ji; Dawei Wang; Qiang Xu; Shangqing Yuan "DC Power Distribution System and Its Protection Configuration published in 6th International Conference on Power and Renewable Energy (ICPRE) 2021
7. HanYu Liu; Long Hua Mu "Integrated protection on shipboard integrated power system" 2011 International Conference on Advanced Power System Automation and Protection
8. IEEE Recommended Practice for Fault Diagnosis and Protection in Smart Distribution System IEEE Std 2748-2023.
9. Jamil Abdo & Tariq Masood "New Technique to develop Intelligent, highly reliable & Efficient Unmanned Power Generation Modular" 22nd International Conference on Renewable Energies and Power Quality (ICREPQ'24) Bilbao (Spain), 26th to 28th June 2024. https://www.icrepq.com/icrepq24/392-24-abdo.pdf
10. Jamil Abdo & Tariq Masood "Operational Reliability- Production Best Practice Pragmatic Case Studies & Improvement" published in Proceedings of the 9th North American Conference on Industrial Engineering and Operations Management Washington D.C., USA, June 4-6, 2020
11. Jamil Abdo & Tariq Masood "Techno Economics Analysis to Reshape Global Clean and Green Energy" 22nd International Conference on Renewable Energies and Power Quality (ICREPQ'24) Bilbao (Spain), 26th to 28th June 2024. https://www.icrepq.com/icrepq24/415-24-abdo.pdf
12. Jane Courtney; Aaron McDonnell "Impact on Distribution System Protection with the Integration of EG on the Distribution Network" published in 2019 54th International Universities Power Engineering Conference (UPEC).
13. Kun Yang; Zhen Xiang; Ruixiong Yang; Lin Jiang "Decentralized Protection Coordination Based on multi-agent system for Active Distribution Networks" published in 2022 7th Asia Conference on Power and Electrical Engineering (ACPEE).
14. Li Jun Qin; Ying Wang; Cuijuan Hao; "Multi-Agent System wide area protection considering distributed generation impact" published in International Conference on Advanced Power System Automation and Protection in 2011

15. M. Szablicki; P. Rzepka; A. Halinka; P. Sowa "Diagnosis of challenges for power system protection—selected aspects of transformation of power systems" published in Modern Electric Power Systems (MEPS) in 2021
16. Qin Lijun; Wang Ying; Zheng Shenrui; Guo Qing " Situation of study on distribution network protection containing distributed generation and application scheme of intelligent protection" 2011 4th International Conference on Electric Utility Deregulation and Restructuring and Power Technologies (DRPT)
17. Qingfa Chai; Jiayin Jiang; Yi Tang; Anning Wang; Congcong Li "High Reliability Relay Protection Setting Scheme of Distribution Network" published in 2023 International Conference on Power System Technology (PowerCon)
18. Qingtao Cao; Yun Jiang; Chunxiu Liu; Xinzhe Ma; Pengpeng Wang "Research on Fault Analysis of Multi source Distribution Network Based on Adaptive Protection" published in 7th Asia Conference on Power and Electrical Engineering (ACPEE) in 2022
19. Rani Kumari; Bhukya K. Naick "Protection Coordination Impact on Reliability of Distribution Systems With Distributed Energy Resources" 2024 Third International Conference on Power, Control and Computing Technologies (ICPC2T).
20. Tariq Masood & Jamil Abdo & "Maintenance Strategy Production Best Practices—Pragmatic Case Studies and Operations Improvement" published in Proceedings of the 9th North American Conference on Industrial Engineering and Operations Management Washington D.C., USA, June 4-6, 20
21. Tariq Masood & Jamil Abdo "New Technique to Enhance Hybrid Renewable Energy Resources' Operational Flexibility and Reliability" 22nd International Conference on Renewable Energies and Power Quality (ICREPQ'24) Bilbao (Spain), 26th to 28th June 2024. https://www.icrepq.com/icrepq24/396-24-masood.pdf
22. Tariq Masood and Jamil Abdo "Review: Uninterrupted Liquid Hydrogen (E-Methanol) Production framework to deliver Clean & Green Energy" Published 22nd International Conference on Renewable Energies and Power Quality (ICREPQ'24) Bilbao (Spain), 26th to 28th June 2024. https://www.icrepq.com/icrepq24/414-24-masood.pdf
23. Tariq Masood, Abdel Aty Edris, Suhail Aftab Qureshi, Muhammad Tajammal, Murtaza Hashmi, DP Kothari, Samer Kareem "SSSC tailored to optimization performance in Between United Arab Emirates and Oman on the GCC Power Grid" published in the 13th IET International conference on AC and DC power Transmission Conference on 14–16 February, 2017. http://ieeexplore.ieee.org/document/7934934/
24. Tariq Masood, Abdel Aty Edris, Suhail Aftab Qureshi, Muhammad Tajammal, Murtaza Hashmi, DP Kothari, Samer Kareem "Smart Grid Operations and Control Challenge by Implementing STATCOM Tailored to Optimization Performance at the GCC Power Grid" published in the 13th IET International conference on AC and DC power Transmission Conference on 14–16 February, 2017. http://ieeexplore.ieee.org/document/7934988/
25. Tariq Masood, Muhammad Tajammal, Samer Kareem, Suhail Aftab Qureshi, Abdul Jabbar Khan, DP Kothari, Nair Jameel "Unified power Flow Controller Modeling and Analysis Technique on the GCC Power Grid Operations and Control" published IEEE Energy Con, 2018 June 7-9 Limassol Cyprus. https://ieeexplore.ieee.org/document/8398834/
26. Tariq Masood, Muhammad Tajammal, Samer Kareem, Suhail Aftab Qureshi, Murtaza Hashmi, Nasir Jameel "Unified power Flow Controller (UPFC) Modeling and Analysis Technique by using Matlab and EMTP" published (ICREQP,18 Salamanca Spain 21–23 March, 2018). http://www.icrepq.com/icrepq18/250-18-masood.pdf
27. Tariq Masood, Muhammad Tajammal, Samer Kareem, Suhail Aftab Qureshi, Murtaza Hashmi, Naser Jameel "STATCOM Controller Performance Enhancement and Optimization Technique" published (ICREQP, 18 Salamanca Spain 21–23 March, 2018). http://www.icrepq.com/icrepq18/249-18-masood.pdf
28. Tariq Masood, Muhammad Tajammal, Samer Kareem, Suhail Aftab Qureshi, Murtaza Hashmi, DP Kothari, "Smart Grid Operational functions and Control Challenges by Implementing SSSC Tailored to Optimize performance in between Qatar and KSA on the GCC Electrical-power

grid" published in the ASTESJ Journal Advances in Science, Technology and Engineering Systems Journal Vol. 2, No. 1, 30.09.2017. https://astesj.com/special-issue-4/

29. Tariq Masood, N Al-Emadi, SA Qureshi, YY Almulla, AJ Khan, SK Shah "FACTS Control Devices (Statcom, SSSC and UPFC) Re-Configuration Techniques By PSIM/MATLAB" Published in: 2007 International Conference on Electrical Engineeringhttps://ieeexplore.ieee.org/abstract/document/4287295
30. Tariq Masood, Raj K Aggarwal, SA Qureshi, RAJ Khan "STATCOM and SVC control operations and optimization during network fault conditions" https://ieeexplore.ieee.org/document/5636868
31. Tariq Masood, Raj K Aggarwal, Suhail Aftab Qureshi, RAJ Khan "STATCOM model against SVC control model performance analyses technique by MATLAB" published in International Conference on Renewable Energies and Power Quality (ICREPQ'10) https://icrepq.com/icrepq'10/655-Masood.pdf
32. Tariq Masood, RK Aggarwal, SA Qureshi, RAJ Khan "Statcom control reconfiguration technique for steady state and dynamic performance optimization during network fault conditions" published in International Conference on Renewable Energies and Power Quality (ICREPQ'10) Granada (Spain) https://www.icrepq.com/icrepq'10/656-Masood.pdf
33. Tariq Masood, Samer Karim Shah, Muhammad Tajammal, "Unified power flow controller modeling and analysis technique on the GCC power grid" published in 2018 IEEE International Energy Conference (ENERGYCON) https://ieeexplore.ieee.org/abstract/document/8398834
34. Tariq Masood, Suhail A Qureshi, RK Aggarwal, DP Kothari "Multiple FACTS devices integration through distributed control system and its application on the GCC interconnection" published in 2013 7th IEEE GCC Conference and Exhibition (GCC) https://ieeexplore.ieee.org/abstract/document/6705750
35. V. S. Sazanov; A. I. Kovalenko; A. A. Voloshin "The Development of System for Automatic"

Overcurrent Relays

3

Contents

3.1 Introduction 24
3.2 Inverse Definite Minimum Time Overcurrent Relays 25
3.2.1 Operating Characteristics 25
3.2.2 Relay Types 27
3.2.3 Overcurrent Relay Setting 31
3.3 Instantaneous Overcurrent Relays 41
3.4 Conclusion 41
3.5 Self-Assessment Questions 41
Bibliography 42

Abstract

Upon completing this chapter, you will be equipped to list the differences between the different types of overcurrent relays and their characteristics. More importantly, you will understand the crucial need for relay coordination of relay characteristics. This understanding will enable you to describe the procedure required to determine relay settings and the use of an overcurrent relay as backup protection. You will also appreciate the need to apply computers for complex and intelligent coordination.

Supplementary Information The online version contains supplementary material available at https://doi.org/10.1007/978-3-031-94312-6_3.

T. Masood et al., *Protection of Electrical Power Transmission Systems*, Power Systems, https://doi.org/10.1007/978-3-031-94312-6_3

Objectives

When you have completed this chapter, you should be able to:

- List the differences between the different types of overcurrent relays and their characteristics.
- Explain the need for relay coordination of relay characteristics.
- Describe the procedure required to determine relay settings.
- Describe the use of an overcurrent relay as back-up protection.
- See the need for the application of computers for complex and smart coordination.

3.1 Introduction

Overcurrent relays are generally used as non-unit protection and therefore provide single-point relaying. In themselves, they only monitor the measured power system current and cannot determine where the fault is on the network. Therefore, their settings, together with the settings of other relays protecting the network, have to be coordinated so that only the relay that is required to trip actually trips. An example of a coordination exercise, or grading, is illustrated in Sect. 3.2.3. Naturally, since this is to be done by hand, the exercise only considers and simplified power system. For more involved networks, computer-based coordination systems are recommended.

Overcurrent relays, in particular, inverse definite minimum time (IDMT) overcurrent relays, are the most widely used relays to protect electricity supply systems. The current–time characteristics of these relays enable them to be used with other similar relays positioned at other points in the network to provide discrimination for faults on adjacent plant.

IDMT overcurrent relays are generally used to protect both high-voltage overhead lines and underground cable feeders from 3.3 kV to 33 kV. They are also used to provide backup protection for subtransmission circuits operating up to 66 kV. They offer single-point, non-unit protection and inherently provide backup for other relays protecting adjacent [1, 2].

Instantaneous overcurrent relays are used in conjunction with IDMT relays and auto-reclose schemes to protect feeder circuits. They are also used as high-set elements for a great variety of applications where their setting corresponds to an excessive fault current. Definite time delayed overcurrent relays are used in conjunction with IDMT relays and auto-reclose schemes for feeder protection, as well as being used alone for earth fault protection. The first overcurrent relays were introduced in the early 1900s. They have proved to be very effective and are now available in electromechanical, static, and digital designs [3]. Their importance to the supply industry in the United Kingdom has led to their current–time characteristics curves being defined by **British standard BS 142**.

3.2 Inverse Definite Minimum Time Overcurrent Relays

3.2.1 Operating Characteristics

The three principal characteristics defined by BS 142 are the STANDARD INVERSE, the VER INVERSE, and the EXTREMELY INVERSE characteristics. These current–time curves shown in Fig. 3.1 were initially derived from the operating characteristics of electromechanical relays and are hence defined empirically. They are generally represented by plotting the operating time in seconds along the Y-axis and input current measured as a multiple of the setting current, along the X-axis. The principal area of interest is from input currents of twice the setting current to 20 times the setting current relays; these curves have been characterized by equations [4]. The equation used for one series of relays to define the British Standard specification is as follows:

STANDARD INVERSE

$$T = \frac{0.14T}{\left[\frac{I_f}{PI_N}\right]^{-0.02} - 1} \tag{3.1}$$

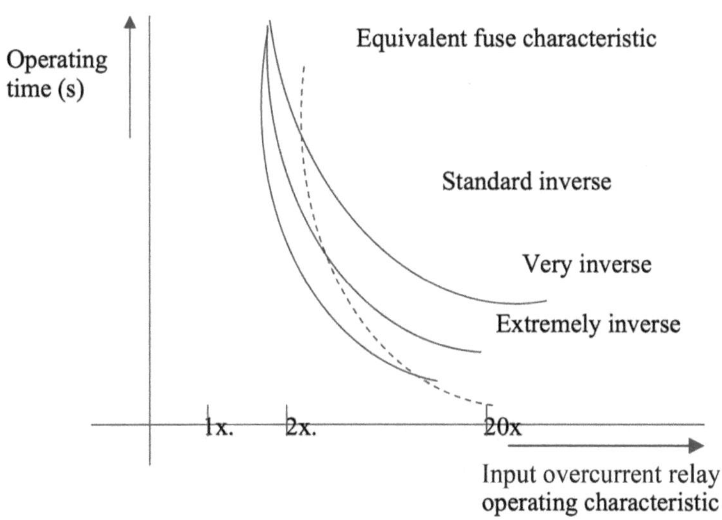

Fig. 3.1 IDMT overcurrent relay operating characteristics. Graph showing the relationship between operating time in seconds and input overcurrent relay operating characteristics. It features four curves: The curves demonstrate how operating time decreases as the input current increases, with each curve representing a different inverse time characteristic

VERY INVERSE

$$T = \frac{13.5T}{\left[\frac{I_f}{PI_N}\right] - 1} \tag{3.2}$$

EXTREMELY INVERSE

$$T = \frac{120T}{\left[\frac{I_f}{PI_N}\right]^{-2} - 1} \tag{3.3}$$

where

- I_f is the fault current as seen by the relay.
- I_N is the nominal relay current.
- T is the time multiplier setting—the relay's time setting.
- P is the plug scale multiple—the relay's current setting.

Exercise 3.1

The operating characteristic of the protection system depends upon the characteristics of the relay and its primary transducers. Assuming that the current transformer (CT) associated with an electromechanical overcurrent relay saturates at ten times the relay current setting, sketch the resulting characteristics for the standard, very, and extremely inverse IDMT relays. For simplicity, assume that the current transformer provides a linear output until it saturates, and thereafter it provides a constant output current.

Figure 3.2 shows the standard inverse and because of its tolerance to current transformer saturation is the most widely used characteristic in the United Kingdom. The very inverse characteristic is an inverse linear characteristic and, in most feeder protection applications, provides an operating time proportional to the distance to the fault position along the feeder. The extremely inverse characteristic, defined by an inverse square law, is similar to the characteristic of a fuse and is the most suitable for applications where part of the network is protected by fuses [5].

Other characteristics exist for specialist applications. These include overload protection and the protection of rectifier circuits. Manufacturers need not comply with British Standard, and several leading international manufacturers supply their own preferred curves. These are generally manufacturers similar to the British Standard curves. A wide range of IDMT characteristics is available, and where a system may contain relays from more than one manufacturer, system analysis can become involved. With the introduction of digital overcurrent relays, customer-defined characteristic curves are also available [6].

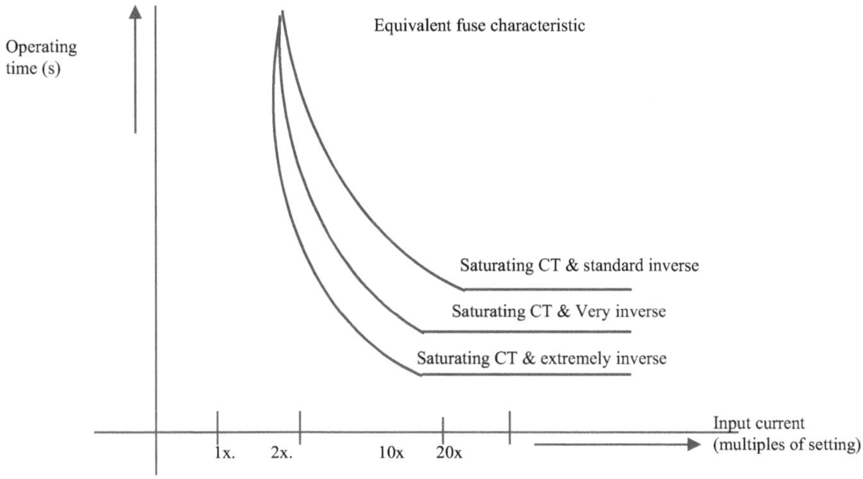

Fig. 3.2 Modified IDMT operating times due to primary CT saturation. Graph illustrating the equivalent fuse characteristic with three curves representing different saturation levels of current transformers (CTs): standard inverse, very inverse, and extremely inverse. The vertical axis shows operating time in seconds, while the horizontal axis indicates input current as multiples of the setting. The curves demonstrate how operating time decreases as input current increases

3.2.2 Relay Types

Electrotechnical Relay

The first IDMT overcurrent relays were developed from single-phase rotating disc watt-hour meters. Since, in the United Kingdom, the saturating standard inverse characteristic has always been the most widely used, these relays were often referred to as bad wattmeters (wattmeters should never saturate!). The majority of electrotechnical designs manufactured today were developed between 1940 and 1960. The main improvements that have been made to these designs have been to reduce costs, to reduce size, to extend to setting range, and to take advantage of modern methods of manufacture [7].

The principal components of an electrotechnical IDMT relay as shown in Fig. 3.3. Characteristic curve shaping is achieved through the magnetic characteristics of the electromagnet and the interaction of the magnetic field produced from the input current and the eddy currents in the disc. When the relay is deenergized, the disc is normally held against the backstop by the restraining spring. The input current generates a force to counter this, and when the current exceeds the pickup setting, i.e., the strength of the spring, the disc rotates. A moving contact is attached to the disc and rotates with it. This moving contact reaches a fixed contact and completes an electrical circuit, thus tripping the relay. The operating time corresponds to the time required for the disc to rotate from the backstop and reach the fixed contact, thus completing the electrical circuit [8]. The relay has two settings, which are the current setting, controlled by varying the number of turns on the

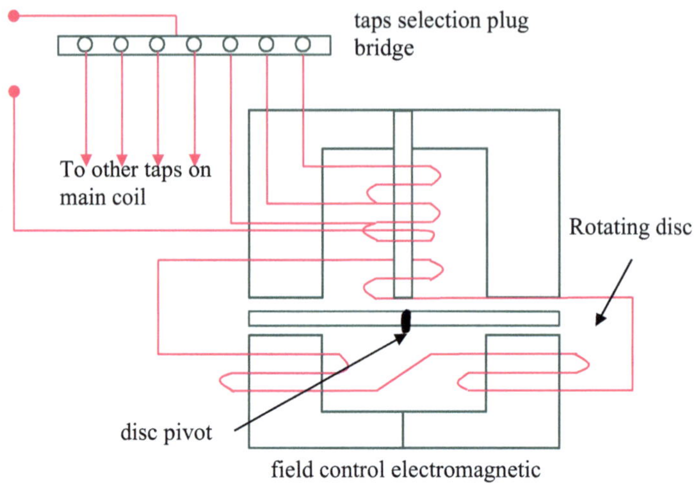

Fig. 3.3 Principal components of an electromechanical IDMT overcurrent relay. Diagram illustrating the components of a tap changer mechanism. At the top, a tap selection plug bridge connects to multiple taps on the main coil. The field control electromagnetic is depicted at the bottom, interacting with the rotating disc. Arrows indicate the flow of connections between components

electromagnet, and the time setting, controlled by adjusting the position of the backstop for the rotating disc. The current setting is referred to as the plug scale (or setting) multiplier (PSM), and the time setting is the time multiplier setting (TMS).

Following the British Standard definition, the relay, "starting current," the input current that will cause the disc to move off the backstop, is set between 3% and 5% above the relay setting current, i.e., nominal current times PSM. Its "closing current," which will result in the relay contacts closing, is set between 15% and 30% above its setting current. If the input current falls below 90% of its setting current, the relay will reset, and the disc will retune to the backstop under the force of the restraining spring. Since the power required to rotate the disc is derived from the input current, these relays have a comparatively high burden. With modern designs, this can vary between 1 VA and 3 VA at the nominal current. As the input current increases, the burden also increases [9].

Exercise 3.2

Assuming that an overcurrent relay has a burden of 1 VA at its setting current, and that its input impedance is constant, calculate the burden at ten times its setting current.

Letting

$$I_s = setting\ current$$
$$Z_s = input\ imedance\, at\, setting\ current$$

Then,

$$I_s^2 \times Z_s = 1VA$$

At the ten times setting current,

$$I_{IN} = 10 \times I_s$$

Z_s remain constant

$$Burden = I_N^2 \times Z_s = (10I_s)^2 \times Z_s = 100\,VA.$$

This relay burden of 100 VA at ten times the setting current places heavy demands not only on the primary current transformer but also on the relay's input circuit. The relay must be able to withstand these input currents and hence power level for the maximum period associated with fault clearance, together with possible auto-reclose onto permanent faults.

Static (Analog) Relays

The first static relays, using an analog electronic circuit, were introduced in the early 1970s. These designs used transistor circuits but were later replaced by operational amplifiers. Compared to electrotechnical relays, they offered the advantages of low burden, improved curve repeatability, reduced frequency errors, reduced temperature errors, and could be housed in smaller cases. The principal disadvantages of these electronic relays were that they were invariably more expensive, they required a power source, they were generally obtained from a d.c. station battery, and early design gained a reputation for poor reliability [10]. A simplified block diagram for a static overcurrent relay is shown in Fig. 3.4. Input isolation is achieved using a current transformer (CT), which, together with a resistance across its secondary, provides the curve-shaping circuit with an input voltage proportional to the

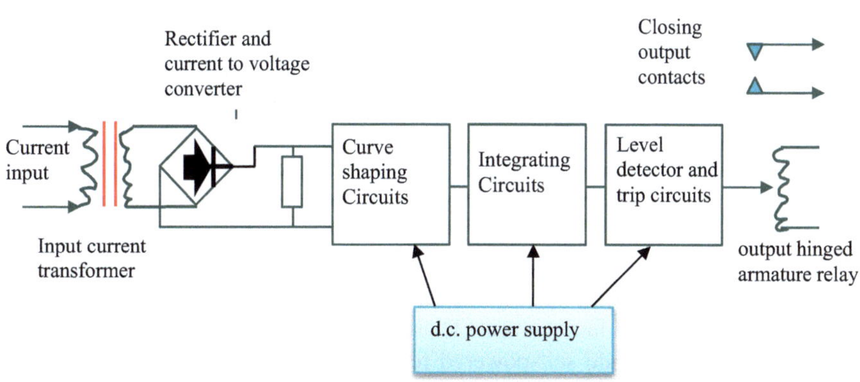

Fig. 3.4 Block diagram for a static IDMT overcurrent relay. Diagram illustrating the process of converting input current to output relay activation. It begins with an input CTcurrent transformer, followed by a rectifier and current-to-voltage converter. The signal then passes through curve shaping circuits, integrating circuits, and level detector and trip circuits. A DC power supply supports these processes

input current. The current setting, PSM, can be adjusted either by adjusting this CT turn ratio or by varying the value of the loading resistor. The time setting, TMS, is an internal setting in the electronic decision-making circuit [11].

The relay output generally uses a hinged armature electromechanical relay, which provides isolated output contacts. The harsh electrical environment of the power system sub-station is hazardous for the well-being of relay electronics, and hence, this method of isolation is very attractive. Solid-state, thyristor-switched outputs are available from some suppliers, but care is required to avoid malfunctioning because of the electrical environment. In general, although static overcurrent relays have enjoyed some success in the industrial market, they have not been successful in the utility market except for certain specialist applications that are difficult to provide using electromechanical technology [12].

Digital Relays

Commercial digital overcurrent relays were first introduced in the late 1970s and early 1980s. The early designs used dedicated digital electronics and erasable programmable read-only memory (EPROMs) to hold the curve-shaping data. These were soon superseded by microprocessor-based relays. The availability of single-chip microcomputers provided designers with a few components and therefore a high tolerance to the harsh electrical environment of the power system. A typical block diagram of microprocess overcurrent relays is shown in Fig. 3.5. As with the static design, a current transformer provides input isolation, and a hinged armature relay provides the output isolation. In this design, the current setting is provided by switching the value of the CT loading resistance, whereas the time setting uses digital inputs into the microprocessor [13].

An immediate advantage of digital technology was the opportunity to provide switch-selectable characteristic curves. This was a particular advantage for manufacture since one design could contain all the standard curves and hence replace a range of electromechanical or static equivalents. Improved manufacturing techniques also reduced the production costs of these designs. Low input burden enabled relays to have wide current setting ranges. These can vary between 2:1, 24:1, and 32:1 depending on relay type. Thus, again, a single digital design could cover a wide range of electromechanical variants [14].

Opportunities to reduce the component count and the physical size of the relays eased environmental problems despite the increased sensitivity of the microelectronics to interference.

Although these relays, as with static relays, have the disadvantage of requiring a power supply, systems have been developed, which can provide power from a protected circuit's load and fault current. These units, when used with the overcurrent relays, effectively provide self-powered relays. More recently, the use of digital designs has enabled complex multi-characteristic curves to be provided and for the overcurrent relaying functions to be complemented by other protection functions and/or control functions, such as an auto-reclose scheme [15].

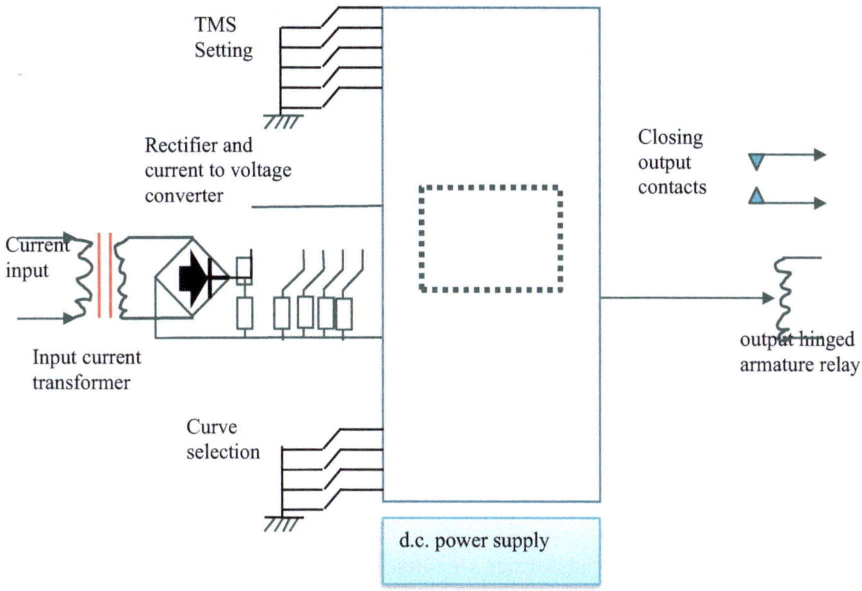

Fig. 3.5 Typical block diagram of a microprocess IDMT overcurrent relay. Diagram illustrating a microcomputer-based system for processing current input. The input current transformer receives the current input, which is then converted from current to voltage by a rectifier. The system includes a TMS setting and curve selection. An on-chip A/D converter processes the signal, and the output is directed to a hinged armature relay with closing output contacts

3.2.3 Overcurrent Relay Setting

When protecting any power system, the operations of the relays should be such that they operate for faults on the plant they are protecting and not for faults on other plant, however close. This discrimination is the principal requirement of any protection system. Relay operations for faults elsewhere on the network and subsequent circuit breaker opening will disconnect the healthy plant from the network.

Naturally, should the protection system for one unit of plant fail for any reason, further protection should be included to isolate that "faulted plant" from the network. There is therefore a need for backup protection and the operation of circuit breakers, disconnecting a larger part of the system. The protection system for a unit of plant is an integral part of that protected plant; hence, if the protection system fails, that unit is considered faulty [16].

The second priority of the protection scheme is the need to break the flow of fault current as quickly as is prudent. The protection system should require the minimum operating time for high fault currents while operating satisfactorily for minimum fault currents. Overload protection is generally covered by other schemes and is not therefore the concern of the main protection system [17].

Protecting Feeder Circuits with IDMT Overcurrent Relays

In order to protect a network of feeder circuits with IDMT relays located in each substation, their settings are chosen so that they correctly discriminate between each other, such that only the relay required to trip will do so. The simplified radial network shown in Fig. 3.6 will be protected by relays located at the input for each feeder at locations A, B, C, and D. The analysis of the protection system is referred to as a grading exercise. Grading exercises are applied to all types of network configurations; however, this radial network of plain feeders provides a useful introduction to the procedures [18].

For an effective power system analysis, a one-line diagram of the power system network is essential, together with source capacities and plant impedances, which are generally expressed in MVA and symmetrical component impedances, respectively. If the system includes motors, and or small- or medium-sized generator, their starting currents and decrement curves under fault conditions will also be required [19].

To evaluate the effectiveness of the protection systems and their settings, details of the initial choices of relays will be required, together with the position and ratios of the primary current transformer's excitation curves, their secondary winding impedance, cabling impedances, and details of other load will be required.

When protecting a network with IDMT relays, it is advisable that the relays used should be of the same type and with similar operating characteristics. This simplifies the matching of curve shapes and ensures that variation in power system frequency and ambient temperatures will introduce the minimum of complications.

At each substation, different relays are used to protect the feeder against phase faults and earth faults. The settings of the phase fault relays are based on three-phase faults and hence consider the system voltages and positive phase sequence impedances. Earth fault relay settings are based on single-phase-to-ground faults and hence require details of the positive-, negative-, and zero-phase sequence impedance components [20].

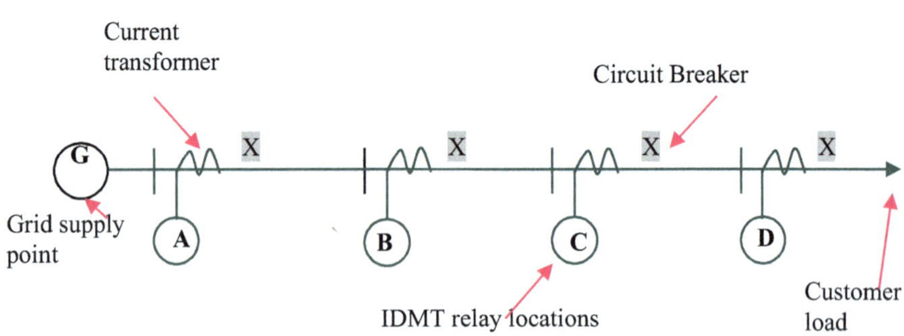

Fig. 3.6 A radial transmission network. Diagram of an electrical distribution system showing a sequence from left to right: a grid supply point, current transformer, circuit breakers, and IDMT relay locations labeled A, B, C, and D. The flow ends at the customer load. Arrows indicate the direction of power flow and label key components

The Relay Current Settings

The first part of the analysis is to select the relay current settings. These are chosen so that the pickup current is well above the maximum load current expected in the circuit, while at the same time it is well below the minimum fault current for which the relay may be required to operate. Invariably, this will lead to the relays furthest from the grid supply point and closest to the ultimate load having the most sensitive settings. As a general rule, the current setting of the relay at a point A in the power system I_{RA} is chosen such that [21]:

$$I_{RA} > 2\left(I_{load}\right)_{\max}$$

and

$$I_{RA} < \frac{1}{2}\left(I_{fault}\right)_{\min}$$

where $(I_{load})_{max}$ is the maximum load current that the relay will experience and $(I_{fault})_{min}$ is the minimum fault current that it will encounter.

The actual choice of the relay setting is a compromise between these limits. Occasionally, these limits conflict, and the choice of settings has to be examined in more detail.

Exercise 3.3

Considering the 11-kV overhead feeder network shown in Fig. 3.7, choose the current setting for IDMT phase fault relays at locations A, B, C, and D, noting that the source capacity can vary between 75 and 150 MVA and the line impedance is as shown in the diagram. Assume that the relays have a nominal rating of 5 A and PSM settings of 50%, 75%, 100%, 125%, 150%, 175%, and 200% are available.

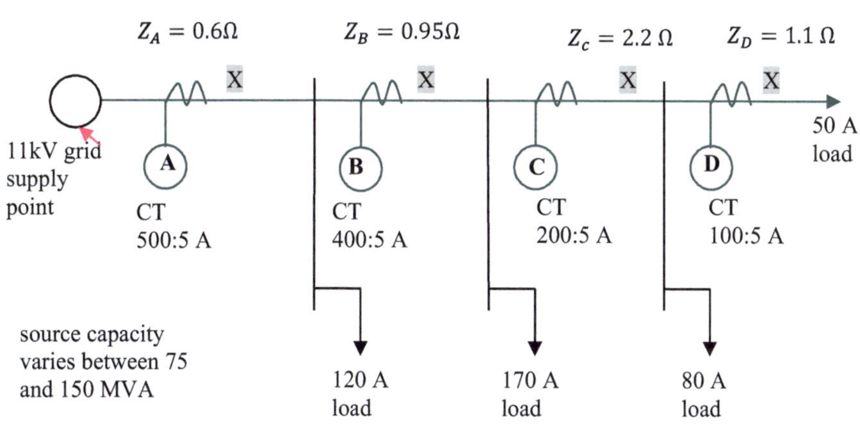

Fig. 3.7 A radial feeder network to be analyzed. Diagram of an electrical grid system showing an 11-kV supply point labeled "G" with source capacity between 75 and 150 MVA. It includes four sections with impedances. Each section has a current transformer (CT) with ratios. Loads are indicated as 120 A, 170 A, 80 A, and 50 A

To choose the current settings for the relays at locations A, B, C, and D in Fig. 3.7, you must first calculate the load and fault currents encountered by the relays. The relay setting is then chosen to be a value that can be set on the relay and that satisfies the relationship:

$$I_{SET} > 2\left(I_{LOAD}\right)_{MAX}$$

and

$$I_{SET} < \frac{1}{2}\left(I_{FAULT}\right)_{MIN}$$

where $(I_{FAULT})_{MIN}$ is the lowest fault current for which the relay must operate.

Since these relays are phase fault relays, their settings are chosen using the fault currents that will flow for balanced three-phase faults.

Considering the load currents [22]:

$$\left(I_{LOAD}\right)_D = 50\,A$$

$$\left(I_{LOAD}\right)_C = 80 + \left(I_{LOAD}\right)_D\, A = 80 + 50A = 130A$$

$$\left(I_{LOAD}\right)_B = 170 + \left(I_{LOAD}\right)_C\, A = 170 + 130A = 300A$$

$$\left(I_{LOAD}\right)_A = 120 + \left(I_{LOAD}\right)_B\, A = 120 + 300A = 420A$$

The analysis of the fault current requires the circuit impedances to be calculated: Instead of defining the source as an ideal voltage with a source impedance, it has been defined by its voltage, 11 kV, and a range of source capacities, which in this case can vary from 75 MVA to 150 MVA. This can be converted into source impedance using:

$$Z_{eq} = \frac{V^2}{M}\Omega \tag{3.4}$$

where

Z_{eq}= equivalent source impedance (Ω)
V = source voltage (kV).
M = source capacity (MVA).

The 75 MVA source can be expressed as follows:

$$\left(Z_s\right)_{75} = \frac{V^2}{M} = \frac{11\times 11}{75}1.6133\Omega$$

$$\left(Z_s\right)_{75} = 1.61\Omega$$

And the 150 MVA source can be expressed as follows:

$$\left(Z_s\right)_{150} = \frac{V^2}{M} = \frac{11\times 11}{150}0.8066\Omega$$

$$\left(Z_s\right)_{150} = 0.81\Omega$$

Returning to the fault currents, the greatest fault current that will be encountered by the relay at location A will be for a fault at the relaying position when the source is at its strongest, i.e., 150 MVA. The circuit impedance will only consist of the source impedance $(Z_s)_{150}$:

$$\therefore \left(I_{FAULT}\right)_{MAX} = \frac{\frac{V}{\sqrt{3}}}{Z}$$

$$\left(I_{FAULT}\right)_{MAX} = \frac{\frac{11000}{\sqrt{3}}}{0.81} A$$

$$\left(I_{FAULT}\right)_{MAX} = 7840\,A.$$

The minimum fault current that the relay at location A will encounter for a fault on its protected feeder corresponds to a fault at the far end of feeder A when the source capacity is at its weakest, i.e., 75 MVA. This will correspond to the maximum circuit impedance for an in-zone fault, Z_{MAX}, where

$$Z_{MAX} = \left(Z_s\right)_{75} + Z_A = 1.61 + 0.6\Omega = 2.21\Omega$$

The fault current will be derived from:

$$\left(I_{FAULT}\right)_{MIN} = \frac{\frac{V}{\sqrt{3}}}{Z_{MAX}}$$

$$\left(I_{FAULT}\right)_{MIN} = \frac{\frac{11000}{\sqrt{3}}}{2.21} A = 2874\,A$$

A similar exercise can be undertaken for the relays at locations B, C, and D. The result is tabulated in Table 3.1 [23].

For the relay at location A, its setting should be at least twice the maximum load current that it will encounter; therefore, its setting I_{RA} must satisfy

Table 3.1 Different location results

Relay location	Minimum in-zone fault impedance	Maximum in-zone fault impedance	Maximum in-zone fault current	Minimum in-zone fault current	Maximum load current
A	0.81	2.21	7840	2874	420
B	1.41	3.16	4504	2010	300
C	2.36	5.36	2691	1185	130
D	4.56	6.46	1393	983	50

$$I_{RA} > 2\left(I_{LOAD}\right)_{MAX\,A}$$
$$I_{RA} > 2 \times 420\,A$$
$$I_{RA} > 840\,A$$

Its setting must also be at most half of the minimum fault current for which it must operate. The relay at location A must operate for all in-zone faults; therefore, its setting I_{RA} must satisfy

$$I_{RA} < \frac{1}{2}\left(I_{FAULT}\right)_{\min A}$$
$$I_{RA} < \frac{1}{2}(2874)\,A$$
$$I_{RA} < 1437\,A$$

Therefore, to operate correctly for all faults on feeder A,

$$840 > I_{RA} > 1437\,A$$

Since this is a generous range, the opportunity for providing backup protection for feeder B can also be considered. For this, the relay at location A must operate for all in-zone faults for feeders A and B, and its setting I_{RA} must satisfy

$$I_{RA} < \frac{1}{2}\left(I_{FAULT}\right)_{\min B}$$
$$I_{RA} < \frac{1}{2}(2010)\,A$$
$$I_{RA} < 1005\,A$$

Therefore, to operate correctly for all faults on feeder A and to provide backup for faults on feeder B

$$840 > I_{RA} > 1005\,A$$

The possible settings on the relay are provided by the nominal rated current, its PSM setting, and the current transformer ratio:

$$\therefore I_{SET} = I_N \times PSM \times CT$$
$$I_{SET} = 5 \times PSM \times \frac{500}{5}\,A$$

Using a PSM of 150%

$$\left(I_{SET}\right)_{150\%} = 5 \times \frac{150}{100} \times \frac{500}{5}\,A$$
$$\left(I_{SET}\right)_{150\%} = 875\,A$$

Using a PSM of 200%

$$(I_{SET})_{200\%} = 5 \times \frac{200}{100} \times \frac{500}{5} A$$

$$(I_{SET})_{200\%} = 1000\,A$$

The preferred setting to satisfy the range of acceptable settings and the settings available is therefore

$$(I_{SET})_{175\%} = 875\,A$$

This is expressed in primary current and can also be expressed in relay current, where

$$(I_{SET})_{175\%} = PSM\, I_N$$

$$(I_{SET})_{175\%} = \frac{175}{100} \times 5A$$

$$(I_{SET})_{175\%} = 8.75\,A$$

A similar exercise can be undertaken for the relays at locations B and C, and the result is tabulated in Table 3.2.

For relay B, there is an immediate conflict. The setting I_{RB} should lie in the range:

$$593 > I_{RB} > 600\,A$$

which cannot be done. However, since 593 is close to 600 A and the relay can be se to 600 A, 600 A is the best compromise. This setting meets the load current criteria to enable it to provide protection for feeder B and backup protection for virtually all possible faults on feeder C [24].

The choice on the setting for the relay at location D is complicated by the lack of information on the system it is supplying. The maximum relay setting data is unavailable. Again, a compromise is required, and a setting is chosen that satisfies the load current criteria, satisfies the fault current criteria for faults on feeder D of 492 A, and provides some back-up protection for the circuit fed from feeder D. A setting of 100 A primary current could have been chosen, but 150 A provides a reasonable compromise [25].

The choice of the settings for the relays at A and B illustrated the compromise that is often required. The lack of information about the ultimate load system leads to the choice of the lowest current setting for relay D, which accommodates the maximum load expected.

Table 3.2 Different location results

Relay location	Maximum relay setting (A) including backup	Minimum relay setting (A) load	Chosen setting (A)	CT Ratio	PSM setting (% PSM)
A	1005	840	875	500:5	175
B	593	600	600	400:5	150
C	492	260	300	200:5	150
D	*	100	150	100:5	150

Relay Time Settings

The relay time settings are chosen to ensure that they operate as quickly as possible with the proviso that the relay closest to the fault, and which sees the fault current, opens its circuit breaker before any other relays have time to trip. It is not sufficient that the correct relay trips before the others in the scheme. To prevent possible malfunctioning, fault clearance times, relay inertia, and relay, transducer, and system errors must all be considered.

To ensure that only the correct relay trips, the minimum time separation required between adjacent relays for the largest fault current that they are likely to experience is set to the grading margin. This is the total time required for the correct circuit breaker to clear the fault current, for overshoot in the relay, which is required not to trip, to accommodate relay and transducer errors, and to provide a safety margin to cover other effects. A breakdown of the grading margin is shown in Fig. 3.8.

The time allowed for the circuit breaker to clear the fault depends upon the type of breaker used. Modern vacuum and SF6 breakers can clear the fault current in under three cycles (60 ms). Other types of modern circuit breakers require a maximum time of about five cycles (100 ms), whereas older switchgear could require up to seven cycles (140 ms) [26].

Relay overshoot occurs when the relay inertia prevents it from stopping rotation immediately after the fault current is removed, and applies to both electromechanical and electronic relays, as well as both analog and digital relays. The time is 50 ms for electromechanical relays and 30 ms for electronic relays.

Relay errors depend on the class of relay used. These are defined by specification **BS-142** and international standard **IEC 255–4.** Typically, the majority of

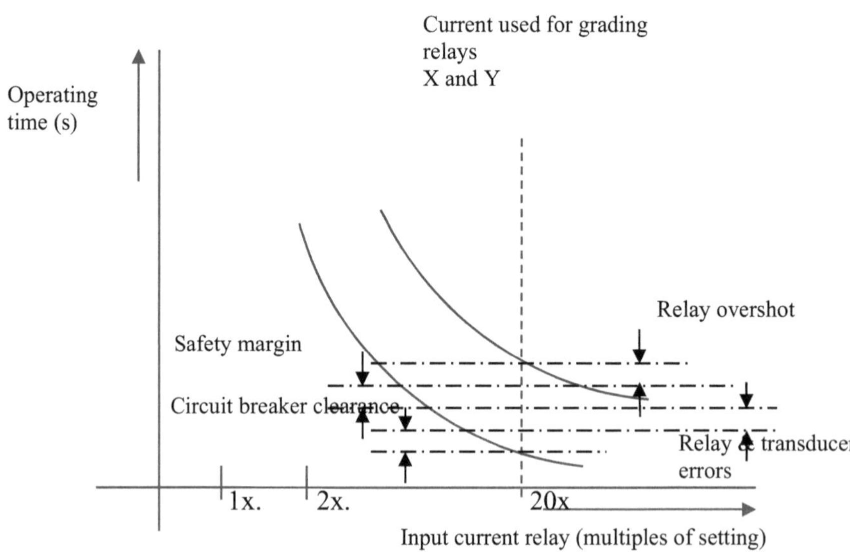

Fig. 3.8 Timings used to form the grading margin. A graph illustrating the relationship between operating time in seconds and input current relay, measured in multiples of the setting. The graph features two curved lines representing the current used for grading relays X and Y. Horizontal lines indicate safety margin, circuit breaker clearance, relay overshoot, and relay and transducer errors

electromechanical relays have a timing error specification of ±7.5%, and electronic relays have a specification of ±5.0%. Transducer errors are treated in a manner similar to relay errors and depend on the class of transducers used. Typical current transformers used for IDMT relays are assumed to have an error equivalent to a relay timing error of ±10.0% [27]. In addition to these, a safety margin is added to ensure that any unforeseen complications do not result in malfunctioning. In general, 100 ms provides a suitable margin.

The equation for the grading margin, t_G, is:

$$t_G = \frac{(2E_R + E_{CT})}{100} t_R + t_{CB} + t_O + t_{SM} \tag{3.5}$$

where

E_R is the relay error (in %).
E_{CT} is the current transformer error (in %).
t_R is the relay operating time.
t_{CB} is the circuit breaker fault clearance time.
t_O is the relay overshoot time.
t_{SM} is the safety margin.

For most applications, this is too involved, and a grading margin of 400 ms is used, consisting of 150 ms for the circuit breaker clearance, 50 ms of that relay overshoot, 100 ms for errors, and 100 ms safety margin [28].

Exercise 3.4
Determine the relay TMS settings for Fig. 3.7 considered above, given that the relay at D has a TMS setting of 0.1, also that the relay has very inverse characteristics, that the TMS settings range from 0.1 to 1.0 in steps of 0.1, and that a grading margin of 0.4 s should be used.

The relay characteristic curve is given by:

$$t = \frac{13.5T}{\left[\frac{I_f}{PI_N}\right] - 1}$$

The first grading exercise considers the response of the relays at C and D to the largest fault current that both relays will experience. This corresponds to a fault close to the relaying point on the feeder taken from D. From Table 3.1, the maximum fault current seen by relay D is 1393 A. Using its PSM setting of 150 A and TMS setting of 0.1, the operating time can be derived from the characteristic equation:

$$t_D = 0.163\,s.$$

For this fault, the operating time of relay C, t_C, should be the operating time of relay D plus the grading margin, i.e.,

$$t_C = 0.163 + 0.4 = 0.563\,s.$$

The current setting for relay C has been chosen to be 300 A; therefore, from the characteristic equation, the TMS setting, *T*, can be found, i.e.:

$$T = 0.152$$

This setting is not available, and the relay must be set to 0.2. The settings for relay C are therefore:

$$PSM_C = 150\%\,and\,TMS_C = 0.2$$

The procedure is tabulated for ease of presentation.

	D	C	B	A
PSM setting (A)	150	300	600	875
Max. Fault current for downstream relay		1393	2691	4504
Trip time for downstream relay		0.163	0.339	0.415
Required trip time for relay including grading margin		0.563	0.739	0.815
Calculated TMS		0.152	0.191	0.25
Chosen TMS	0.1	0.2	0.2	0.3

The chosen PSM and TMS settings define the relay characteristic curves, which can be presented as time–current curves as shown in Fig. 3.9 [29].

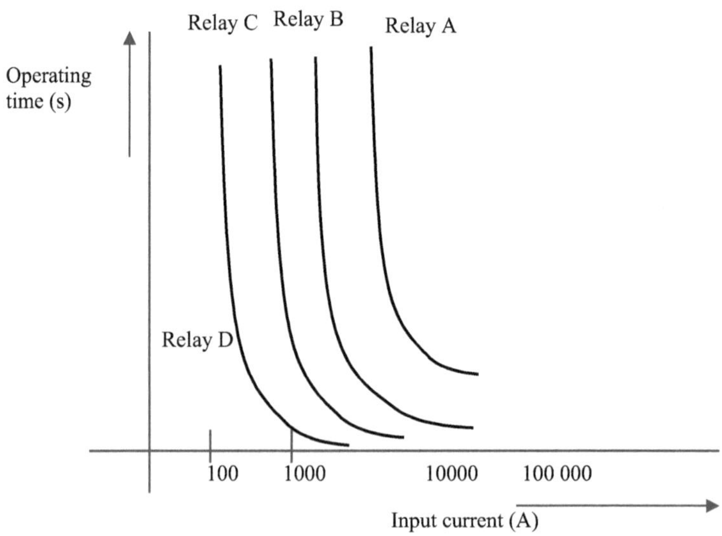

Fig. 3.9 Characteristic curves for IDMT relays to protect radial feeder in Fig. 3.7. A graph depicting the relationship between input current in amperes (x-axis) and operating time in seconds (y-axis) for four relays: A, B, C, and D. Each relay is represented by a curve, showing that, as the input current increases, the operating time decreases. Relay A has the steepest curve, indicating the fastest response time, while Relay D has the least steep curve, indicating the slowest response time

3.3 Instantaneous Overcurrent Relays

Instantaneous overcurrent relays provide high-speed tripping whenever the fault current exceeds the pick-up current and are most frequently used as high-set units for detecting excessive fault currents. They have limited discrimination capabilities and are generally used in conjunction with other types of protection. For feeder circuits, they can improve discrimination by providing high-speed response for close-in faults or where the protected system impedance is greater than the source impedance, such as with long feeders or circuits containing transformers [29].

Care must be taken with the settings of instantaneous relays and consideration given to their overreach under transient conditions. General practice is to set the instantaneous relays to 20% or 30% higher than the largest fault current that would be produced by a fault at the far end of the protected feeder. This provides a margin to accommodate transducer errors, impedance errors, and transient overreach problems [30, 31].

3.4 Conclusion

Overcurrent relaying provides a comparatively simple non-unit protection for feeder circuits and hence is the most widely used protection for electricity supply systems. The main disadvantages are that discrimination is difficult with instantaneous relays and that IDMT relays, which overcome this problem, do so at the expense of tripping times. This is particularly undesirable for a fault close to the point of generation or bulk supply. As power supply networks become more interconnected, other difficulties arise in the use of overcurrent relaying, and other types of protection have to be used.

3.5 Self-Assessment Questions

? Self-assessment question 3.1

Determine the optimum settings for the very inverse characteristic IDMT relays positioned at A, B, C, and D used to protect the feeder network shown in Fig. 3.10, given the power system data shown in the diagram and that the relays have very inverse characteristic, are nominally rated at 5 A, and have PSM settings of 50%, 75%, 100%, 125%, 150%, 175%, and 200% and TMS settings from 0.1. Use a grading margin of 0.4 s.

Computer-Based Overcurrent Relay Grading

The above exercise has considered a radial network of plain feeders with a single grid supply point. Although these systems do exist, they are becoming the minority as supply companies are increasingly using interconnected network to enhance the

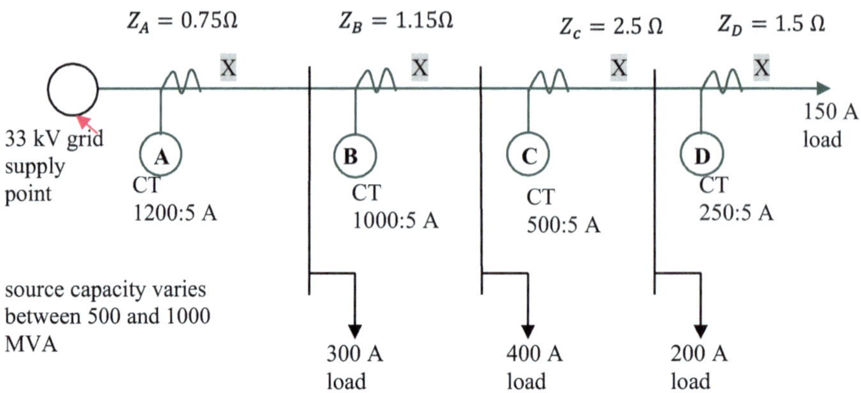

Fig. 3.10 Feeder network model for SAQ 3.1. Diagram of an electrical grid system with a 33-kV supply point. The system includes four sections labeled A, B, C, and D, each with a current transformer (CT) and associated impedance values. Current transformers have ratios of 1200:5 A, 1000:5 A, 500:5 A, and 250:5 A, respectively. Loads are 300 A, 400 A, 200 A, and 150 A

security of supply. These networks can be analyzed by hand, but with the advent of readily available computer systems and power system fault analysis software, this job is now best handled by a computer. Most relay coordination software has been developed from fault analysis packages where the characteristics of the appropriate protection relays are included. Details of the types of protection are input into the software along with the system parameters. Most programs accept suggested relay settings, and some can progress through an extensive grading exercise to determine the desired relay settings. These packages are extremely useful with interconnected systems since they complicate the choice of appropriate settings. The first complication is that the fault current can be supplied from more than one point in the network and/or by more than one route. In order to protect these systems, this has led to the need for protection systems to be located at both ends of the feeder and for directional elements to assist in ensuring proper discrimination. Even with these, however, the grading exercise can highlight the impossibility of providing an effective protection scheme and the need to use other types of relaying.

Bibliography

1. Qingtao Cao; Yun Jiang; Chunxiu Liu; Xinzhe Ma; Pengpeng Wang "Research on Fault Analysis of Multi source Distribution Network Based on Adaptive Protection" published in 7th Asia Conference on Power and Electrical Engineering (ACPEE) in 2022
2. Feng Deng; Xiang Ji; Guo-Fang Huang A centralized distribution protection system published in China International Conference on Electricity Distribution in 2012
3. Congbo Wang; Ke Jia; Bohan Liu; Jiankang Zhang "Coordination Control and Protection for Photovoltaic DC Distribution System" published in 2nd International Conference on Smart Power & Internet Energy Systems (SPIES) in 2020
4. Bin Li; Chao Li; Fengrui Guo; Ying Xin "Overcurrent Protection Coordination in a Power Distribution Network With the Active Superconductive Fault Current Limiter" Published in IEEE Transactions on Applied Superconductivity

5. Bruno Fazio; Sanja Bajic; Serge St-Antoine "Relay Protection Coordination Analysis using Fault Current Measurements:" published in 2022 IEEE Power & Energy Society General Meeting (PESGM).
6. Dong Chen; Bing Hu; Mingjun Xue; Yinian Qi; Chang Song "Research and Application of the System of Relay Protection and Self-Healing for Distribution Network with Mixed Use of Circuit Breaker and Load Switch" published in 2023 International Conference on Power System Technology (PowerCon)
7. Hana A. Halim; Melaty Amirruddin; Nur S. Noorpi; Nurhakimah M. Mukhtar "An Improved Protection Scheme for Smart Distribution Grid published in 2013 1st International Conference on Artificial Intelligence, Modelling and Simulation.
8. IEEE Recommended Practice for Fault Diagnosis and Protection in Smart Distribution System IEEE Std 2748-2023.
9. Jamil Abdo & Tariq Masood "New Technique to develop Intelligent, highly reliable & Efficient Unmanned Power Generation Modular" 22nd International Conference on Renewable Energies and Power Quality (ICREPQ'24) Bilbao (Spain), 26th to 28th June 2024. https://www.icrepq.com/icrepq24/392-24-abdo.pdf
10. Jamil Abdo & Tariq Masood "Operational Reliability- Production Best Practice Pragmatic Case Studies & Improvement" published in Proceedings of the 9th North American Conference on Industrial Engineering and Operations Management Washington D.C., USA, June 4-6, 20
11. Jamil Abdo & Tariq Masood "Techno Economics Analysis to Reshape Global Clean and Green Energy" 22nd International Conference on Renewable Energies and Power Quality (ICREPQ'24) Bilbao (Spain), 26th to 28th June 2024. https://www.icrepq.com/icrepq24/415-24-abdo.pdf
12. Jane Courtney; Aaron McDonnell "Impact on Distribution System Protection with the Integration of EG on the Distribution Network" published in 2019 54th International Universities Power Engineering Conference (UPEC).
13. Kun Yang; Zhen Xiang; Ruixiong Yang; Lin Jiang "Decentralized Protection Coordination Based on multi-agent system for Active Distribution Networks" published in 2022 7th Asia Conference on Power and Electrical Engineering (ACPEE).
14. Qin Lijun; Wang Ying; Zheng Shenrui; Guo Qing "Situation of study on distribution network protection containing distributed generation and application scheme of intelligent protection" 2011 4th International Conference on Electric Utility Deregulation and Restructuring and Power Technologies (DRPT)
15. Qingfa Chai; Jiayin Jiang; Yi Tang; Anning Wang; Congcong Li "High Reliability Relay Protection Setting Scheme of Distribution Network" published in 2023 International Conference on Power System Technology (PowerCon)
16. Rani Kumari; Bhukya K. Naick " Protection Coordination Impact on Reliability of Distribution Systems With Distributed Energy Resources" 2024 Third International Conference on Power, Control and Computing Technologies (ICPC2T).
17. Tariq Masood & Jamil Abdo & "Maintenance Strategy Production Best Practices—Pragmatic Case Studies and Operations Improvement" published in Proceedings of the 9th North American Conference on Industrial Engineering and Operations Management Washington D.C., USA, June 4–6, 2020
18. Tariq Masood & Jamil Abdo "New Technique to Enhance Hybrid Renewable Energy Resources' Operational Flexibility and Reliability" 22nd International Conference on Renewable Energies and Power Quality (ICREPQ'24) Bilbao (Spain), 26th to 28th June 2024. https://www.icrepq.com/icrepq24/396-24-masood.pdf
19. Tariq Masood and Jamil Abdo "Review: Uninterrupted Liquid Hydrogen (E-Methanol) Production framework to deliver Clean & Green Energy" Published 22nd International Conference on Renewable Energies and Power Quality (ICREPQ'24) Bilbao (Spain), 26th to 28th June 2024. https://www.icrepq.com/icrepq24/414-24-masood.pdf
20. Tariq Masood, Abdel Aty Edris, Suhail Aftab Qureshi, Muhammad Tajammal, Murtaza Hashmi DP Kothari, Samer Kareem "SSSC tailored to optimization performance in Between United Arab Emirates and Oman on the GCC Power Grid" published in the 13th IET International

conference on AC and DC power Transmission Conference on 14–16 February, 2017. http://ieeexplore.ieee.org/document/7934934/
21. Tariq Masood, Abdel Aty Edris, Suhail Aftab Qureshi, Muhammad Tajammal, Murtaza Hashmi DP Kothari· Samer Kareem "Smart Grid Operations and Control Challenge by Implementing STATCOM Tailored to Optimization Performance at the GCC Power Grid" published in the 13th IET International conference on AC and DC power Transmission Conference on 14-16 February, 2017. http://ieeexplore.ieee.org/document/7934988/
22. Tariq Masood, Muhammad Tajammal, Samer Kareem, Suhail Aftab Qureshi, Abdul Jabbar Khan, DP Kothari, Nair Jameel "Unified power Flow Controller Modeling and Analysis Technique on the GCC Power Grid Operations and Control" published IEEE Energy Con, 2018 June 7-9 Limassol Cyprus. https://ieeexplore.ieee.org/document/8398834/
23. Tariq Masood, Muhammad Tajammal, Samer Kareem, Suhail Aftab Qureshi, Murtaza Hashmi, Nasir Jameel "Unified power Flow Controller (UPFC) Modeling and Analysis Technique by using Matlab and EMTP" published (ICREQP,18 Salamanca Spain 21–23 March, 2018). http://www.icrepq.com/icrepq18/250-18-masood.pdf
24. Tariq Masood, Muhammad Tajammal, Samer Kareem, Suhail Aftab Qureshi, Murtaza Hashmi, Naser Jameel "STATCOM Controller Performance Enhancement and Optimization Technique" published (ICREQP, 18 Salamanca Spain 21–23 March, 2018). http://www.icrepq.com/icrepq18/249-18-masood.pdf
25. Tariq Masood, Muhammad Tajammal, Samer Kareem, Suhail Aftab Qureshi, Murtaza Hashmi, DP Kothari, "Smart Grid Operational functions and Control Challenges by Implementing SSSC Tailored to Optimize performance in between Qatar and KSA on the GCC Electrical-power grid" published in the ASTESJ Journal Advances in Science, Technology and Engineering Systems Journal Vol. 2, No. 1, 30.09.2017. https://astesj.com/special-issue-4/
26. Tariq Masood, N Al-Emadi, SA Qureshi, YY Almulla, AJ Khan, SK Shah "FACTS Control Devices (Statcom, SSSC and UPFC) Re-Configuration Techniques By PSIM/MATLAB" Published in: 2007 International Conference on Electrical Engineeringhttps://ieeexplore.ieee.org/abstract/document/4287295
27. Tariq Masood, Raj K Aggarwal, SA Qureshi, RAJ Khan "STATCOM and SVC control operations and optimization during network fault conditions" https://ieeexplore.ieee.org/document/5636868
28. Tariq Masood, Raj K Aggarwal, Suhail Aftab Qureshi, RAJ Khan "STATCOM model against SVC control model performance analyses technique by MATLAB" published in International Conference on Renewable Energies and Power Quality (ICREPQ'10) https://icrepq.com/icrepq'10/655-Masood.pdf
29. Tariq Masood, RK Aggarwal, SA Qureshi, RAJ Khan "Statcom control reconfiguration technique for steady state and dynamic performance optimization during network fault conditions" published in International Conference on Renewable Energies and Power Quality (ICREPQ'10) Granada (Spain) https://www.icrepq.com/icrepq'10/656-Masood.pdf
30. Tariq Masood, Samer Karim Shah, Muhammad Tajammal, "Unified power flow controller modeling and analysis technique on the GCC power grid" published in 2018 IEEE International Energy Conference (ENERGYCON) https://ieeexplore.ieee.org/abstract/document/8398834
31. Tariq Masood, Suhail A Qureshi, RK Aggarwal, DP Kothari "Multiple FACTS devices integration through distributed control system and its application on the GCC interconnection" published in 2013 7th IEEE GCC Conference and Exhibition (GCC) https://ieeexplore.ieee.org/abstract/document/6705750

Distance Protection

4

Contents

4.1 Introduction 46
4.2 Impedance Measurement 47
4.3 Three-Phase Distance Protection 63
4.3.1 Relay Input Measurands 64
4.3.2 Load Encroachment and Power Swings 72
4.4 Three-Phase Distance Protection Schemes 78
4.4.1 Three-Zone Stepped Distance 78
4.4.2 Full and Switched Distance Protection Scheme 83
4.4.3 Carrier-Assisted Distance Protection 85
4.5 Directional Earth Fault Protection Schemes 90
4.5.1 The Basic Directional Earth Fault Protection Scheme 91
4.5.2 Directional Earth Fault Protection Schemes Integrated with Distance Relaying .. 94
Self-Assessment Questions 97
Bibliography 98

Abstract

When you have finished this chapter, you should be able to: List the main types of distance protection characteristics and their advantages and disadvantages. List the measurements used for distance relay comparators. Describe the central electrotechnical distance relays. Explain the operation of the block average comparator. Describe the use of three-zone stepped distance protection. Give the derivation of the residual compensation factor. Describe the operation of differ-

Supplementary Information The online version contains supplementary material available at https://doi.org/10.1007/978-3-031-94312-6_4.

T. Masood et al., *Protection of Electrical Power Transmission Systems*, Power Systems, https://doi.org/10.1007/978-3-031-94312-6_4

ent polarization methods and their response to close-up faults. Describe the operation of the standard communication to enhance distance protection schemes. Show how the operation of a directional earth fault protection scheme is used to strengthen a distance protection scheme's performance. Describe the performance of a distance relay to power swings and the need to inhibit its operation.

Objectives

When you have finished this chapter, you should be able to:

- List the main types of distance protection characteristics and their advantages and disadvantages.
- List the measurands used for distance relay comparators.
- Describe the main electrotechnical distance relays.
- Explain the operation of the block average comparator.
- Describe the use of three-zone stepped distance protection.
- Give the derivation of the residual compensation factor.
- Describe the operation of different polarization methods and their response to close-up faults.
- Describe the operation of the common communication enhance distance protection schemes.
- Show how the operation of a directional earth fault (DEF) protection scheme is used to enhance a distance protection scheme's performance.
- Describe the performance of a distance relay to power swings and the need to inhibit its operation.

4.1 Introduction

The term distance protection is used to describe a family of relays that use the impedance of the protected circuit as measured by the relay to determine whether or not the plant is faulted. For a transmission line circuit, the impedance measured or seen by the relay under fault conditions is: roughly proportional to the distance to the fault. If this "distance" seen by the relay is less than the known length of the line, which corresponds to the relay's setting, the relay will trip, and if it is greater, the relay will restrain. Traditionally, distance relays are associated with extra-high voltage (EHV) overhead transmission line circuits. They are the standard form of protection used for both main and sub transmission lines operating at voltages from 33 kV to 765 kV. In addition to these circuits, distance relays are used to protect EHV cable circuits, and with the growing complexity of the distribution network, they are increasingly being applied to protect both overhead and underground HV feeders operating at voltages as low as 6.6 kV. The principal advantages of distance relays are their speed of operation and the simplification of the grading process. Their disadvantage is cost, not only the cost of the relay but also the cost of the voltage transducer required in addition to those monitoring the currents [1, 2].

4.2 Impedance Measurement

Distance relays use both power system voltage and current measurements derived from primary transducers to decide whether or not the protected plant is faulted. The relaying point is defined by the position of the voltage transformer. For the immediate zone of protection, zone 1, time is not required for discrimination, and therefore the relay operates as quickly as it can and therefore is defined as being "instantaneous." Using enlarged operating characteristics, distance relays also provide backup protection, zones 2 and 3, for which time delays are required for grading.

A simplified representation of a distance relay is shown in Fig. 4.1. The relay's inputs are the voltage transformer (VT's) secondary voltage measured at the relaying point and the line current transformer (CT's) secondary current. In this simplified case, the relay will trip when the ratio of the measured voltage and current is less than the known impedance of the protected transmission line. This decision-making process can be reduced to a comparison technique by comparing the amplitude of the measured voltage with that of the voltage generated by the *replica impedance* through which the measured current passes, rather than actually measuring the impedance of the circuit [3].

The above technique formed the basis for the first time-independent distance relays using the electromechanical balanced beam amplitude comparator, as illustrated in Fig. 4.2. Under normal (upfaulted) conditions, the voltage across the restraint coil exceeds that across the operate coil, and the beam is attracted to the restraint arm holding the relay trip contact open. When, however, the voltage across the operate coil exceeds that across the restraint coil, the beam is attracted to the

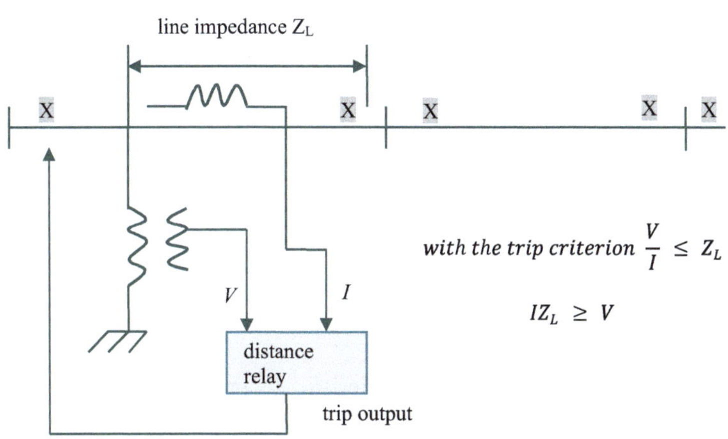

Fig. 4.1 Basic distance protection. Diagram illustrating a power line protection system using a distance relay. The line impedance is labeled as ZL, with multiple fault points marked by "X" along the line. The system measures voltage (V) and current (I) to determine the trip output. The trip criterion is given as the absolute value of V divided by I being less than or equal to the absolute value of ZL

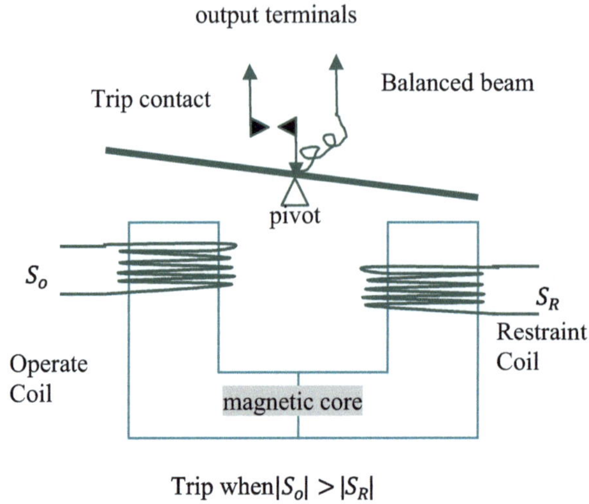

Fig. 4.2 Balanced beam electromechanical distance relay comparator. Diagram of a balanced beam relay system. The system includes an operating coil & restriaining coil, both wrapped around a magnetic core. A balanced beam pivots at the center, with a trip contact mechanism above it. The output terminals are positioned at the top. The system trips when the absolute value is greater than its operating limits

operate arm and causes the trip contact to close, hence tripping the circuit breaker. The relay resets after the fault circuit has been broken and the current falls to zero [4].

Balanced beam relays were manufactured from the 1920s until the 1950s and are still in use today for specialist applications, in particular 25-Hz single-phase railway systems. In general, they were superseded by more sensitive devices introduced in the 1940s, which gave an improved protection capability [5].

Exercise 4.1
For the trip criterion shown in Fig. 4.1, sketch the relay's operating characteristic in the $(R + jX)$ impedance diagram. Comment on the principal limitation of the characteristics.

See Fig. 4.4 for the impedance characteristic. Its principal limitation is that it is a non-directional characteristic and will trip equally for faults in front of and behind the relaying point on the system fed from both ends.

The first complication with applying distance protection is that impedance is not a scalar quantity. Impedance is a vector and has both amplitude and argument. The impedance seen by the relay when there is a fault on a transmission circuit consists of the impedance of the line and return path from the relaying point to the fault arc. The combination for all likely values of fault impedance lies within the speckled

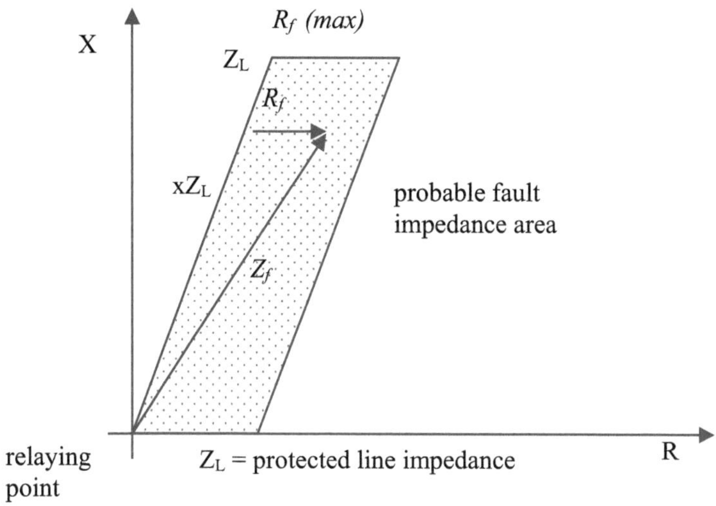

Fig. 4.3 Probable fault impedances seen by a distance relay protecting a transmission line. A graph illustrating the probable fault impedance area in an electrical system. The X and R axes represent reactance and resistance, respectively. A shaded, dotted parallelogram indicates the fault impedance area, labeled with "R_f (max)" at the top

area, which is the probable fault impedance area as shown in Fig. 4.3 [6]. In Fig. 4.3, the impedance of the protected line is shown as Z_L, and hence, for a fault at x along that line, the fault impedance includes xZ_L. The fault impedance Z_f comprises xZ_L and the fault's arc resistance R_f. The probable fault impedance area includes all possible values of Z_f and hence is a quadrilateral defined by all values along Z_L and all values of R_f from zero to $R_{f(max)}$ [7].

The distance relay's operating characteristic is required to cover the probable fault impedances for the portion of the line to be protected and must avoid the probable fault impedance areas for faults outside the tripping zone. The characteristic generated by the relay illustrated in Fig. 4.1 produces the impedance characteristic as shown in Fig. 4.4. This circular characteristic is centered on the relaying point with a radius equal to the replica impedance Z_R used for the relay's setting. The major limitation of this characteristic is that it is non-directional and will trip equally for faults in front of as well as behind the relaying point is moved to the circumstance of the circle produces the directional characteristic shown in Fig. 4.5. This offsetting of the circle is achieved by moving its center away from the origin to the point corresponding to $\frac{1}{2}Z_R$ and reducing its diameter to Z_R. Both the operating and restraining phasors are modified by subtracting $\frac{1}{2}IZ_R$,thus producing:

$$S_O = \frac{1}{2}IZ_R \tag{4.1}$$

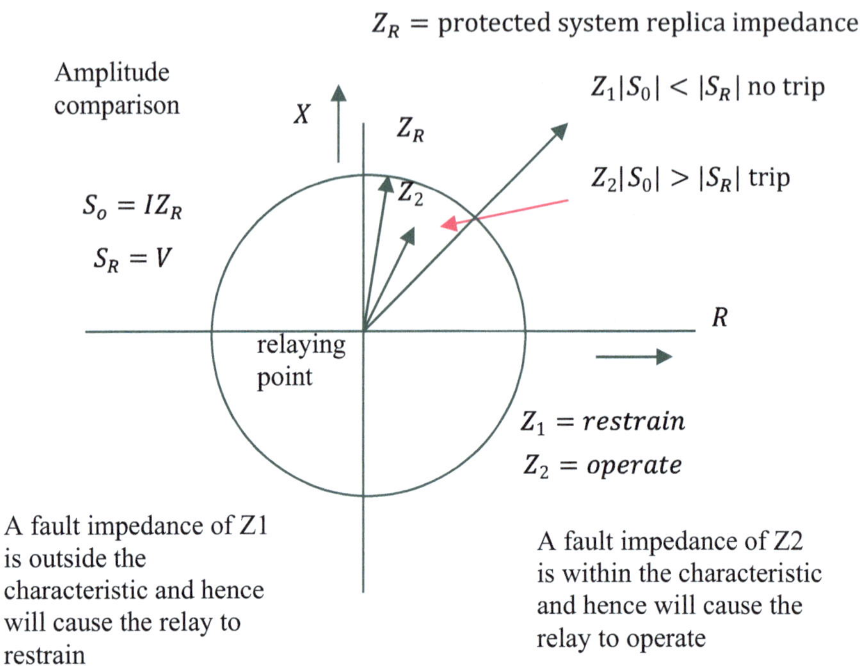

Fig. 4.4 The impedance characteristic. Diagram illustrating a relay protection system with a circle centered at the relaying point on an X and R axis. Amplitude comparison, "A fault impedance is outside the characteristic and hence will cause the relay to restrain," and "A fault impedance is within the characteristic and hence will cause the relay to operate"

$$S_R = V - \frac{1}{2} IZ_{R.} \tag{4.2}$$

The characteristic produced is the unit of conductance (MHO) circle, which was first described in the early 1920s and has become the industry standard for distance protection. It provides both a well-defined and limited forward reach, together with an inherent directional capability. Both the IMPEDANCE characteristic shown in Fig. 4.4 and the MHO circle characteristic shown in Fig. 4.5 have been obtained using amplitude comparison. The relay trips when the amplitude of the operating phasor, S_O is greater than the amplitude of the restraining phasor, S_R. This concept provided the basis of the first distance relays used, and several designs are still in service today. However, it was quickly realized that this approach, together with the technology used in early relays, had an inherently high burden, which restricted the number of applications where they could be used [8].

Exercise 4.2

Plot the probable fault impedance as seen by a distance relay for faults on a transmission line together with the IMPEDANCE and MHO circle characteristics on an $(R + jX)$ impedance diagram. Assume that the transmission line has an impedance of 1.0 *at* 75° and that the maximum arc resistance is 0.25.

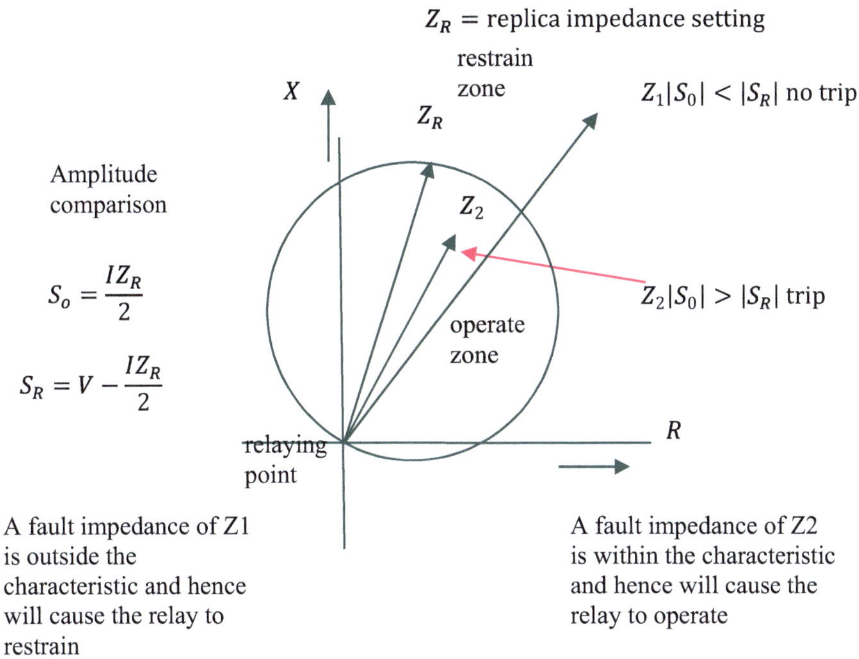

Fig. 4.5 The MHO circle characteristic. The image is a diagram illustrating impedance relay operation zones on an R-X plane. It features a circle centered at the relaying point, dividing the plane into restrain and operate zones. Text explains that a fault impedance of Z_1 is outside the characteristic, causing the relay to restrain, while Z_2 is within, causing the relay to operate. The replica impedance setting \(Z_R\) is noted

The impedance diagram showing the probable fault impedance area together with the IMPEDANCE and MHO circle characteristic is shown in Fig. 4.6.

This diagram illustrates the improved selectivity of the MHO circle characteristic compared to the IMPEDANCE characteristic since it provides a closer match to the probable fault impedance area. The MHO circle characteristic is a directional characteristic, whereas the IMPEDANCE characteristic is not. It provides a greater resistive reach than is required for the probable fault impedance area, but its droop at the reactive reach point limits its ability to detect high resistance faults at the far end of the line. A more detailed examination of the impedance seen during fault conditions shows that this droop is advantageous [9].

Phase angle comparison techniques, where the operations of the relay depended upon the phase angle between the relaying measurands and not the comparison of the signal amplitudes, provided an improved sensitivity, and hence were easier to apply.

The most successful electromechanical phase angle comparator was the induction-cup relay, illustrated in Fig. 4.7. This design provided a long-lasting principle for a generation of distance relays, which was manufactured by most of the leading suppliers from the 1940s and is still in manufacture today. For this relay, the

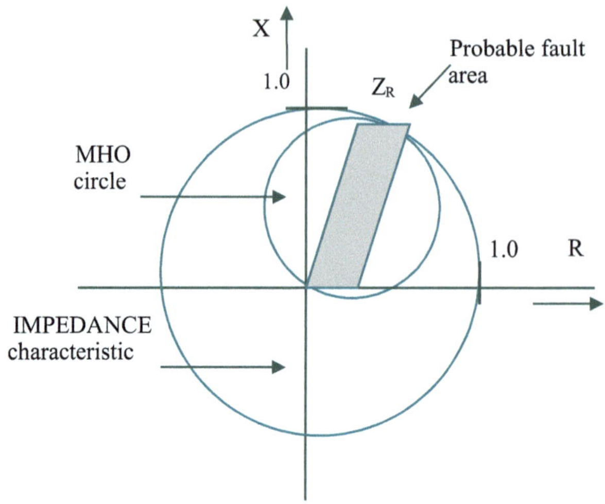

Fig. 4.6 Probable fault impedance area as seen by a distance relay together with the IMPEDANCE and MHO circle characteristics. Diagram illustrating an impedance characteristic graph with X and R axes. A large circle labeled "MHO circle" intersects a smaller circle. A shaded area within the smaller circle is marked as "Probable fault area." The axes are marked at 1.0 on both X and R

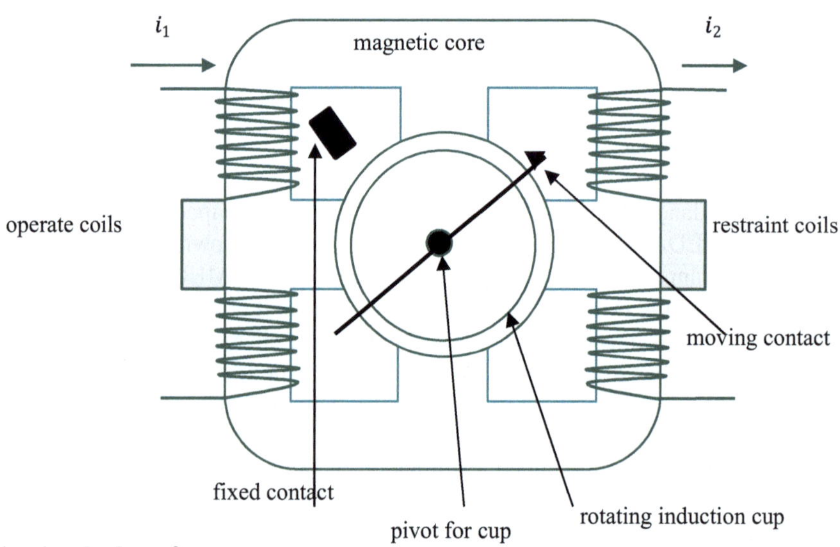

trip when I_1 , $I_2 \cos \theta_0$

Fig. 4.7 The induction cup electromechanical distance relay comparator. Diagram of an induction cup relay showing a magnetic core with operate coils on the left and restraint coils on the right. The central rotating induction cup has a pivot and is connected to a moving contact. The fixed contact is positioned near the operate coils. Arrows indicate current flow, labeled as i_1 and i_2. The relay trips when i_1 and $i_2 \cos \theta$ equals zero

decision to trip was performed by a light metal cup suspended in the field generated by the operate and restraint coils. When restrained, the cup was held against an end-stop. When the phase angle between the two input phasors was between the trip limits, the cup would rotate away from the end-stop. Eventually, the cup would rotate sufficiently such that a contact attached to the cup would meet a fixed contact mounted on the relay assembly and complete the electrical trip circuit. If the fault persisted, the cup would be restrained against further rotation by the fixed contact. Only when the fault condition was removed, did the cup rotate back to the restraining end-stop, resetting the relay. Since the cup required a very small force to rotate, the relay was very sensitive and could be applied in a wide variety of situations. These included conditions where the relay's input signals were well below the nominal values [10].

The light metal cup of the induction cup relay operates as a small motor and rotates in the field generated by the input currents i_1 *and* i_2 *shown in* Fig. 4.7. By convention, its rotation results in tripping when:

$$I_1 I_2 cos\theta > 0$$

where I_1 and I_2 are the r.m.s. values of i_1 and i_2, respectively, and θ is the phase angle between these inputs; referring to Fig. 4.7, this would result in an anticlockwise rotation of the cup, causing the moving contact to complete the circuit with the fixed contact [11].

Since the tripping of the induction cup relay is controlled by the cosine of the angle between the two relay input signals, θ, it is commonly referred to as a cosine comparator. The relay trips when cosine θ is positive and therefore when the angle between its input θ is between the values −90° and +90°.

Cosine comparator techniques have become the standard for most designs of phase angle comparators and are used in designs based on electromechanical static and microcomputer technologies [12].

The use of a phase angle comparator to generate an MHO circle characteristic is illustrated in Fig. 4.8. The boundary restraining and operating conditions are shown in the voltage diagrams shown in Fig. 4.8 (i), (ii), and (iii), respectively. The voltage or phasor diagram is derived from the impedance diagram by multiplying it by the fault current *I*. The relay measurands for the MHO circle are:

$$S_1 = IZ_R - V \tag{4.3}$$

and

$$S_2 = V$$

where Z_R is the relay's setting impedance often referred to as the replica impedance since it represents the impedance of the protected system.

For a fault on the relay boundary of the characteristic, the angle between S_1 and S_2 is seen to be 90°, since the fault is not really inside the characteristic, the relay will not operate (actually, it will operate in an infinite time). Similarly, the relay will not operate if the fault lies outside the characteristic since the angle between S_1 and

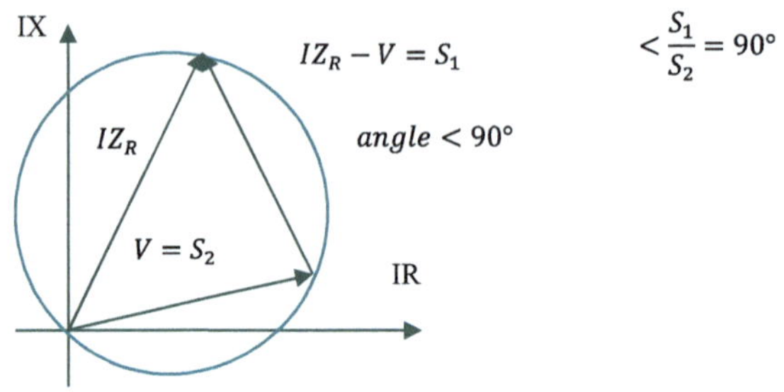

(i) Fault on relay boundary- A phasor diagram illustrating electrical components in a circuit. The vectors form an angle less than 90 degrees. The vertical axis is labeled IX, and the horizontal axis is labeled IR. indicating the relationship between the vectors.

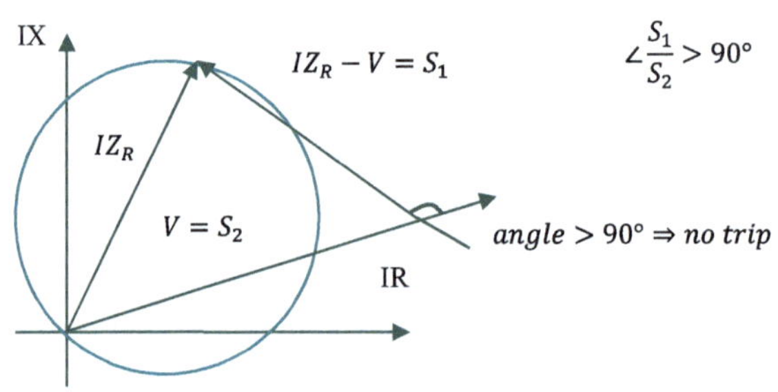

(ii) Fault outside boundary (Diagram illustrating phase angle comparison with a circle on an IX and IR coordinate plane. Two vectors, originate from the center. Equations shown.

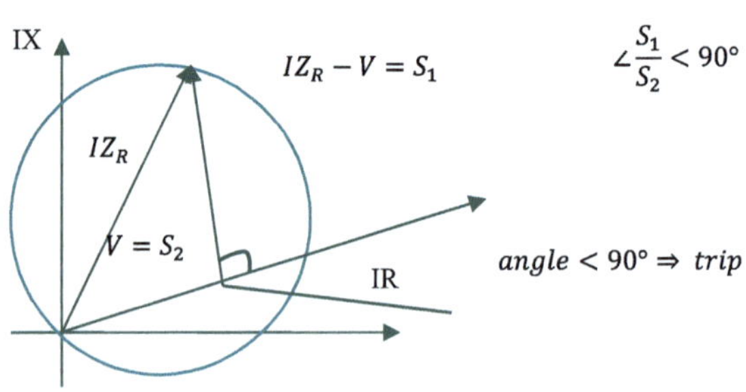

(iii)Fault inside boundary- Diagram illustrating a circle with vectors and mathematical expressions. The circle is centered at the origin of an IX coordinate system.

Fig. 4.8 Boundary restraining conditions

S_2 is greater than 90°. If the fault lies within the characteristic, then relay operation occurs since the angle between S_1 and S_2 is less than 90° [13].

The application of the cosine phase angle comparator to generate the MHO, OFFSET MHO, and IMPEDANCE characteristic is shown in Fig. 4.9 (i), (ii), and (iii), respectively. The OFFSET MHO characteristic is widely used to provide backup protection since it has independently controllable forward and reverse reach. For transmission line protection, this reverse reach enables it to provide protection for the substation to which the transmission line is connected [14].

The input measurands for the OFFSET MHO relay are:

$$S_1 = IZ_{R1} - V$$

and

$$S_2 = IZ_{R2} - V$$

For the IMPEDANCE characteristic, the input measurands are:

$$S_1 = IZ_R - V$$

and

$$S_2 = IZ_R - V$$

For the IMPEDANCE characteristic is therefore a special case of the OFFSET MHO where:

$$Z_{R1} = Z_R$$

and

$$Z_{R2} = Z_R$$

Exercise 4.3

Explain how the input measurands for the MHO characteristic are related to those for the OFFSET MHO characteristic.

The input measurands for the OFFSET MHO characteristic are:

$$S_1 = IZ_{R1} - V$$

$$S_2 = V$$

These can be derived from the input measurands for the OFFSET MHO characteristic, where:

$$Z_{R1} = Z_R$$

The forward reach of the relay and

$$Z_{R2} = 0$$

That is, the characteristic is forced through the origin and therefore has no reverse reach. It is therefore a directional characteristic.

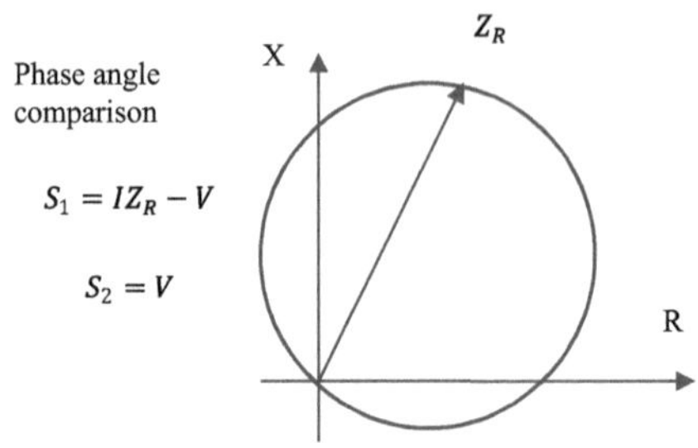

(i) The MHO characteristic- Diagram illustrating phase angle comparison with a circle centered at the origin of an X-R coordinate system. The circle represents impedance, labeled \(Z_R \). Two vectors originate from the center: one along the X-axis and another at an angle within the circle.

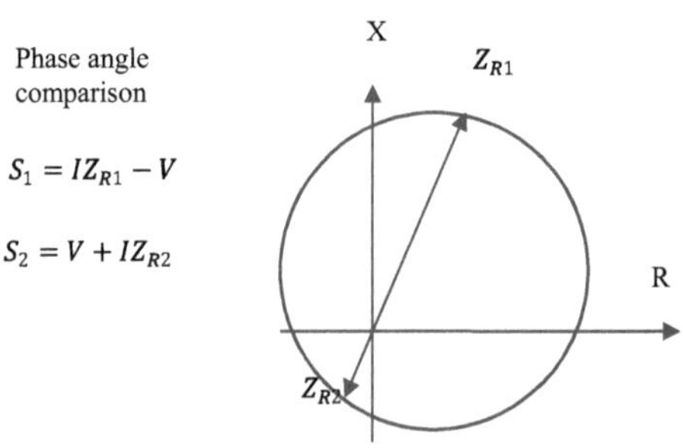

(ii) The offset MHO characteristic- Diagram illustrating phase angle comparison with a circle on an X and R coordinate plane. Two vectors, originate from the center. \(Z_{R1}}\) points upwards and \(Z_{R2}}\) points downwards. Equations shown.

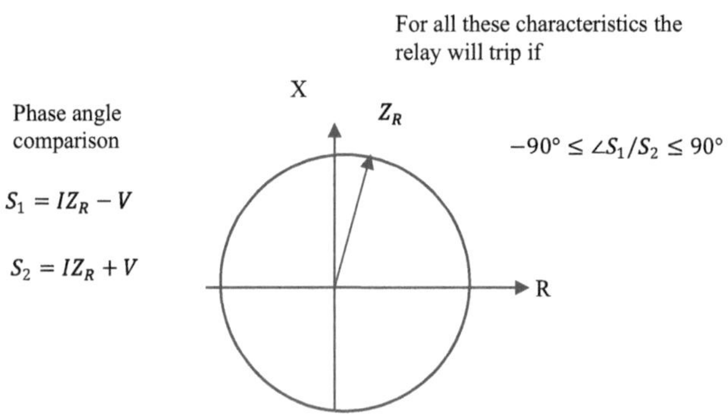

(iii) The IMPEDANCE characteristic- Diagram illustrating phase angle comparison for relay tripping. A circle with horizontal and vertical axes is shown, labeled R and X, respectively. Two vectors are depicted within the circle. The condition for relay tripping is given.

Fig. 4.9 Circular characteristics obtained using phase angle comparison

The MHO circle characteristic offers a direction characteristic with a constrained forward reach. The replica impedance magnitude determines how far along the protected feeder the relay will trip for a fault condition. The replica impedance angle is generally set slightly power than the protected line's angle to enhance the resistive reach [15].

The OFFSET MHO characteristic includes a deliberate reverse reach, and hence the relay will trip if a fault occurs that the relay sees in that area. This is attractive for several applications since the relay will trip for faults just behind the relaying point, i.e., in the feeder's substation. The reverse reach is defined by the reverse replica impedance ZR2 and fault further behind that will not result in tripping.

The IMPEDANCE characteristic trips for all faults where the magnitude of the fault impedance is less than the magnitude of the replica impedance used for the relay's setting. It is therefore a non-directional characteristic and trips equally for faults behind the relay as well as for those in front of it. An advantage of the IMPEDANCE relay is that it provides a greater resistive reach for faults close to the relay point than the MHO circle. In order to overcome the non-directional feature of the IMPEDANCE characteristic, a composite characteristic can be generated by combining the outputs of a comparator with an IMPEDANCE characteristic and one with a characteristic that has a directional feature [16].

Exercise 4.4
Suggest how a composite characteristic can be provided where the non-directional feature of the impedance characteristic can be overcome.

The MHO circle characteristic is inherently directional. Where an MHO characteristic is used for its directional properties, the forward reach is set much greater than the impedance of the protected line. This is illustrated in Fig. 4.10, which

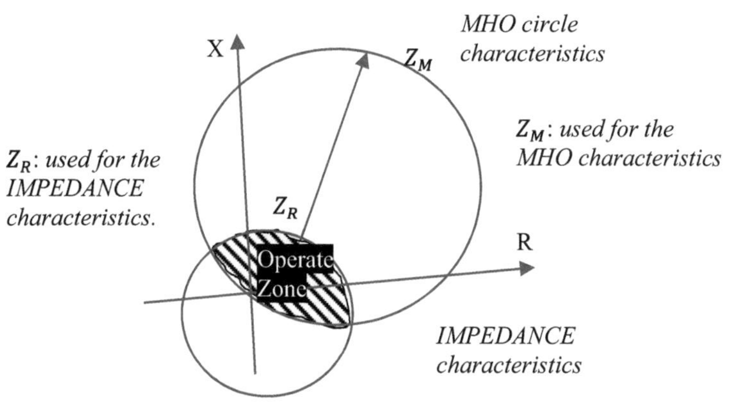

Fig. 4.10 Combined IMPEDANCE and MHO characteristics. Diagram illustrating impedance characteristics with two intersecting circles is used for impedance characteristics, while the Z_M circle represents MHO circle characteristics. The intersection area is marked as the "Operate Zone." Axes are labeled X and R

shows the combination of an IMPEDANCE characteristic to give the forward reach and an MHO circle characteristic to provide directional discrimination.

Considering the case where the reach of the MHO characteristic is increased to infinity, a straight-line characteristic is produced. This DIRECTIONAL characteristic and other linear characteristics can be generated in a manner similar to the production of circular characteristics [17].

The combination of the IMPEDANCE characteristic and a DIRECTIONAL characteristic is shown in Fig. 4.12a. In addition to providing circular characteristics, both amplitude and phase angle comparators can be used to provide linear characteristics [18].

The DIRECTIONAL characteristic, shown in Fig. 4.11a, provides a convenient method of determining whether the direction of current flow is in the forward or the reverse direction. In general, this indicates the direction of the fault as seen from the relaying point [19].

The REACTANCE characteristic, shown in Fig. 4.11b, provides an indication of whether or not the reactance of the system as seen from the relaying point is greater than the setting impedance or not. Since the impedance of a transmission line is predominantly reactive, this gives an indication as to where the fault is on the system [20].

The LOAD BLINDER characteristic, shown in Fig. 4.11c, is particularly important where high load conditions on a power system may cause the impedance seen by the relay to encroach into the trip characteristic and hence can be used to inhibit relay tripping. Singe-phase railway systems often have this problem, and hence a LOAD BLINDER characteristic relay can be used in conjunction with the main distance to prevent false trips [21]. In addition to the diagram for the three linear characteristics, Fig. 4.11 also shows the input measurands used to produce them when using cosine phase angle comparators. The DIRECTIONAL characteristic, for example, is produced using

$$S_1 = V$$

$$S_2 = IZ_R$$

and the trip criterion,

$$-90° \leq \angle S_1 / S_2 \leq 90°$$

The DIRECTIONAL characteristic can also be produced using an amplitude comparator with the input measurands:

$$S_O = V + IZ_R \tag{4.4}$$

$$S_R = V - IZ_R \tag{4.5}$$

and the trip criterion,

$$|S_O| > |S_R|.$$

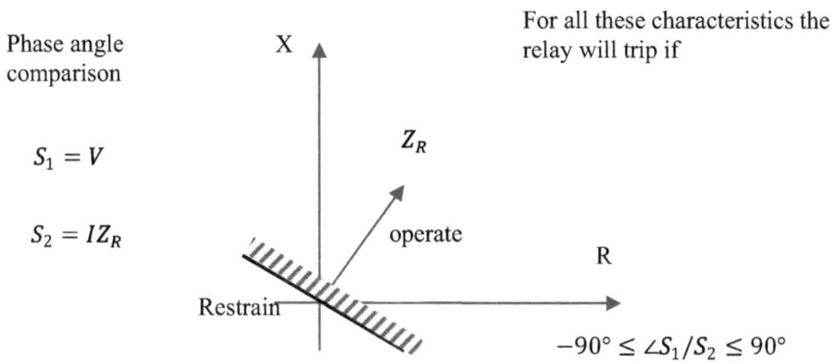

(i) The DIRECTIONAL characteristic - Diagram illustrating phase angle comparison with two equations: indicating a boundary between "Restrain" and "Operate" zones.

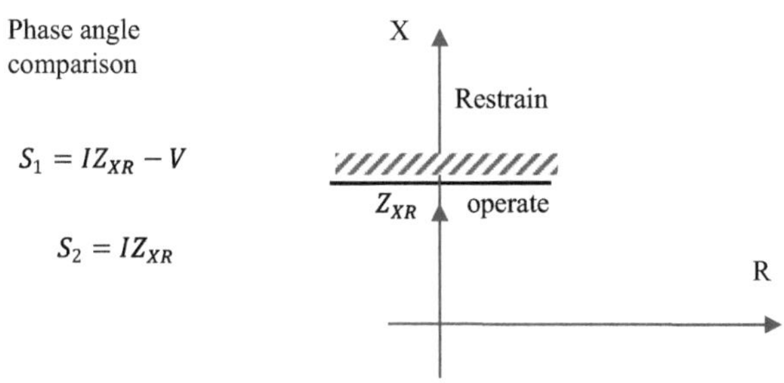

(ii) The REACTANCE characteristic- Diagram illustrating phase angle comparison with two equations: The graph features X and R axes, with a shaded region labeled ZXR. The area above the shaded region is marked "Restrain," and below is labeled "Operate."

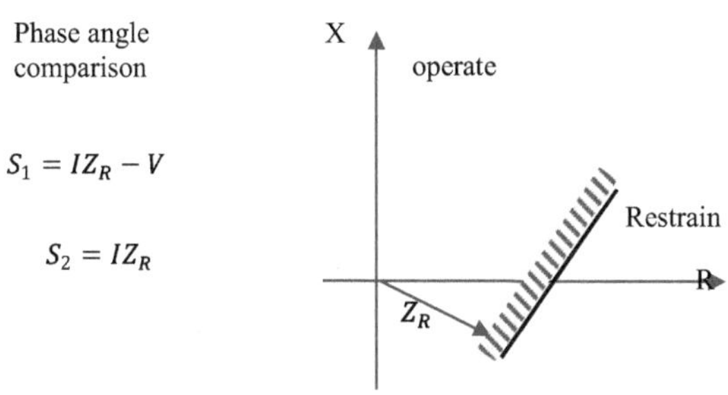

(iii) The LOAD BLINDER characteristic- Diagram illustrating phase angle comparison with two equations: It features an X and R axis, labeled "operate" and "restrain" respectively. A vector \(ZR\) is shown at an angle, indicating the restrain region.

Fig. 4.11 Linear characteristics obtained using phase angle comparison

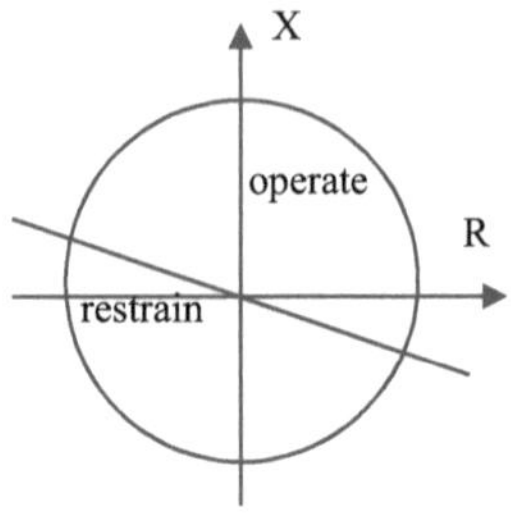

(i) Combined IMPEANCE and DIRECTIONAL characteristic

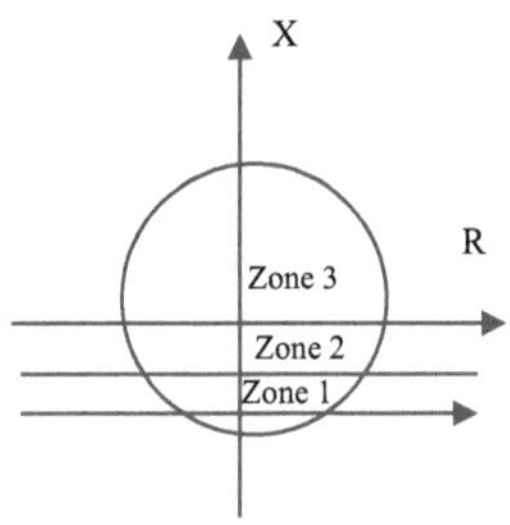

(ii) Three zone distance scheme using combined MHO and REACTANCE characteristic

Fig. 4.12 Examples of composite characteristics for distance relays

An immediate parallel can be drawn with the IMPEDANCE characteristic shown in Fig. 4.4. For an IMEPDANCE characteristic, the input measurands used with an amplitude comparator are:

$$S_O = IZ_R$$

$$S_R = V$$

and with a cosine phase angle comparator, they are:

$$S_1 = V - IZ_R$$

$$S_1 = V + IZ_R$$

Linear and circular characteristics are related, as are amplitude and cosine phase angle comparators.

In addition to using a single characteristic, a composite characteristic can be constructed, which builds on the advantages of some while overcoming their disadvantages. Figures 4.10 and 4.12 illustrate three of these composite characteristics, where in Fig. 4.10, a combination of IMPEDANCE and MHO characteristics is shown, in Fig. 4.12a, a combination of IMPEDANCE and DIRECTIONAL characteristics is shown, and in Fig. 4.12b, a combination of a MHO circle and several REACTANCE characteristics is shown, providing a three-zone distance relay configuration. This is described in more detail in Sect. 4.3 [22]. With the introduction of static relays in the 1960s, phase angle comparison techniques continued to be developed and provided the basis for most modern distance protection. Static relays offer the advantages of low burdens on the primary transducer, no moving parts, higher reliability and flexibility when compared with electromechanical designs, and, above all, reduced cost [23].

Work on static relays started in the early 1930s with valves, but these proved unsuitable for protection systems. With the introduction of transistors, static relays could offer the life required for protection systems, and the development of static distance relays began in the early 1950s.

For many years, the industry standard static phase angle comparator was the block average comparator, the block diagram of which is illustrated in Fig. 4.13. Using this comparator, the operating time in response to a fault for which the relay measurands were in phase was reduced to half a cycle of the power system frequency, while at the same time providing an infinite time to trip for faults on the characteristic's boundary. This was a result of integrating the feature of the comparator, which also ensured a high noise tolerance [24].

The squaring circuits convert the sinusoidal input phasors into square waves containing only polarity information. When the inputs to the coincidence circuit are of the same polarity, its output is high; when they are of differing polarity, its output is zero. The output is therefore a square wave of twice the power system frequency with a mark–space ratio dependent upon the phase angle between the two input phasors. This mark–space ratio is unity when the phase angle is ±90°. If the phase angle is less than 90°, the mark–space ratio will be greater than unity [25].

The integrator is biased such that if the input is high, it will ramp up, and if zero, it will ramp down. If the mark–space ratio is greater than unity, the integrator's output will ramp to the upper supply rail. If this is less than unity, the output will ramp to the lower supply rail. The reset and trip levels are commonly located at one-third and two-thirds of the range between the supply rail; typical internal waveforms are shown in Fig. 4.14.

When two phasors are in phase, the tripping time is a minimum at half a power frequency cycle. At the boundary of the relay characteristic where the phase angle

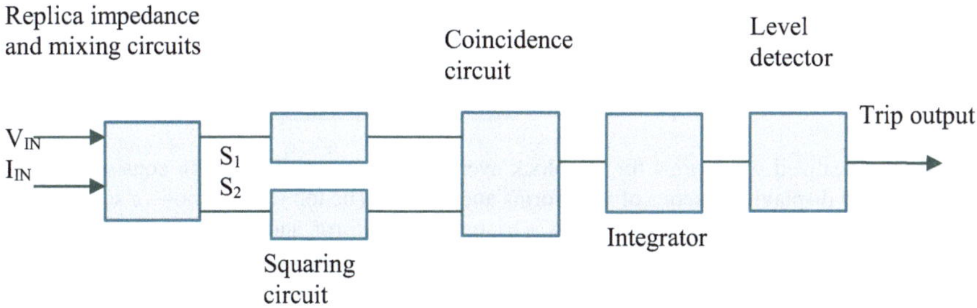

$$trip\ when -90° \le \angle\frac{S_1}{S_2} \le +90°$$

Fig. 4.13 The block average static distance relay comparator. Diagram of an electrical circuit system with labeled components. Enter the "Replica impedance and mixing circuits"

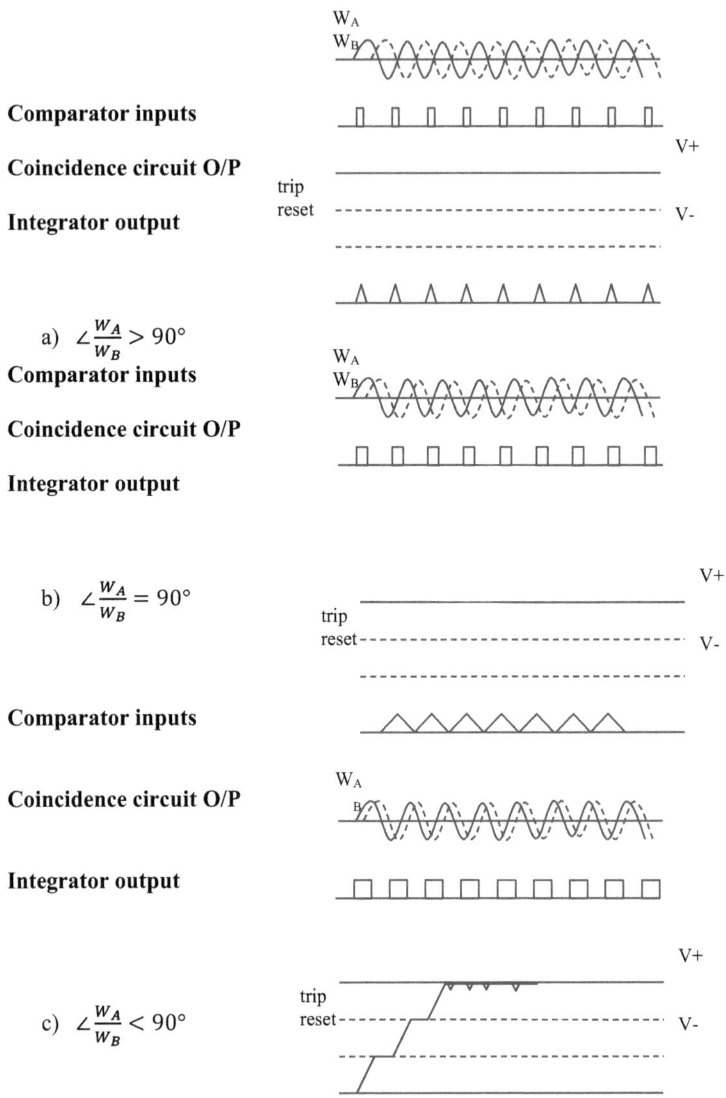

Fig. 4.14 Idealized waveforms for the block average comparator. The image consists of three sections, each displaying a series of waveforms and signals. The top section shows a square wave signal with constant "trip" and "reset" lines, a triangular waveform, and two sinusoidal waves. The bottom section includes a stepped waveform with increasing steps, a triangular waveform, and sinusoidal waves. Each section is labeled with "V+" on the right side

is 90°, the block average comparator requires an infinite time to trip. This ensures that there is the greatest security against malfunctioning, and hence the comparator provides excellent discrimination between faults on the protected plant and elsewhere on the network [26].

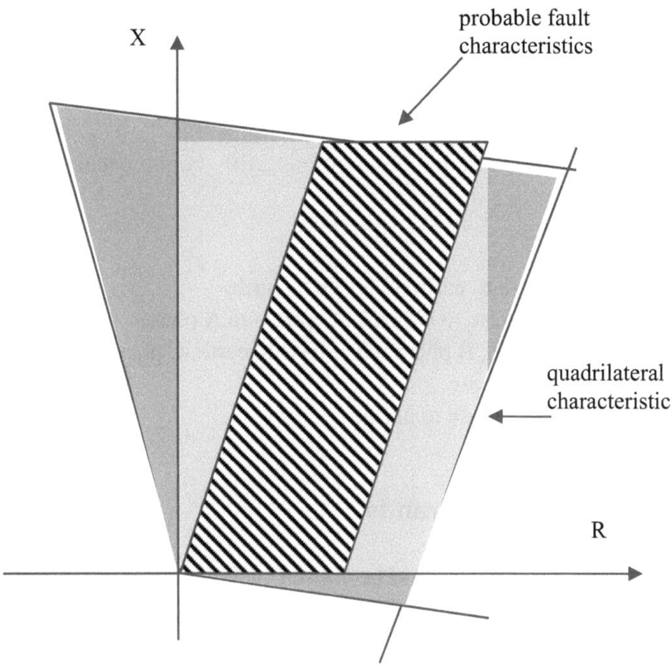

Fig. 4.15 The quadrilateral characteristic. Diagram showing a geometric representation with axes labeled X and R. A shaded quadrilateral area is labeled "quadrilateral characteristic," and a striped area within it is labeled "probable fault characteristics." The diagram illustrates relationships between these elements

The introduction of digital techniques and the desire for shorter trip times have led to the development of several other comparators. Also, the desire to produce a closer match between the relay's trip characteristic and probable fault impedance area has led to the introduction of quadrilateral characteristics. These can be realized in a variety of ways using modified comparators and combinations of linear and circular characteristics. The quadrilateral characteristic is shown in Fig. 4.15. To ensure high-speed tripping for faults on the replica impedance, coverage on both sides of the protected feeder impedance is desirable.

The impedance of distance relays using microprocessor-based hardware is discussed in Chap. 7.

4.3 Three-Phase Distance Protection

In the foregoing, distance protection has been presented using a single-phase representation. However, most power supply systems are three-phase. For these and other poly-phase systems, separate measurements are used to detect faults involving the different possible fault combinations.

With a three-phase system, there are 11 possible short circuit fault conditions, and each presents a different scenario for the relay to assess [27].

Exercise 4.5
Identify the 11 possible fault conditions with could occur on a three-phase system.

A phase to earth, B phase to earth, C phase to earth.
A phase to B phase, B phase to C phase, C phase to A phase.
A phase to B phase to earth, B phase to C phase to earth, C phase to A phase to earth.
A phase to B phase to C phase.
A phase to B phase to C phase to earth.

4.3.1 Relay Input Measurands

Ideally, the input measurands would be chosen such that, irrespective of the type of fault, the impedance seen by the distance relay will be dependent upon the line parameters and the position of the fault along the transmission line.

Unfortunately, this cannot be done by a single device. However, the 11-fault combination above can be covered by six combinations of input measurands, three for phase-to-earth faults. The comparator directly involved with the fault will see an impedance representing the impedance of the system between the relaying point and the fault.

The three combinations of input measurands used to monitor for phase-to-phase faults, phase-to-phase-to-earth faults, and three-phase faults are as follows:

Phase a to phase b faults:

$$V_{ab} = V_a - V_b \quad I_{ab} = I_a - I_b$$

Phase b to phase c faults:

$$V_{bc} = V_b - V_c \quad I_{bc} = I_b - I_c$$

Phase c to phase a fault:

$$V_{ca} = V_c - V_a \quad I_{ca} = I_c - I_a$$

The impedance derived from these measurands is the positive phase sequence impedance of the system between the relaying point and the fault.

The three combinations of input measurands used to monitor for phase to earth faults are as follows:

Phase a to earth faults

$$V_{aN} = V_a \quad I_{aN} = I_a + K_N I_N$$

Phase b to earth faults

$$V_{bN} = V_b \quad I_{bN} = I_b + K_N I_N$$

Phase c to earth faults

$$V_{cN} = V_c \quad I_{cN} = I_c + K_N I_N$$

The current I_N (the residual current) is the effective earth current derived from the summation of the three-phase currents:

$$I_N = I_a + I_b + I_c$$

The value of K_N (the residual compensation factor) is chosen so that the impedance derived from the phase-to-earth measurands is also the positive phase sequence impedance of the system between the relaying point and the fault.

Exercise 4.6

Derive the equation for the residual compensation factor KN such that the impedance seen, Z_{aN} for an "a" phase-to-earth fault is the positive phase sequence impedance of the transmission line to the fault. Use symmetrical components to analyze this fault condition and the impedance diagram shown in Fig. 4.16.

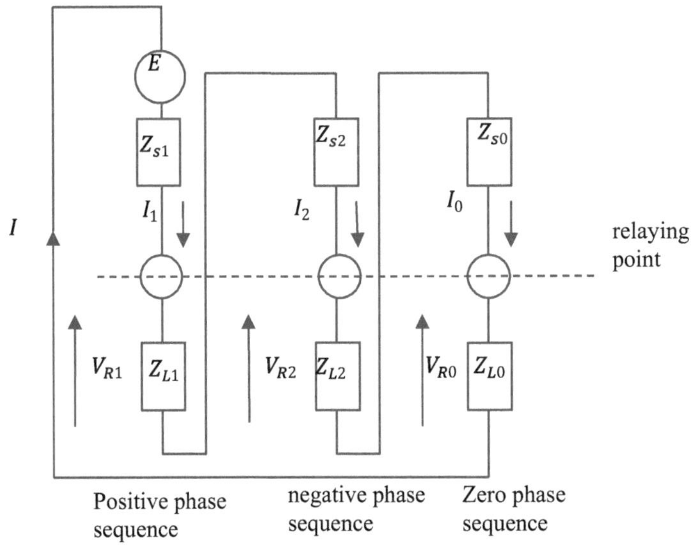

Fig. 4.16 Symmetrical components diagram for a single-phase to earth fault. Diagram illustrating a three-phase power system with positive, negative, and zero phase sequences. The positive phase sequence includes three phasors that are equal in magnitude and displaced by 120 Deg with the same phase sequence as the original system. The negative phase sequence shows unbalanced in three phases. The zero-phase sequence contains three components that are equal in magnitude and in phase. A relaying point is indicated by a dashed line. Arrows depict the direction of current flow

Using symmetrical components to analyze a single-phase to ground fault gives the network diagram shown in Fig. 4.16:

$$I_1 = I_2 = I_0 = I \text{ and } I_0 = I_N$$

$$I = E / \left(Z_{S1} + Z_{L1} + Z_{S2} + Z_{L2} + Z_{S0} + Z_{L0} \right)$$

$$V_{R1} = IZ_{L1}$$

$$V_{R2} = IZ_{L2}$$

$$V_{R0} = IZ_{L0}$$

From symmetrical component theory, the voltage at the relaying point V_a is given by:

$$V_a = V_{R1} + V_{R2} + V_{R0}$$

Hence

$$V_a = I_1 Z_{L1} + I_2 Z_{L2} + I_0 Z_{L0}$$

For a transmission line,

$$Z_{L1} = Z_{L2}$$

Hence

$$V_a = I_1 Z_{L1} + I_2 Z_{L2} + I_0 Z_{L0}$$

$$V_a = I_1 Z_{L1} + I_2 Z_{L1} + I_0 Z_{L1} - I_0 Z_{L1} + I_0 Z_{L0}$$

$$V_a = Z_{L1} \left(I_1 + I_2 + I_0 \right) + I_0 \left(Z_{L0} + Z_{L1} \right)$$

$$V_a = I_a Z_{L1} + I_0 \left(Z_{L0} + Z_{L1} \right)$$

where I_a is the current measurand at the relaying point, i.e.,

$$I_a = I_1 + I_2 + I_0$$

$$V_a = I_a \, Z_{L1} + \frac{I_N}{3} \times Z_{L1} \left(\frac{Z_{L0}}{Z_{L1}} - 1 \right)$$

Thus, defining the residual compensation factor, $K_{N,}$ as

$$K_N = \frac{1}{3} \left(\frac{Z_{L0}}{Z_{L1}} - 1 \right) = \frac{1}{3} \left(\frac{Z_{L0-} Z_{L1}}{Z_{L1}} \right)$$

And substituting into the above equation for Va gives

$$V_a = I_a Z_{L1} + I_N Z_{L1} \times K_N$$

i.e.,

$$Z_{L1} = \frac{V_a}{I_a + K_N I_N}.$$

Using these measurands, the impedance seen by the different relaying elements can be expressed as follows:

$$Z_{ab} = \frac{V_{ab}}{I_{ab}} = \frac{V_a - V_b}{I_a - I_b}$$

$$Z_{bc} = \frac{V_{bc}}{I_{bc}} = \frac{V_b - V_c}{I_b - I_c}$$

$$Z_{ca} = \frac{V_{ca}}{I_{ca}} = \frac{V_c - V_a}{I_c - I_a}$$

$$Z_{aN} = \frac{V_{aN}}{I_{aN}} = \frac{V_a}{I_a + K_N I_N}$$

$$Z_{bN} = \frac{V_{bN}}{I_{bN}} = \frac{V_b}{I_b + K_N I_N}$$

$$Z_{cN} = \frac{V_{cN}}{I_{cN}} = \frac{V_c}{I_c + K_N I_N}$$

To provide full protection for a three-phase system requires six comparators, three monitoring for phase-to-phase fault and three monitoring for phase-to-earth fault. The operating characteristics for each comparator are defined by the choice of comparator input voltage and current phasors and the mixing circuits used to define the trip characteristic's shape [28]:

$$Z_1 = 0.177 + j0.402\,\Omega / km$$

$$Z_2 = 0.177 + j0.402\,\Omega / km$$

$$Z_0 = 0.354 + j1.022\,\Omega / km$$

$$CT\ ratio\ 600\,A / 1A$$

$$VT\ ratio\ 132kV / 110\,V$$

Exercise 4.7

Calculate the residual compensation setting for a distance relay used to protect a 70 km, 132 kV transmission line with primary impedance and transformer ratio.

The residual compensation setting is derived from the positive sequence Z_1 and the zero-sequence impedance Z_0 using the equation:

$$K = \frac{(Z_0 - Z_1)}{3Z_1}$$

$$K = \frac{(0.354 + j1.022) - (0.177 + j0.402)}{3(0.177 + j0.402)}$$

$$K = 0.485 + j0.067$$

$$K = 0.49\angle 8^\circ$$

The residual compensation setting is therefore o.49. Its phase angle is generally neglected for overhead line circuits since it is small. For cable circuits, where the angle is greater, an angular setting is advantageous.

The calculation of the residual compensation setting relies on the ratio of sequence impedance and hence is independent of the line length, its operating voltage, and, assuming that ideal transformers produce the same value irrespective of whether primary or secondary impedances are used.

Relay Input Measurands to Provide the MHO Circle Characteristic

For the phase-to-phase fault comparators, the choice of comparator inputs can be illustrated by considering the "a" phase to "b" phase comparator. To provide an MHO circle characteristic using phase angle comparison, as shown in Fig. 4.9, the comparator inputs would be:

$$S_1 = I_{ab} Z_R - V_{ab}$$

$$S_2 = V_{ab}$$

For the phase-to-earth fault comparators and considering the "a" phase-to-earth comparator, the MHO circle characteristic is obtained using inputs:

$$S_1 = I_{aN} Z_R - V_{aN}$$

$$S_2 = V_{aN}$$

Note that I_{aN} includes the residual compensation. Since using these inputs, the impedance seen by the comparators is the positive phase sequence of the faulted system, the reach of the MHO circle Z_R can also be referred to the positive phase sequence impedance of the protected system [33].

Close-Up Fault Protection

For faults close to the relaying point, difficulties may arise with the MHO circle characteristic when the relay's input voltage falls to zero or nearly to zero. The MHO circle, correctly referred to as the self-polarized MHO circle characteristic, relies on the voltage signal to provide a reference to determine whether or not the fault is inside the trip characteristic. In the absence of the voltage input, the comparator cannot decide whether or not to trip and may malfunction. The risk of false operations for a fault close to the relaying point, but behind it, is generally not acceptable.

Similar difficulties also exist for the directional and idealized quadrilateral characteristic passes through the origin of the impedance diagram; it does not arise, however, for characteristics that either have a reverse reach and hence include the origin and will trip for voltage collapse, or those that are offset away from the origin and will therefore restrain for a zero voltage fault. For those characteristics that suffer from this problem, the difficulties can be overcome using several methods, either alone or in combination [29]. These techniques include:

- The stability notch.
- Healthy phase cross-polarization.
- Memory polarization.

The Stability Notch

The stability notch is the simplest technique to ensure that a relay does not trip incorrectly for voltage collapse. By inhibiting the comparator's operation whenever the input voltage falls below a present value, any indecision within the comparator is avoided. This, however, introduces the problem that the relay will not trip for a fault at or near the relay point. The setting of the threshold value is therefore kept as low as possible and is typically between 0.5% and 1% of the relay's nominal voltage input level. The effect of the stability notch when used with an MHO circle characteristic is shown in Fig. 4.17. The stability notch has been used on several commercial relays, a particularly low-cost relay.

Healthy Phase Cross-Polarization

A more effective technique, which avoids the need to inhibit the comparator when the faulted phase input voltage collapses, is to use a measured value derived from healthy phase or phases that are not by definition involved with the fault. This

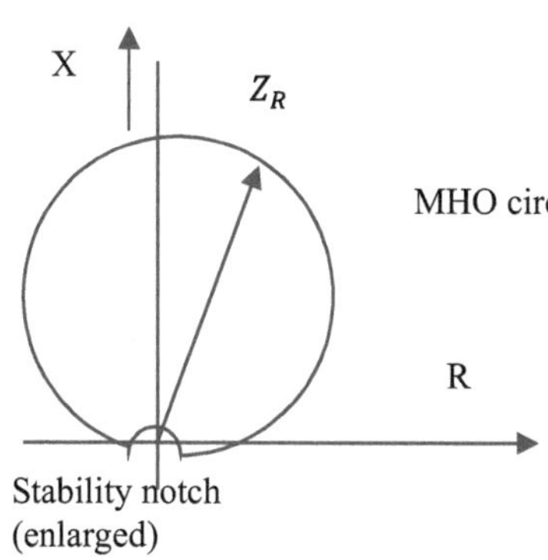

Fig. 4.17 Stability notch applied to MHO circle characteristic. Diagram illustrating an MHO circle with labeled axes. The horizontal axis is marked as R, and the vertical axis is marked as X. An arrow labeled \(Z_R\) points diagonally within the circle. A small section at the bottom of the circle is labeled "Stability notch (enlarged)"

technique may use the voltage derived from the unfaulted phases alone to provide to reference input, referred to as fully cross-polarized, or a percentage of the health phase input together with the voltage from the faulted phase, referred to as partially cross-polarized. Using phase comparison and the MHO circle characteristic, the full cross-polarized MHO characteristic is produced by using the following inputs to the comparator [30]:

$$S_1 = IZ_R - V$$

$$S_2 = V_{POL}$$

where V_{POL} is the polarizing voltage derived from the health phases or phases. The polarizing voltage is chosen such that under nonfaulted conditions, it is in phase with the protected phase or phases' voltage input. The choice of polarizing voltage can be illustrated by considering an "a" phase-to-earth comparator, where the polarizing voltage would be:

$$V_{POLaN} = V_{bc}\angle 90^\circ$$

Exercise 4.8
Suggest which polarizing voltages could be used for the six comparators used to provide full protection for a three-phase transmission system.

Comparator → 'aN'

$$V_{POL} = V_{bc}\angle 90^\circ$$

Comparator → 'bN'

$$V_{POL} = V_{ca}\angle 90^\circ$$

Comparator → 'cN'

$$V_{POL} = V_{ab}\angle 90^\circ$$

Comparator → 'ab'

$$V_{POL} = V_{cN}\angle 90^\circ$$

Comparator → 'bc'

$$V_{POL} = V_{aN}\angle 90^\circ$$

Comparator → 'ca'

$$V_{POL} = V_{bN}\angle 90^\circ$$

Healthy phase cross-polarization ensures correct comparator operations for faults close to the relaying point by expanding the MHO circle characteristic to include the origin when the fault is in front of the relay. In addition to reinforcing the directional capabilities of the MHO circle, it also expands the circle to provide additional resistive reach. This helps overcome a second criticism of the MHO circle, particularly for applications involving short transmission lines. The expansion

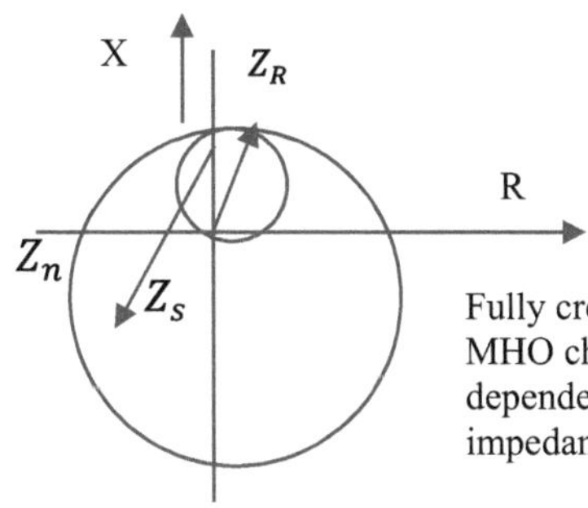

Fig. 4.18 MHO circle characteristic expansion produced by full cross-polarization. Diagram illustrating MHO characteristics with two circles on an X and R plane. The left text reads "Self-polarized MHO characteristic," and the right text states "Fully cross-polarized MHO characteristic Zn is dependent on source impedance"

of the MHO circle using full cross-polarization compared to the self-polarized MHO circle is illustrated in Fig. 4.18.

When the fault is behind the relaying point, the reinforcement of the characteristic by the polarizing input is changed to a reduction of the characteristic, and in effect, the tripping area collapses. The expansion shown in Fig. 4.18 therefore only exists for forward faults.

On occasion, the expansion of the MHO circle can lead to other difficulties when the expanded characteristic overlaps the possible load area, or where single-pole tripping is required, it can include the impedances seen by other comparators not directly involved with the fault and hence result in healthy phase tripping. For this situation, partial cross-polarization is more appropriate. For partial cross-polarization, the polarizing input is derived from the self-polarized input together with a percentage of the cross-polarized input.

The level of polarization used depends on the relay being used and has generally been chosen empirically. Since less than 100% polarization is used, the expansion of the characteristic is reduced accordingly when compared to the fully cross-polarized relay [31].

The expansion of the characteristic is governed by the source impedance of the system feeding the protected transmission line. If the source impedance were zero, there would be no expansion of the characteristic, and the cross-polarized characteristic, whether full or partial, would be the same as the self-polarized characteristic. There is also no expansion of the characteristic for three-phase faults where there is comparable voltage collapse on all phases. The use of memory polarization overcomes this difficulty.

Memory Polarization

Memory techniques use the pre-fault voltage to provide a polarizing voltage, which can be used as a reference signal after voltage collapse. It has been used on many distance relays, including electromechanical, static, digital, and microcomputer designs. The early memory elements used tuned circuits, tuned to the power system frequency and driven by the input voltage signal. The input voltage acted as a forcing function, which, when removed, enabled the memory circuit to provide an output at the power system frequency and in phase with the pre-fault phasor. When used as a polarizing input to the comparator, they produced an expansion of the MHO circle of forward-looking faults similar to that produced by health phase cross-polarization.

With digital and microcomputer design, memory circuits have been used, which provide a square-wave output synchronized to the pre-fault input voltage. These are more successful at tracking any changes in the power system frequency while at the same time retaining the phase relationship that existed before the fault [32].

4.3.2 Load Encroachment and Power Swings

Distance relays rely on the impedance seen by the relay for their operation. If the impedance that is seen lies within the relay trip characteristic, the relay will trip whether or not it is a result of a fault condition or not. The load fed by the transmission line will appear to the relay as a non-infinite impedance, and care must be taken to ensure that the load presents to the relay. Power swing, resulting from generators swinging with respect to each other to achieve a new stability, also presents impedance to the relay, which can cause incorrect operations of the protection.

Load Encroachment

Load encroachment describes the condition where the impedance seen by the relay due to high load levels encroaches into the trip characteristic and causes tripping. This is a particular difficulty for back-up protection for long transmission line systems, where the reactive impedance of the system requires a high reach setting. Since the power system is operated as a balanced system, this difficulty is a problem for the phase-to-phase comparators and not for those concerned with phase-to-earth faults. To reduce losses in the transmission line, the power factor of the load current is corrected such that it is close to a unity power factor. The power system is typically operated at power factors between ±0.85, corresponding to impedance angles varying approximately between ±30°. At unity power factor, the impedance seen by the relay is purely resistive. For a power transmission line carrying a power of M *MVA* and operating at a line voltage of V volts, the magnitude of the primary system impedance will be:

$$\left|Z_{primary}\right| = \frac{V^2}{M}$$

The impedance seen by the relay is the secondary impedance given by:

$$\left|Z_{secondary}\right| = \frac{V^2}{M}\frac{CT}{VT}$$

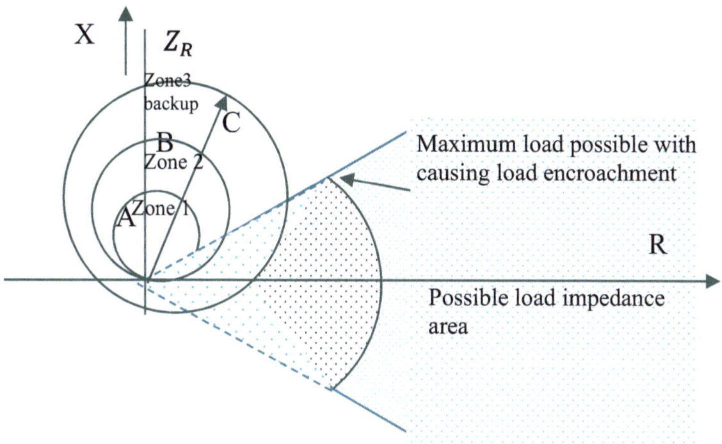

Fig. 4.19 Load encroachment into the distance relay characteristic. Diagram illustrating load impedance zones on a graph with X and Z_R axes. Three concentric circles labeled zone 1, zone 2, and zone 3 backup are centered at the origin. A shaded area extends from the origin, labeled "Possible load impedance area." An arrow points to the boundary of this area, indicating "Maximum load possible with causing load encroachment"

where the primary current transformer ratio, CT, is defined as the ratio of primary current to secondary current, and the primary voltage transformer ratio, VT, is defined as the ratio of the primary voltage to secondary voltage.

The angle, ∅, of this impedance is derived from the power factor, pf, i.e.,

$$\varnothing = arc\cos\left(pf\right)$$

The effects of the load impedance are illustrated in Fig. 4.19, which shows that the maximum load possible before load encroachment will cause tripping for a system using an OFFSET MHO zone three characteristic for back-up protection [33].

Exercise 4.9

Calculate the maximum load which can be carried by a 132 kV transmission line at a power factor of 0.866 before load encroachment problems could occur with an MHO circle zone 3 characteristic se to 50 $\Omega \angle 70°$ given that the primary transducer ratios are:

$$CT : 600\,A / 1A$$

$$VT : 132\frac{kV}{110}V$$

The position described in this question is similar to that shown in Fig. 4.19, except that the zone 3 characteristic is an MHO circle. The angle θ, corresponding to the maximum possible load without causing load encroachment, is given by the power factor where:

$$\theta = \cos^{-1}(pf)$$

$$\theta = \cos^{-1}(0.866)$$

$$\theta = \pm 30^\circ$$

The boundary of the MHO circle is defined by the impedance Z_B where:

$$Z_B = Z_R \cos(\theta_B - \theta_R)$$

$$Z_R = reach\ setting\,(\Omega)$$

$$\theta_R = angle\ setting\,(^\circ)$$

$$\theta_B = angle\ at\ the\ boundary\,(^\circ)$$

Therefore, at 30°

$$Z_B = Z_R \cos(30^\circ - \theta_R)$$

$$Z_B = 50\cos(30^\circ - 70^\circ)$$

$$Z_B = 50\cos(-40^\circ)$$

$$Z_B = 38.3\Omega$$

The maximum load that can be carried by the line is therefore equivalent to an impedance as seen by the relay of:

$$Z_B = 38.3\Omega\angle 30^\circ$$

This impedance appears on the secondary side of the primary transducer. To convert it to a primary equivalent impedance:

$$Z_{primary} = \frac{VT}{CT} Z_{secondary}$$

$$\therefore (Z_B)_{primary} = \frac{VT}{CT} Z_B$$

$$(Z_B)_{primary} = \frac{132\,000}{110} \times \frac{1}{600} \times 38.3\Omega$$

Converting this to an equivalent power level M:

$$M = \frac{V^2}{Z}$$

$$M = \frac{132\,000 \times 132\,000}{76.6} VA$$

$$M = 227.5\,MVA$$

The maximum load that the line can carry at a power factor of 0.866 and avoid load encroachment with the zone 3 characteristic is 227.5 MVA.

Power Swings

Power swings are caused by system disturbances where the generators swing with respect to each other to achieve a new equilibrium point. They produce abnormal phase shifts between the voltage connected to the transmission system and impedance seen by the distance relays, which can cut across the trip characteristic.

For a distance relay, it is important that it can distinguish between a power swing and a fault condition. Power swings are operating phenomena, and although they can confuse a distance relay, relay operations may cause further problems.

The relay's response to a power swing can assist in operating the system. For intertie circuits connecting two major utilities, it can be advantageous for the link to be broken during periods of instability, and hence the relay's malfunctioning may be welcome since it provides an automatic means to break the circuit. On other occasions, breaking the system could be unfortunate, and relay operations should be blocked. Power swing detection is often available in a distance protection scheme, and since it can be used to block the relay's operation, it is generally referred to as power swing blocking.

The effects of a power swing on the impedance seen by a distance relay can be illustrated by considering the simple network shown in Fig. 4.20. Since a power swing is a three-phase balanced phenomenon, only the positive phase sequence network needs to be considered [34]. Analyzing the network shown in Fig. 4.20

$$E_A - E_B = I_R\left(Z_{SA1} + Z_L + Z_{SB1}\right)$$

$$V_R = E_A - I_R Z_{SA1}$$

The impedance seen by all of the comparators in the relay reduces to:

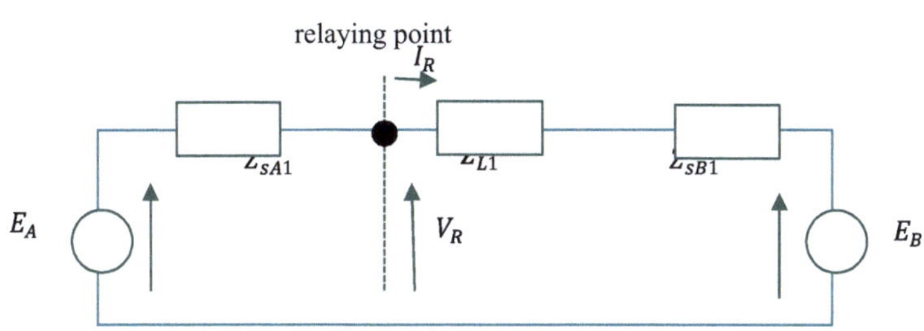

Fig. 4.20 Positive phase sequence network to illustrate power swings and the impedance they present to a distance relay. Diagram of an electrical circuit showing two voltage sources connected in series with three impedance blocks. A relaying point is marked with the current indicated by arrows

$$Z_R = \frac{V_R}{I_R}$$

Therefore

$$Z_R = \frac{V_R}{I_R} - Z_{SA1}$$

Defining the system impedance Z_s as

$$Z_S = Z_{SA1} + Z_L + Z_{SB1}$$

and substituting I_R from.

$$E_A - E_B = I_R Z_{S,}$$

we obtain

$$Z_R = \frac{\left(\frac{E_A}{E_B}\right) Z_s}{\left(\frac{E_A}{E_B}\right) - 1}$$

This defines a family of orthogonal circles focused on the sources of supply E_A and E_B and defined by the ratio of the amplitudes of the two generator voltages and the phase angle between them. Considering the situation illustrated in Fig. 4.20, when the amplitudes of the generator voltages are the same and their source impedances are similar, the impedance seen by the relay would correspond to that for a valid fault at the midpoint of the transmission line when the two generators are 180°apart. Under this condition, a distance relay protecting that line would trip. Illustration of the possible impedance trajectories seen by the distance relay located at the relaying point of Fig. 4.20 is shown in Fig. 4.21.

Power Swing Detection and Blocking

Power swing detection and blocking is a frequently used and optional feature of a distance protection scheme. The key difference between a fault condition and a power swing is that, during a fault, the post-disturbance impedance is relatively fixed, whereas, during a power swing, the impedance seen by the relay follows that swing, which is a relatively slow phenomenon. The impedance trajectory of the impedance seen by a distance relay is shown in Fig. 4.22.

The method used to differentiate a power swing from a fault condition is: to use two similar characteristics, where one is larger than and includes the other, and measure the difference between the times when they trip. This time is then compared with a preset timer. For a fault condition, they should almost trip together; however, for a power swing, there will be a time delay between the tripping of the outer and inner characteristics. In Fig. 4.22, a typical situation is shown where the outer characteristic is a dedicated offset MHO and the inner characteristic is the scheme's zone 3 characteristic. The choice of time setting is not critical, provided that it is greater than the operating times of the comparators and is typically between

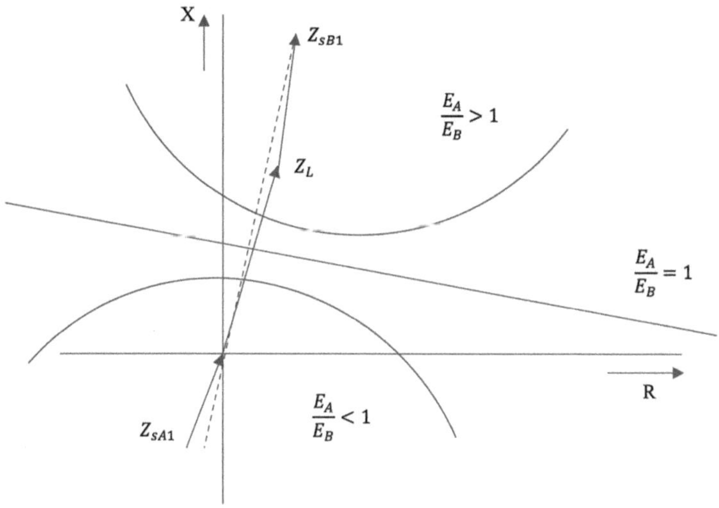

Fig. 4.21 Possible impedance trajectories seen by a distance relay during power swings. A graph with two axes labeled X and R, featuring curves and lines. There are vectors intersecting the curves. The graph illustrates relationships between these variables

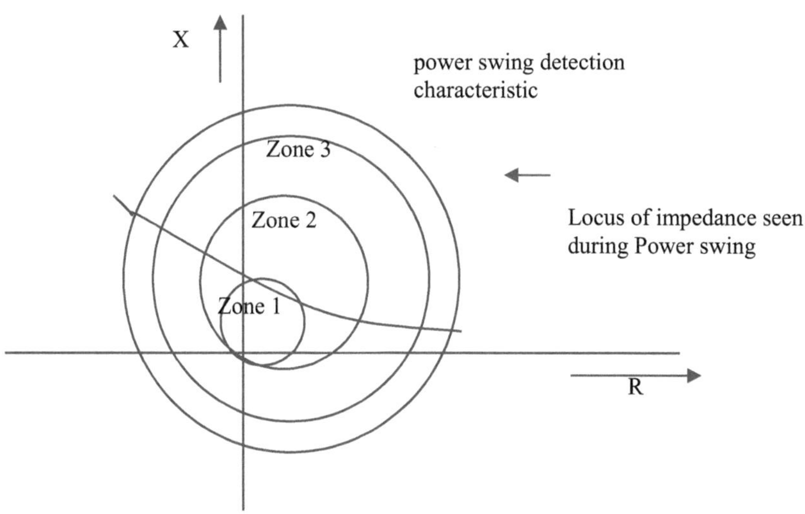

Fig. 4.22 Power swing detection using distance relays. Diagram illustrating power swing detection characteristics with three concentric circles labeled zone 1, zone 2, and zone 3. The circles are centered on a graph with X and R axes. A curved line crosses through the zones, labeled as the "Locus of impedance seen during Power swing." An arrow points to the right, indicating the direction of the power swing detection characteristic

50 and 100 ms. The response to a power swing is a matter for the power system operators since it is not a fault condition. The relay is therefore fitted with a setting to allow the tripping to be inhibited, i.e., blocked or not, when a power swing is detected. The normal settings are as follows:

Block zone one during a power swing	OR	NOT
Block zone two during a power swing	OR	NOT
Block zone three during a power swing	OR	NOT

Traditionally, these settings have been provided using links or switches in the relay; however, with the introduction of microprocessor relays, software switches are also used.

4.4 Three-Phase Distance Protection Schemes

4.4.1 Three-Zone Stepped Distance

For a fault inside the relay's operating characteristic, a time-independent distance relay will trip "instantaneously." The reach of the relay will be defined by the operating characteristic and by the setting of the replica impedance.

In order to protect the immediate feeder of a multi-feeder transmission system properly, as shown in Fig. 4.23, the relay's characteristic and setting should be chosen such that the relay trips only for faults on the protected feeder and restrains for all faults on other parts of the system. Ideally, the relay would be set to protect 100% of the immediate feeder, but, because of errors, this value must be reduced to prevent malfunctioning.

The measurement used for the decision-making depends upon the accuracy of the VTs and CTs, the relay, and the knowledge of the power system impedance, allowing for 5% error in each of these; the setting of the relay has to be reduced to

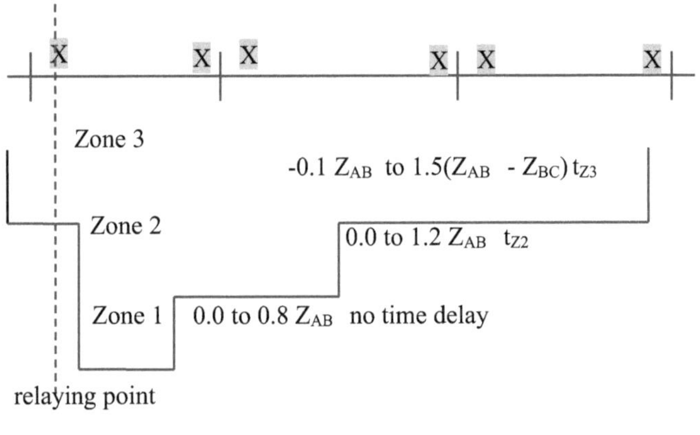

Fig. 4.23 Three-zone steeped distance protection. Diagram illustrating three protection zones for a relay system. The relaying point is marked on the left

80% of the protected feeder's reach, thus preventing possible relay tripping for faults beyond that feeder. The relay's zone one reach is therefore set to 80% of the positive phase sequence impedance of the protected line [33].

Exercise 4.10

Assuming that the errors in the VTs, CTs, the relay, and the power system parameters are all ±5%, determine the range of the relay's actual zone-one reach when sent to protect 80% of the immediate feeder. Use a simplified straight-line concept of system impedances and reach and neglect the two-dimensional nature of impedances

Taking a simplistic approach to the situation, the relay will trip when the percentage of the protected line impedance seen by the relay is less than the percentage reach setting of the relay, i.e.:

$$P_{SEEN} < P_{REACH\ EFFECTIVE}$$

Ideally, the percentage of the protected line impedance seen by the relay is equal to the actual percentage of the protected line to the fault position. In reality, it is affected by the errors introduced by the primary voltage and current transformers, E_V and E_C, i.e.:

$$P_{SEEN} = P_{ACTUAL} \frac{1+E_V}{1+E_C}$$

Ideally, the effective percentage reach setting of the relay is derived from the ratio of the relay's impedance setting to the calculated impedance of the protected line. In reality, this is affected by the errors in the relay and the line data, E_R and E_L, i.e.:

$$P_{REACH\ EFFECTIVE} = Z_{RELAY} \frac{1+E_V}{(1+E_L)Z_{LINK}}$$

$$P_{REACH\ EFFECTIVE} = P_{REACH} \frac{1+E_R}{1+E_L}$$

Therefore, returning to the equation for the trip decision and substituting the above:

$$P_{SEEN} < P_{REACH\ EFFECTIVE}$$

$$P_{ACTUAL} \frac{1+E_V}{1+E_C} < P_{REACH} \frac{1+E_R}{1+E_L}$$

$$P_{ACTUAL} < P_{REACH} \frac{(1+E_R)(1+E_C)}{(1+E_L)(1+E_V)}$$

With a reach setting of 80% and errors of 5% for EV, EC, ER, and EL, the range of the percentage of the line protected by the relay varies between:

$$P_{ACTUAL\,MIN} = \frac{80 \times 0.95 \times 0.95}{1.05 \times 1.05} = 65.5\%$$

and

$$P_{ACTUAL\,MAX} = \frac{80 \times 1.05 \times 1.05}{0.95 \times 0.95} = 97.7\%$$

In the worst case, only 65.5% of the line is protected, and at most 97.7%. The relay will therefore not overreach; however, zone 1 tripping may only apply for nearly two-thirds of the line. Protection for the remaining 20% of the feeder, the end zone, is provided by using a deliberately time-delayed zone 2 characteristic. Allowing for the same errors as before, this characteristic is typically set to 120% of the protected feeder's reach. The time delay is chosen such that if the fault were on the adjacent feeder, the protection for that feeder would have sufficient time to operate and clear the fault before the local relay trips. This time delay is typically set in the range from 0.05 to 2.0 s, with 0.2 s being a common choice. Overall, back-up protection is provided by a time-delayed zone 3 characteristic. Depending upon the system and the type of relay used, this is typically set to provide a reverse reach of 10% of the protected feeder and a forward reach equivalent to 150% of the impedance of both the protected and the adjacent feeder. The time delay is chosen to allow sufficient time for all the relevant protection to operate before the backup protection trips and is typically set in the range 0.1–5.0 s.

The operating characteristic for this type of protection based on MHO and OFFSET MHO relays together with those for the relays protecting the adjacent feeder is shown in Fig. 4.24. Using these overlapping zones ensures the rapid and accurate clearance of faults on the protected system. The only limitation of the scheme is the delayed tripping for faults in the end zone of the protected feeder.

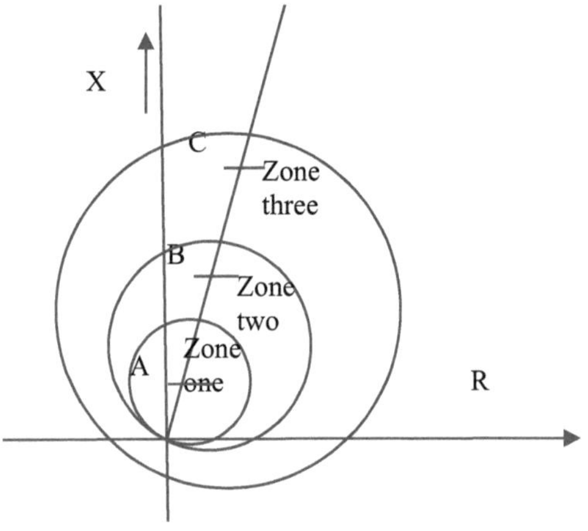

Fig. 4.24 .Relay characteristic for three-zone stepped distance protection. Diagram illustrating three concentric circles labeled as zone 1, zone 2, and zone 3, intersected by two lines. The horizontal axis is labeled R, and the vertical axis is labeled X. Points A, B, and C are marked on the circles, corresponding to each zone

This, however, is a cost incurred due to the errors associated with the measurement involved.

Exercise 4.11

For the 132 kV network shown in Fig. 4.25, calculate the zone 1, zone 2, and zone 3 settings for a three-zone distance relay, R, using MHO, MHO, and OFFSET MHO characteristics, respectively.

The settings required for the phase-to-phase fault protection are determined by the percentage reach requirements and the line's positive phase sequence impedance. For phase-to-earth protection, the residual compensation coefficient is also required and is derived from the line's positive and zero sequence impedances.

The impedances seen by the relay are referred to the secondary side of the primary transducers. The impedances given refer to the primary system. In the following, subscript "p" refers to primary values and subscript "s" to secondary values:

$$\textit{Current transformer ratio}: CT = 600:1$$

$$\textit{Voltage transformer ratio}: VT = 132\,000:110$$

$$CT = \frac{I_P}{I_S} \quad I_P = CT\, I_S$$

$$VT = \frac{V_P}{V_S} \quad V_P = CT\, V_S$$

$$Z_P = \frac{V_P}{I_P}$$

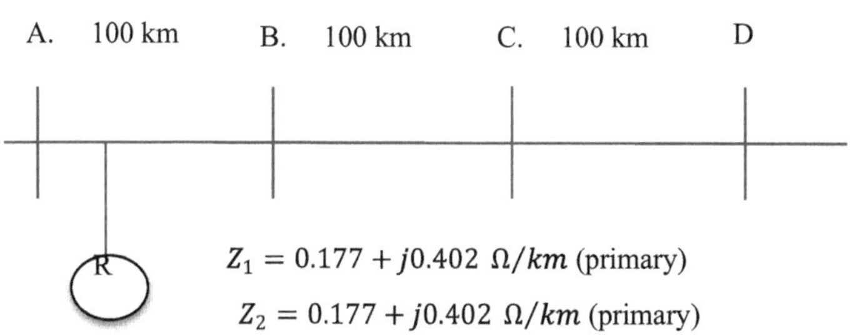

$$Z_1 = 0.177 + j0.402\ \Omega/km \text{ (primary)}$$

$$Z_2 = 0.177 + j0.402\ \Omega/km \text{ (primary)}$$

$$Z_0 = 0.354 + j1.022\ \Omega/km \text{ (primary)}$$

$$CT\ ratio = 600{:}\,1$$

$$VT\ ratio = 132\,000{:}\,110$$

Fig. 4.25 132 kV network for exercise. Diagram of a transmission line with four segments labeled A, B, C, and D, each 100 km long. A circle labeled "R" is connected to the line. Impedance values are given

$$Z_P = \frac{VT}{CT}\frac{V_S}{I_S}$$

$$Z_P = Z_S \frac{VT}{CT}$$

$$Z_S = Z_P \frac{CT}{VT}$$

The positive phase sequence impedance of the protected line is:

$$Z_{L1(P)} = 100(0.177 + j0.402) = 17.7 + j40.2.$$

Referring this to the secondary, from the above:

$$Z_{L1(s)} = Z_{L1(P)}\left(CT/_{VT}\right) = (17.7 + j40.2)\left(\frac{600 \times 110}{132\,000}\right) = 8.85 + J20.10$$

The zone 1 reach is set to 80% of this, i.e.:

$$Z_{R1} = 0.8Z_{L1(s)} = 7.08 + j16.08 = 17.57\angle 66.2°$$

The zone 2 reach is set to 120% of $Z_{L1(s)}$, i.e.:

$$Z_{R2} = 1.2Z_{L1(s)} = 10.62 + j24.12 = 26.36\angle 66.2°$$

The zone 3 forward reach is set to 150% of the impedance of the protected feeder and the next adjacent feeder. In this example, line BC is similar to line AB with the same impedances, i.e.:

$$Z_{R3F} = 1.5 \times 2Z_{L1(s)} = 26.55 + j60.3 = 65.89\angle 66.2°$$

The zone 3 reverse reach is set to 10% of the impedance of the protected feeder, i.e.:

$$Z_{R3R} = 0.1Z_{L1(s)} = 0.885 + j2.01 = 2.2\angle 66.2°$$

For the earth fault protection, the setting of the residual compensation factor is required, i.e.:

$$K = \frac{Z_0 - Z_1}{3Z_1}$$

This can be calculated using the primary system impedance, the secondary system impedances, or the primary system impedances per unit length, i.e.:

$$K = \frac{(0.354 + j1.022) - (0.177 + j0.402)}{3(0.177 + j0.402)}$$

$$K = \frac{0.177 + j0.62}{0.531 + j1.206}$$

$$K = \frac{0.842 + j0.116}{1.736}$$

$$K = 0.485 + j0.067$$

$$K = 0.49\angle 8^\circ$$

Neglecting the small angle concerned,

$$K = 0.49$$

4.4.2 Full and Switched Distance Protection Scheme

The practical implementation of a three-phase, three-zone distance relay can take one of several configurations. As explained in Sect. 4.3.1, the 11 fault combinations associated with a three-phase system can be reduced to six combinations of input measurands to provide full fault coverage. In the simplest scheme, therefore, six independent comparators can be used for each zone of the relay to ensure that the relay accurately detects each fault condition. This is described as full scheme distance protection.

For each zone of the relay, there are three comparators used for phase-to-phase fault coverage and three for phase-to-earth fault combinations. For three-zone relaying, this requires a total of 18 comparators in the configuration shown in Fig. 4.26. The costs of having 18 comparators in the scheme have led to the development of switched schemes, which use a lower number of comparators. These schemes are particularly important with electromechanical relays since not only are the comparators expensive to manufacture in their own right, but the panel space and housing required for a full scheme distance relay is itself expensive [34].

In a totally switched scheme as illustrated in Fig. 4.27, only one comparator is used and this is switched to examine a particular fault condition in response to starters are frequently overcurrent relay elements or voltage controlled overcurrent relay elements which provide a basic impedance type characteristic; once these detect the possible presence of a fault, the comparator will determine in which zone the fault is located.

Exercise 4.12
Suggest why the relay's size is important.

The size of a relay governs the size of the panel required to house it. As the amount of panel space increases, so does the size of the substation and the civil engineering costs.

For retrofit projects, where the substation has already been built, panel space can be at a premium, and any need to increase the size of the panels will inhibit development.

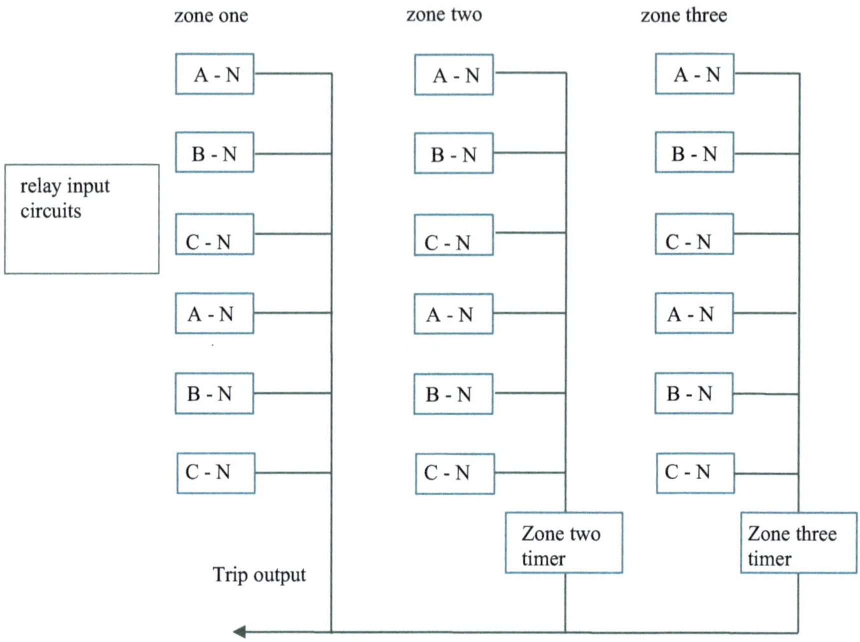

Fig. 4.26 Full scheme three-zone, three-phase distance relay. Diagram of relay input circuits divided into three zones. Zone 1 includes circuits A-N, B-N, and C-N, each connected vertically. Zone 2 has similar circuits, with an additional connection to a "Zone two timer." Zone 3 also features circuits A-N, B-N, and C-N, linked to a "Zone three timer." All zones lead to a "Trip output"

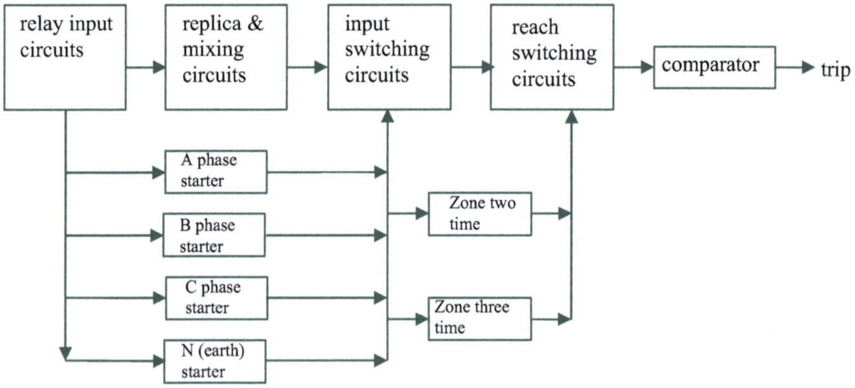

Fig. 4.27 Fully switched distance relay configuration. A flowchart diagram illustrating the process of electrical circuit operation. It begins with "relay input circuits" leading to "replica & mixing circuits," followed by "input switching circuits" and "reach switching circuits." The flow continues to a "comparator" and ends with "trip output"

4.4.3 Carrier-Assisted Distance Protection

The slow tripping times for faults in the end zone of the protected feeder can be reduced by enhancing the non-unit standalone distance scheme with some of the features of unit protection.

Where a fault on a protected feeder is fed from both ends of the feeder, a second set of relays and associated circuit breakers will be needed at the remote end of the feeder to break that fault circuit. The response of these relays, used with the response of the local relay, provides the basis for the unit protection scheme. A communications system is required to transfer the status of selected relaying elements between the relays at each end of the feeder (Fig. 4.28) [35].

The various carrier-assisted distance protection schemes can be illustrated using the three distance zones for a feeder protected by relays at both the local and remote terminals, as shown in Fig. 4.29. Without carrier assistance, faults in the end zones will be cleared from the infeed from the opposite terminal in zone 2 times.

The simplest type of carrier-assisted scheme uses a direct transfer tripping scheme, in which whenever a relay trips on zone 1, the trip signal is also used to trip the breaker at the opposite end of the feeder. This ensures that faults in the end zone are tripped in no more than the zone tripping time plus the time delays incurred in the communication channel. Unfortunately, interference on the communication channel could cause malfunctioning of the scheme and incorrect tripping of the circuit breakers.

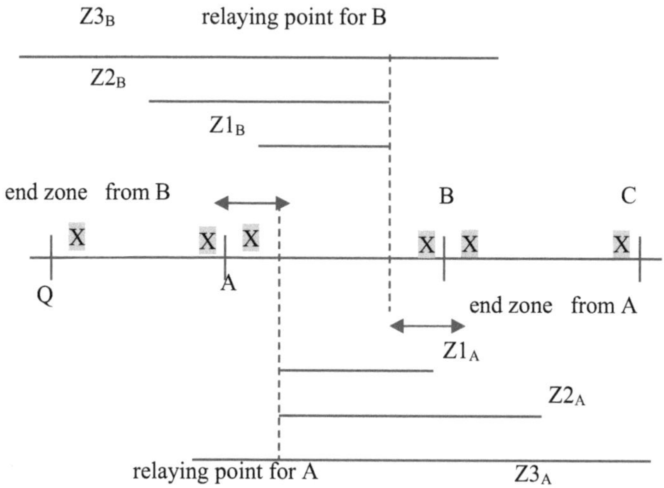

Fig. 4.28 Three-zone stepped distance protection applied to feeder fed from both ends of the line. Diagram illustrating relaying points and zones for entities A and B. The top section shows the relaying point for B with zones $Z3_B$, $Z2_B$, and $Z1_B$. The bottom section shows the relaying point for A with zones $Z1_A$, $Z2_A$, and $Z3_A$. The central line features positions Q, A, B, and C, with end zones. Arrows indicate directional flow between points and zones

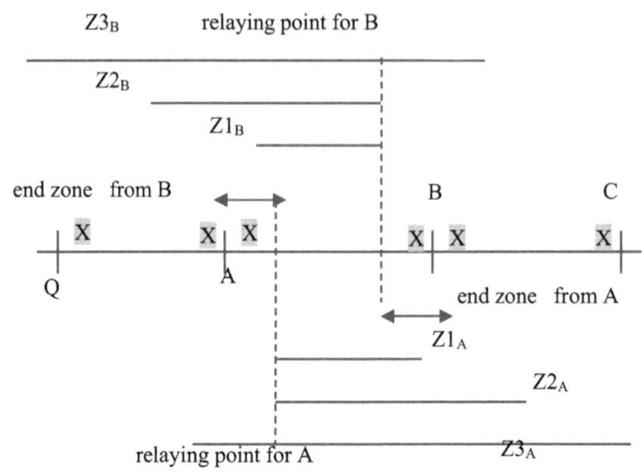

(a) Direct transfer scheme

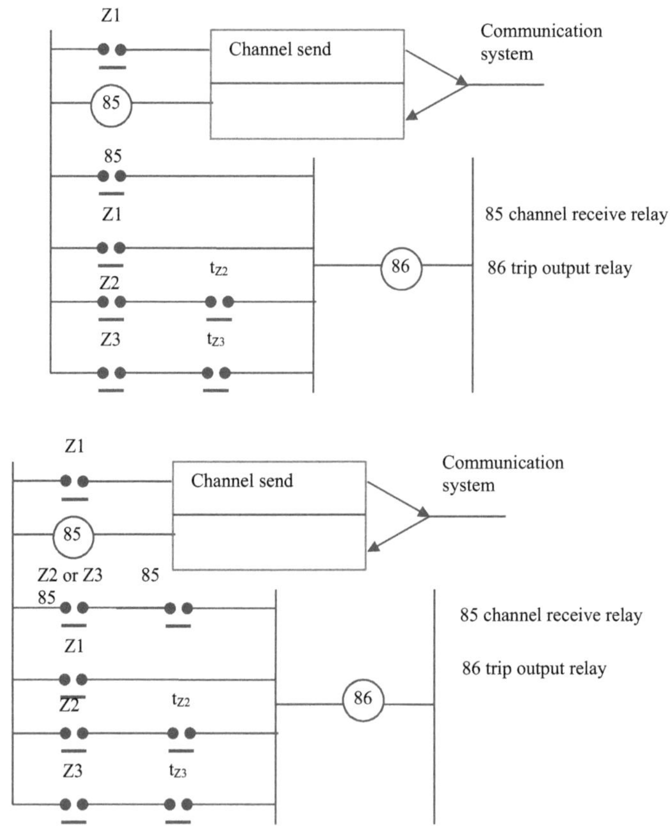

(b) Permissive underreaching transfer trip scheme

Fig. 4.29 Direct transfer tripping and permissive underreaching transfer tripping schemes. Diagram illustrating a communication system for relaying points A and B. The top section shows zones Z1, Z2, and Z3 for both A and B, with arrows indicating end zones and relaying points. The bottom section features a schematic with labeled components: Channel send and receive connected to a communication system, and relays labeled 85 and 86

The possibility of malfunctioning of the scheme can be dramatically reduced by checking the validity of the inter-tripping signal with the operation of comparators in the receiving scheme's relays. This produces the permissive underreaching transfer trip scheme by which the transfer trip is permitted only when the fault is also seen by the receiving relay's zone 2 or zone 3 comparators. The operations of the direct transfer tripping and permissive under-reaching transfer tripping schemes are shown in Fig. 4.29.

An alternative to the permissive underreaching scheme is to provide two independent, non-time delayed protection schemes using the distance relays at each end of the feeder. The direct tripping scheme uses the zone 1 comparators, which will instantaneously trip their associated breakers whenever they detect a fault. The carrier-assisted scheme is designed to rely on the operations of the zone 2 comparators and, since these are over-reaching comparators, provides a permissive overreaching transfer tripping scheme. The operation of this scheme is illustrated in Fig. 4.30.

The operation of the zone 2 comparator in a distance scheme initiates the transfer tripping communication signal. Receipt of this by the protection scheme at the opposite end of the line, coupled with the operation of the zone 2 comparator at that end of the line, results in instantaneous tripping. The operation of the zone 2 comparators is therefore used both to initiate and to check the validity of the transfer trip. Since these can be independent of the zone 1 comparators, they can be used to provide an independent protection scheme.

For both permissive schemes, inter-tripping is produced by the receipt of the communications signal. An alternative philosophy is to use the loss of communications as the basis for the scheme operations, where the presence of the communications blocks the inter-trip. The operations of the classical and practical blocking schemes are illustrated in Fig. 4.31.

In the classical blocking scheme, a reverse-looking distance comparator is used to initiate the carrier. Whenever this comparator is operating, the communication carrier is transmitted to the scheme at the opposite end of the protected feeder. However, when it is restrained, the carrier is blocked. Failure to receive the communications from the relaying scheme at the opposite end of the feeder, coupled with the operations of the scheme's zone 2 comparator, indicates that the fault is on the protected feeder and initiates transfer tripping [36].

The disadvantage of the classical blocking scheme is the need for a fourth set of distance comparators. This can be overcome in a practical blocking scheme by using a standard set of three-zone comparators by combining the responses of the scheme's zone 2 and offset zone 3 comparators. In this scheme, the blocking signal is transmitted whenever the scheme's zone 2 comparators restrain, but the offset zone 3 comparators operate for a fault condition. As with the classical blocking scheme, failure to receive the blocking signal together with the operations of the scheme's zone 2 comparator initiates tripping.

Exercise 4.13

Describe how distance relays can be used to provide the characteristics of both UNIT and NON-UNIT protection.

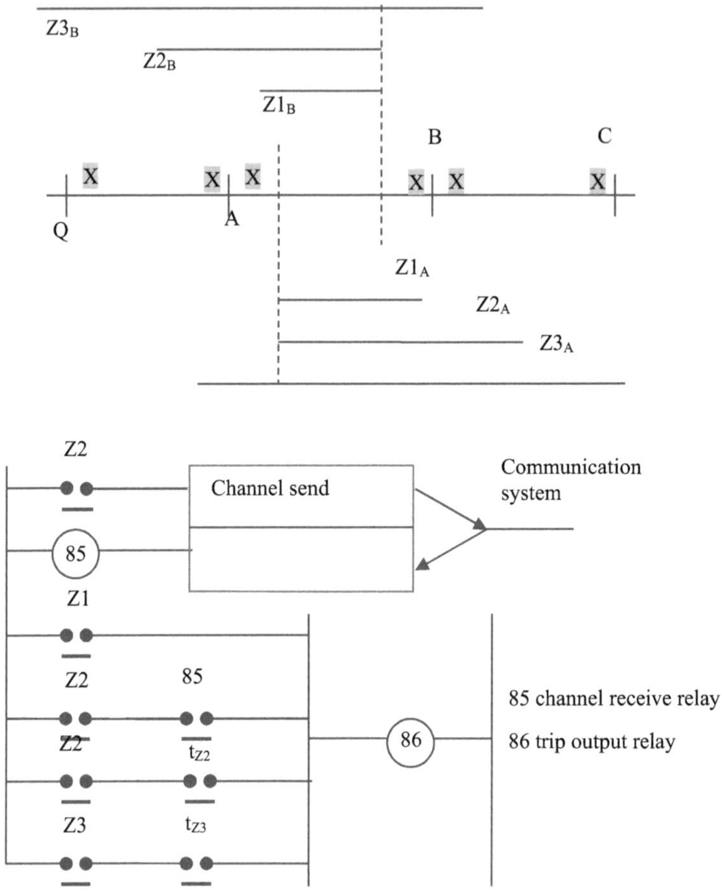

Fig. 4.30 Permissive overreaching transfer tripping scheme. The image depicts a schematic diagram of a power system protection setup. The top section shows a transmission line with labeled zones: impedances marked between points Q, A, B, and C. There are multiple "X" symbols indicating fault locations. The bottom section illustrates a communication system with relays and channels. It includes labeled components

Distance protection is inherently a NON-UNIT type of protection. It uses inputs derived from a single relaying point, namely, the voltage and currents measured at that point. It provides distinct zones of protection defined by the impedances seen by the relays, and it provides inherent backup protection using an item to delay the operation of zones of protection that extend beyond the protected feeder.

The most widely used configuration for a distance protection is the three-zone, stepped scheme shown in Fig. 4.22. The zone 1 characteristic provides instantaneous protection for 80% of the protected feeder. The zone 2 provides time-delayed protection for the end zone and backup protection for the remote substation. The zone 3 characteristic provides overall backup protection for the protected feeder, its substation, and the adjacent feeder.

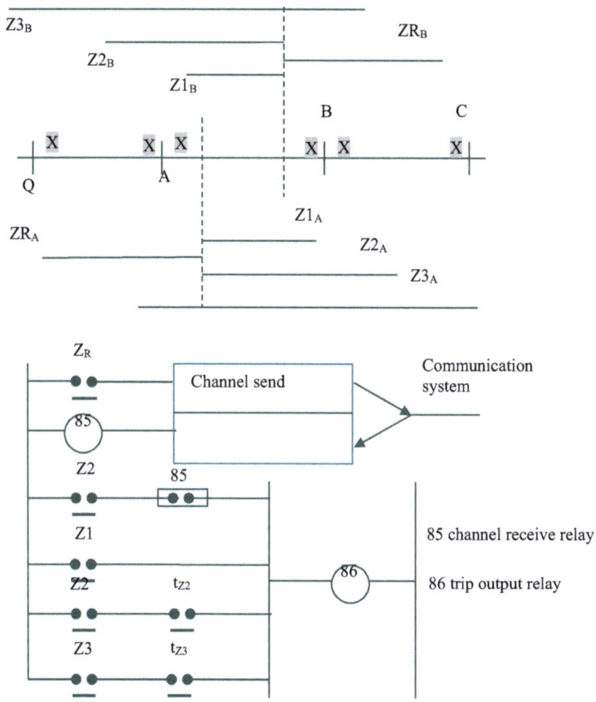

(a) Classic blocking scheme- Diagram illustrating a communication system for a relay protection scheme. The top section shows a series of lines labeled Z impedance with points marked Q, A, B, and C, and several 'X' symbols indicating fault locations. The bottom section features a schematic with components labeled ZR, Z2, Z1, ZZ, Z3, and relays 85 and 86. Arrows connect "Channel send" and "Channel receive" to a "Communication system," indicating signal flow. The 85 channel receive relay and 86 trip output relay are highlighted.

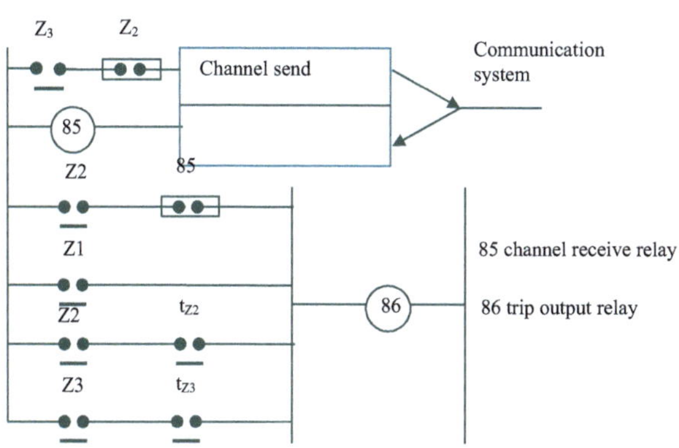

(b) Practical blocking scheme

Fig. 4.31 Carrier-assisted blocking schemes. Diagram illustrating a communication system with relays and channels. The system includes "Channel send" and "Channel receive" connected to a "Communication system." Relays labeled "85" and "86" are shown, with "85 channel receive relay" and "86 trip output relay" indicated. Zones Z1, Z2, and Z3 are depicted with corresponding time delays t_{Z2} and t_{Z3}

When the transmission line can be supplied from both ends, separate distance relays are required at each end of the line to provide adequate protection. This produces the situation shown in Fig. 4.28, where the relay at A is responsible for protecting the feeder when the source is connected to substation A and the relay at B is responsible for its protection when the source is connected to substation B. The relaying scheme shown in Fig. 4.28 can be converted to a UNIT protection scheme by adding a communication link between the two relays. This can be done using the carrier-assisted distance protection schemes described in Sect. 4.4.3. The advantage of the UNIT protection scheme is that it provides absolute discrimination for the protected feeder and will trip instantaneously for a fault on that feeder. It also has a high security against tripping for faults not on the protected feeder [37].

The limitation of the UNIT protection scheme is that in itself it does not provide backup protection. However, since the scheme is based on NON-UNIT protection distance relays, these provide the back-up facilities.

4.5 Directional Earth Fault Protection Schemes

High-impedance earth faults present particular problems for transmission line protection since the fault currents are invariably low and the impedances they present to the protection cannot be covered by the distance relay characteristic without introducing load encroachment. These types include fallen conductors, particularly where the transmission line runs over desert or rocky areas, burning grassland faults where the line runs over savannah areas, and tree-top faults, where the line runs over forested areas. The fault impedances associated with these faults can be as high as several hundred ohms and are resistive. Directional earth fault (DEF) schemes were introduced to protect transmission lines against high impedance earth faults and are now frequently offered as an independent protection or an optional feature of a distance protection scheme. When used with distance relaying, they are generally used in conjunction with both phase and earth fault distance protection schemes, although some utilities prefer to dispense with the earth fault distance protection and use phase fault distance and DEF alone. The scheme can be provided using discrete relays or integrated into the main distance relay protection package [37].

Exercise 4.14
Explain why high impedance earth faults on transmission lines create difficulties for distance type protection.

The distance relay's comparators are required to restrain when the transmission line is carrying heavy loads and only to trip when there is no doubt about the presence of a fault. All of the relay's comparators see system load as an impedance in the impedance plane. Their characteristics are therefore chosen into to trip for these impedances, i.e., to avoid load encroachment. Unfortunately, high-impedance earth faults present a resistive impedance to the appropriate earth fault comparator, which is similar to that presented by possible loads. Since the distance relay must restrain for load impedance, to will also restrain for high-impedance earth fault.

4.5.1 The Basic Directional Earth Fault Protection Scheme

The directional earth fault protection scheme, as associated with transmission line protection, is a unit protection scheme. It includes relay functions at both ends of the protected transmission line with a communication channel. The relay functions involve the detection of the earth current and the direction in which it is flowing. The communication system is required to communicate this information to the other end of the line. The protection trip trips when earth current is flowing into the line from both ends, indicating an internal shunt fault on flowing through the line at one end but out of it at the other. A simple representation of the basic permissive directional earth fault scheme is shown in Fig. 4.32.

The overcurrent elements at both the local and remote substation are set such that they will operate for the zero-sequence current corresponding to the maximum fault impedance for which the scheme is required to trip. The fault current is flowing and is set to operate when its flowing channel is activated, when both of the relays at one terminal of the line have operated and hence have communicated that fact to the other terminal. In the basic scheme, separate channels are required for communications in each direction. A scheme trip is initiated when communication is received from the other end, and the local overcurrent and directional relay functions have operated.

This scheme is described as a permissive scheme since tripping is only permitted when the communication signals are received from the other terminals of the

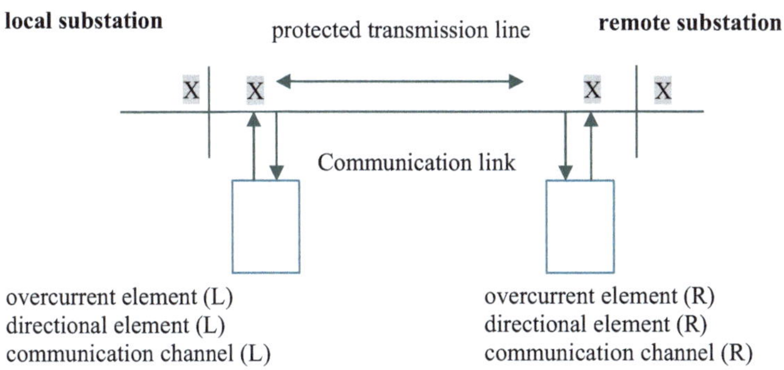

Fig. 4.32 Basic permissive directional earth fault protection scheme. Diagram illustrating a protected transmission line between a local substation and a remote substation. The line includes overcurrent and directional elements at both ends, labeled as (L) for local and (R) for remote. Communication links connect these elements, facilitating data exchange. The diagram highlights the importance of communication channels in managing the transmission line's protection system

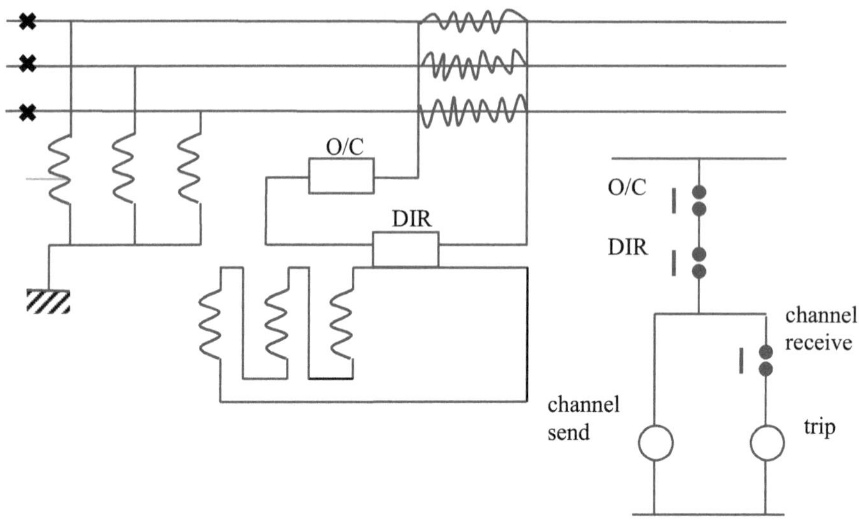

Fig. 4.33 Basic directional earth fault protection scheme using discrete relay elements. A schematic diagram of an electrical protection system. It includes three parallel lines with coils and transformers, labeled "O/C" for overcurrent and "DIR" for directional protection. The diagram shows connections to a channel send and receive system, with a trip mechanism. The setup illustrates the flow and control of electrical signals for protection purposes

transmission line and the local relays indicate the presence of a fault. The communications system can use a basic ON/OFF keyed carrier, since the information required to be transferred relates to whether or not the relaying scheme at that terminal has operated [38]. In the simplest scheme, instantaneous overcurrent elements can be used to detect the earth current. To prevent possible malfunctioning due to a mismatch in the line charging currents or magnetizing inrush currents of transformers on the system, either definite time or inverse definite minimum time overcurrent relays may prove more effective.

The choice of inputs for the directional relays is also dictated by the power system configuration. With single-circuit transmission lines, zero sequence voltage offers an immediate polarizing quantity. With double-circuit lines, where mutual coupling between the lines can create difficulties using zero sequence quantities, negative sequence measurements have been found to be more effective.

Directional earth fault protection cannot discriminate between faults on different phases and therefore is used for three-pole tripping only. The three-phase relaying scheme shown in Fig. 4.33 is based on an atypical DEF scheme using discrete relays. The earth current is derived from the summation of the three-phase current, and the zero-sequence voltage polarizing signal is derived from an open delta connection of the VT secondaries. The voltage output of this voltage transformer connection is the residual voltage, which is three times the zero-sequence voltage.

Exercise 4.15

For the 70 km 132 kV transmission line used in the exercise on pages 4–28, calculate the relay's current setting required for it to trip for a high-impedance earth fault of impedance 200 Ω (resistive). Assume that the system appears as an infinite bus connected to the line.

The symmetrical components diagram for this type of fault is shown in Fig. 4.35. The setting for the earth fault relay would correspond to I'_N for a fault at the far end of the line (Fig. 4.34).

$$I_N = I_1 + I_2 + I_0 = 3I_0$$

$$\therefore I_0 = \frac{V}{Z_{pps} + Z_{nps} + Z_{zps} + 3R_g}$$

$$I_0 = \frac{\frac{132,000}{\sqrt{3}}}{70(0.177 + j0.402 + 0.177 + j0.402 + 0.354 + j1.022) + 600} A$$

$$I_0 = \frac{\frac{132,000}{\sqrt{3}}}{649.56 + j127.82} A$$

$$I_0 = 115.12\angle -11.13° A$$

The earth current I_N is then:

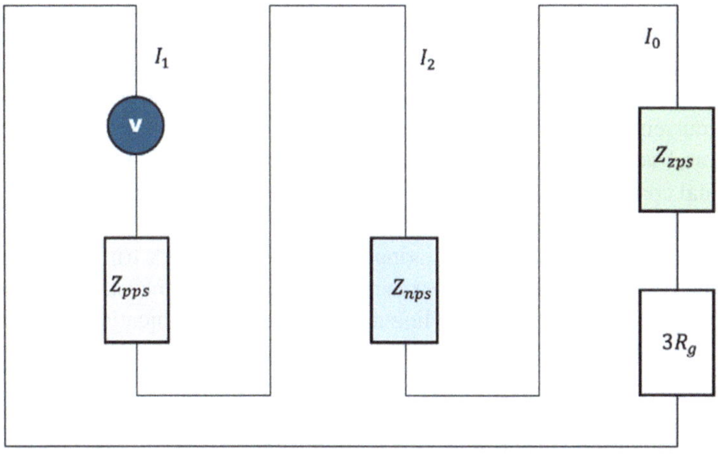

Fig. 4.34 Calculated relay settings. The image is a circuit diagram featuring three loops. The first loop includes a voltage source labeled "V" and impedance "Zpps," with current "I_1" flowing through it. The second loop contains impedance "Znps" with current "I_2." The third loop has impedance "Zzps" and "3Rg," with current "I_0." Each component is represented by a rectangle, and the diagram shows the connections between these elements

$$I_N = 3I_0$$

$$I_N = 345.36\angle -11.13°A$$

And the relay current I'_N is:

$$I'_N = \frac{I_N}{CT}$$

$$I'_N = \frac{345.36\angle -11.13°}{\frac{600}{1}} A$$

$$I'_N = 0.576\,A$$

The calculated relay setting so that the relay trips for an earth fault with 200 Ω fault resistance is therefore 0.576.

4.5.2 Directional Earth Fault Protection Schemes Integrated with Distance Relaying

The use of directional earth fault schemes with modern distance protection schemes has led to this function being integrated into the distance protection package. Functions common to both the distance protection and the directional earth fault scheme, such as the current and voltage input circuits, the power supply, and the scheme logic, can be shared. The costs of adding directional earth fault protection can therefore be minimized. Various communication philosophies can be accommodated by using either permissive or blocking schemes.

The Basic Permissive Directional Earth Fault Scheme

The basic permissive DEF scheme used with distance protection is similar to the scheme described above. The operation of the scheme is illustrated in Fig. 4.35.

Overcurrent elements, referred to as level detectors, monitor the zero-sequence current and activate the scheme when that current exceeds the pick-up point. A directional comparator is used for the directional elements and determines when the current is flowing into the protected transmission line. Communication is activated when both functions operate, initiating the channel send. A trip output signal is given when the DEF scheme at either location detects sufficient zero-phase sequence current flowing into the protected line and receives communication from the other terminal to indicate that it is also detecting zero sequence current flowing into the line [39].

The Basic Blocking Directional Earth Fault Scheme

The blocking scheme, the presence of other communication signal from the relay at the other terminal of the protected transmission line, inhibits tripping of the directional earth fault scheme. Since the block signal can be initiated by either terminal

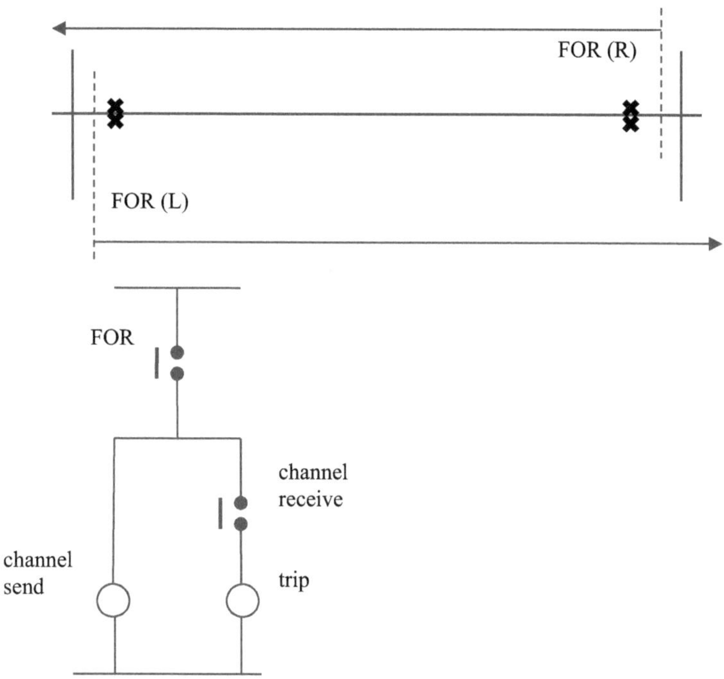

Fig. 4.35 Permissive directional earth fault protection scheme. Diagram illustrating a circuit with two main sections. The top section shows a horizontal line labeled "FOR (L)" on the left and "FOR (R)" on the right, with two crosses marking positions along the line. The bottom section depicts a circuit schematic with a vertical line labeled "FOR" connected to two parallel paths. The left path is labeled "channel send," and the right path is labeled "channel receive" and "trip"

of the line, a common signaling channel can be used. The operation of the scheme is illustrated in Fig. 4.36.

Two sets of overcurrent level detectors are used together with two directional comparators. One of these directional overcurrent functions is used to detect earth fault current flowing out of the protected feeder, REV, i.e., the reverse direction, and the other is used to detect earth fault current flowing into the line, FOR, i.e., for the forward direction. The reverse-looking directional overcurrent functions control the channel send. The channel is activated by the presence of sufficient zero sequence current flowing in the transmission line and indicates that the scheme should not trip. Operations of the forward-looking directional earth fault function and, therefore, by default, the restraint of the associated reverse-looking directional earth fault function produce a trip output if the communication channel is off. While the channel is active, either from the local or remote terminal, tripping is inhibited. To avoid possible malfunctioning of the scheme, the overcurrent level detectors are used with the forward-looking comparator. This ensures that is in the marginal case, where zero-sequence current is flowing through the line, the line's capacitive elements do not shunt sufficient zero elements at the remote terminal [40].

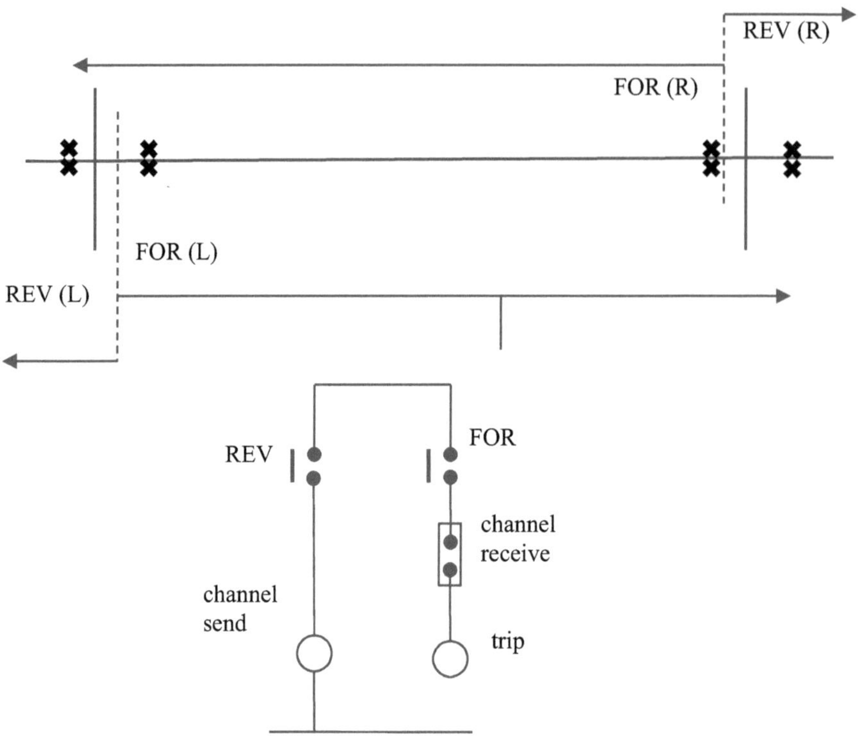

Fig. 4.36 Blocking directional earth fault protection scheme. Diagram illustrating a communication channel with forward (FOR) and reverse (REV) directions. The top section shows a horizontal line with labeled arrows: "FOR (L)" and "REV (L)" pointing left, and "FOR (R)" and "REV (R)" pointing right. Cross marks indicate signal points. The bottom section depicts a circuit diagram with labeled components: "channel send," "channel receive," and "trip," connected by lines with "REV" and "FOR" switches

Grading and Backup Protection

Directional earth fault protection provides an alternative protection to earth fault distance relaying. Both schemes can be operated without introducing deliberate delays, or the distance protection can be given priority by introducing a definite time delay into the directional earth fault protection scheme. A limitation of the directional earth fault protection is that it is inherently three-phase and cannot discriminate between faults on different phases. The directional earth fault scheme can also be used to provide time delay backup protection to other power system plants beyond the end of the protected line. To provide this feature, the operation of the relaying functions at the different substations must be graded to ensure that the protection closest to the fault is given every opportunity to trip first.

Self-Assessment Questions

? Self-assessment question 4.1

Suggest a possible combination of relay characteristics that could be used to produce a quadrilateral characteristic, and comment on any precautions required to avoid malfunctioning.

? Self-assessment question 4.2

Calculate the zone 1, zone 2, and zone 3 reach settings together with the residual compensation setting for the three-zone distance relay at location R for the 400 kV transmission line system shown in Fig. 4.37. The relay uses the MHO characteristic for zones 1 and 2 and the OFFSET MHO for zone 3.

The line data and transformer ratio are:

$$Z_1 = 0.0383 + j0.320\,\Omega / km$$
$$Z_2 = 0.0383 + j0.320\,\Omega / km$$

$$Z_0 = 0.1474 + j0.837\,\Omega / km$$

$$CT\ ratio = 2000A / 1A$$

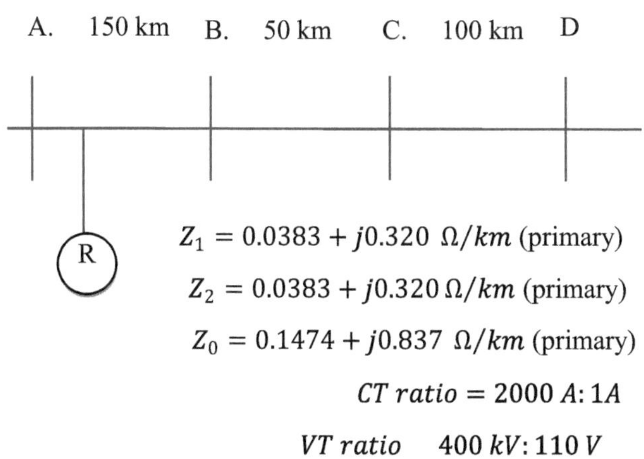

Fig. 4.37 kV network for exercise. Diagram of a transmission line with distances marked: A at 150 km, B at 50 km, C at 100 km, and D. A branch labeled "R" connects downward from the main line. Impedance values are given

$$VT\ ratio = 400\frac{kV}{110}V$$

? Self-assessment question 4.3

Briefly describe how a carrier-assisted distance protection scheme using a three-zone, full scheme distance relay at each end of a protected feeder can be configured to provide instantaneous MAIN A and MAIN B relaying for the feeder.

Bibliography

1. Juan Noh; Wookyu Chae; Woohyun Kim; Sungyun Choi "A Study on Mesh Distribution system and Protection Coordination Using HILS System" published in 13th International Conference on Information and Communication Technology Convergence (ICTC) in 2022
2. Congbo Wang; Ke Jia; Bohan Liu; Jiankang Zhang "Coordination Control and Protection for Photovoltaic DC Distribution System" published in 2nd International Conference on Smart Power & Internet Energy Systems (SPIES) in 2020
3. Tara Khayyamim; Bita Nouparvar; Arman Safaei "A Novel Scheme for Grid Interconnection and Protection of EV Charging stations" published in 15th International Conference on Protection and Automation of Power Systems (IPAPS) in 2020
4. Chong Tong; Yunfeng Cai; Ziyang Zhang; "Artificial intelligence-based lightning protection of smart grid distribution system" published in International Symposium on Lightning Protection (XIV SIPDA) in 2017
5. Harsha Ravindra; M. Omar Faruque; "Impact of PV on distribution protection system" Published in North American Power Symposium (NAPS) in 2012
6. Hesam-Edin Hayati Soloot; Edris Agheb; Amir Hayati Soloot; Shahab Moghadam "A SWOT Analysis of Two Protection Strategies Due to the Expansion of Renewable Distributed Generation Distribution Network" in published in 15th International Conference on Protection and Automation of Power Systems (IPAPS) in 2020
7. Bin Li; Chao Li; Fengrui Guo; Ying Xin "Overcurrent Protection Coordination in a Power Distribution Network With the Active Superconductive Fault Current Limiter" Published in IEEE Transactions on Applied Superconductivity
8. Bruno Fazio; Sanja Bajic; Serge St-Antoine "Relay Protection Coordination Analysis using Fault Current Measurements:" published in 2022 IEEE Power & Energy Society General Meeting (PESGM).
9. Dong Chen; Bing Hu; Mingjun Xue; Yinian Qi; Chang Song "Research and Application of the System of Relay Protection and Self-Healing for Distribution Network with Mixed Use of Circuit Breaker and Load Switch" published in 2023 International Conference on Power System Technology (PowerCon)
10. Feng Deng; Xiang Ji; Guo-Fang Huang "A centralized distribution protection system" published in 2012 China International Conference on Electricity Distribution.
11. Hana A. Halim; Melaty Amirruddin; Nur S. Noorpi; Nurhakimah M. Mukhtar "An Improved Protection Scheme for Smart Distribution Grid published in 2013 1st International Conference on Artificial Intelligence, Modelling and Simulation.
12. Jamil Abdo & Tariq Masood "New Technique to develop Intelligent, highly reliable & Efficient Unmanned Power Generation Modular" 22nd International Conference on Renewable Energies and Power Quality (ICREPQ'24) Bilbao (Spain), 26th to 28th June 2024. https://www.icrepq.com/icrepq24/392-24-abdo.pdf

13. Jamil Abdo & Tariq Masood "Operational Reliability- Production Best Practice Pragmatic Case Studies & Improvement" published in Proceedings of the 9th North American Conference on Industrial Engineering and Operations Management Washington D.C., USA, June 4–6, 2020
14. Jamil Abdo & Tariq Masood "Techno Economics Analysis to Reshape Global Clean and Green Energy" 22nd International Conference on Renewable Energies and Power Quality (ICREPQ'24) Bilbao (Spain), 26th to 28th June 2024. https://www.icrepq.com/icrepq24/415-24-abdo.pdf
15. Jane Courtney; Aaron McDonnell "Impact on Distribution System Protection with the Integration of EG on the Distribution Network" published in 2019 54th International Universities Power Engineering Conference (UPEC).
16. Kun Yang; Zhen Xiang; Ruixiong Yang; Lin Jiang "Decentralized Protection Coordination Based on multi-agent system for Active Distribution Networks" published in 2022 7th Asia Conference on Power and Electrical Engineering (ACPEE).
17. M.Saeed Ullah, Tariq Masood, Umair Zahid, H. Abuzar Ahsan "Electrochemical Oxidation of Dissolved organic Dye in Aqueous Solution' Published in IMASCON-2019
18. Nasser Alemadi, Tariq Masood, R Aggarwal, Suhail A Qureshi, D Kothari "Novel selection approach of STATCOM device tailored and optimized engineering configuration to implement in between Qatar and Bahrain" published in Int. J. Autom. Power Eng. https://doi.org/10.14355/ijape.2014.0302.01
19. Qin Lijun; Wang Ying; Zheng Shenrui; Guo Qing "Situation of study on distribution network protection containing distributed generation and application scheme of intelligent protection" 2011 4th International Conference on Electric Utility Deregulation and Restructuring and Power Technologies (DRPT)
20. Qingfa Chai; Jiayin Jiang; Yi Tang; Anning Wang; Congcong Li "High Reliability Relay Protection Setting Scheme of Distribution Network" published in 2023 International Conference on Power System Technology (PowerCon)
21. Rani Kumari; Bhukya K. Naick "Protection Coordination Impact on Reliability of Distribution Systems with Distributed Energy Resources" 2024 Third International Conference on Power, Control and Computing Technologies (ICPC2T).
22. Tariq Masood & Jamil Abdo & "Maintenance Strategy Production Best Practices—Pragmatic Case Studies and Operations Improvement" published in Proceedings of the 9th North American Conference on Industrial Engineering and Operations Management Washington D.C., USA, June 4-6, 2020
23. Tariq Masood & Jamil Abdo "New Technique to Enhance Hybrid Renewable Energy Resources' Operational Flexibility and Reliability" 22nd International Conference on Renewable Energies and Power Quality (ICREPQ'24) Bilbao (Spain), 26th to 28th June 2024. https://www.icrepq.com/icrepq24/396-24-masood.pdf
24. Tariq Masood and Jamil Abdo "Review: Uninterrupted Liquid Hydrogen (E-Methanol) Production framework to deliver Clean & Green Energy" Published 22nd International Conference on Renewable Energies and Power Quality (ICREPQ'24) Bilbao (Spain), 26th to 28th June 2024. https://www.icrepq.com/icrepq24/414-24-masood.pdf
25. Tariq Masood, Abdel Aty Edris, Suhail Aftab Qureshi, Muhammad Tajammal, Murtaza Hashmi, DP Kothari, Samer Kareem "SSSC tailored to optimization performance in Between United Arab Emirates and Oman on the GCC Power Grid" published in the 13th IET International conference on AC and DC power Transmission Conference on 14–16 February, 2017. http://ieeexplore.ieee.org/document/7934934/
26. Tariq Masood, Abdel Aty Edris, Suhail Aftab Qureshi, Muhammad Tajammal, Murtaza Hashmi, DP Kothari, Samer Kareem "Smart Grid Operations and Control Challenge by Implementing STATCOM Tailored to Optimization Performance at the GCC Power Grid" published in the 13th IET International conference on AC and DC power Transmission Conference on 14–16 February, 2017. http://ieeexplore.ieee.org/document/7934988/
27. Tariq Masood, Muhammad Tajammal, Samer Kareem, Suhail Aftab Qureshi, Abdul Jabbar Khan, DP Kothari, Nair Jameel "Unified power Flow Controller Modeling and Analysis

Technique on the GCC Power Grid Operations and Control" published IEEE Energy Con, 2018 June 7–9 Limassol Cyprus. https://ieeexplore.ieee.org/document/8398834/
28. Tariq Masood, Muhammad Tajammal, Samer Kareem, Suhail Aftab Qureshi, Murtaza Hashmi, Nasir Jameel "Unified power Flow Controller (UPFC) Modeling and Analysis Technique by using Matlab and EMTP" published (ICREQP,18 Salamanca Spain 21–23 March, 2018). http://www.icrepq.com/icrepq18/250-18-masood.pdf
29. Tariq Masood, Muhammad Tajammal, Samer Kareem, Suhail Aftab Qureshi, Murtaza Hashmi, Naser Jameel "STATCOM Controller Performance Enhancement and Optimization Technique" published (ICREQP,18 Salamanca Spain 21–23 March, 2018). http://www.icrepq.com/icrepq18/249-18-masood.pdf
30. Tariq Masood, Muhammad Tajammal, Samer Kareem, Suhail Aftab Qureshi, Murtaza Hashmi, DP Kothari· "Smart Grid Operational functions and Control Challenges by Implementing SSSC Tailored to Optimize performance in between Qatar and KSA on the GCC Electrical-power grid" published in the ASTESJ Journal Advances in Science, Technology and Engineering Systems Journal Vol. 2, No. 1, 30.09.2017. https://astesj.com/special-issue-4/
31. Tariq Masood, N Al-Emadi, SA Qureshi, YY Almulla, AJ Khan, SK Shah "FACTS Control Devices (Statcom, SSSC and UPFC) Re-Configuration Techniques By PSIM/MATLAB" Published in: 2007 International Conference on Electrical Engineeringhttps://ieeexplore.ieee.org/abstract/document/4287295
32. Tariq Masood, Jamil Abdo, Atif Iqbal "Protection of Electrical Power Distribution Systems: Smart-grid, Microgrid, AI and Cybersecurity" ISBN: 978-1-394-32764-5, published by Wiley-IEEE Press 2025
33. Tariq Masood, Raj K Aggarwal, SA Qureshi, RAJ Khan "STATCOM and SVC control operations and optimization during network fault conditions" https://ieeexplore.ieee.org/document/5636868
34. Tariq Masood, Raj K Aggarwal, Suhail Aftab Qureshi, RAJ Khan "STATCOM model against SVC control model performance analyses technique by MATLAB" published in International Conference on Renewable Energies and Power Quality (ICREPQ'10) https://icrepq.com/icrepq'10/655-Masood.pdf
35. Tariq Masood, RK Aggarwal, SA Qureshi, RAJ Khan "Statcom control reconfiguration technique for steady state and dynamic performance optimization during network fault conditions" published in International Conference on Renewable Energies and Power Quality (ICREPQ'10) Granada (Spain) https://www.icrepq.com/icrepq'10/656-Masood.pdf
36. Tariq Masood, Samer Karim Shah, Muhammad Tajammal, "Unified power flow controller modeling and analysis technique on the GCC power grid" published in 2018 IEEE International Energy Conference (ENERGYCON) https://ieeexplore.ieee.org/abstract/document/8398834
37. Tariq Masood, Suhail A Qureshi, RK Aggarwal, DP Kothari "Multiple FACTS devices integration through distributed control system and its application on the GCC interconnection" published in 2013 7th IEEE GCC Conference and Exhibition (GCC) https://ieeexplore.ieee.org/abstract/document/6705750
38. V. S. Sazanov; A. I. Kovalenko; A. A. Voloshin "The Development of System for Automatic Adaptive Change of Relay Protection Settings in Distribution Networks" published 2022 5th International Youth Scientific and Technical Conference on Relay Protection and Automation (RPA)
39. Vassilis C. Nikolaidis; Evangelos Papanikolaou; Anastasia S. Safigianni "A Communication-Assisted Overcurrent Protection Scheme for Radial Distribution Systems With Distributed Generation" IEEE Transactions on Smart Grid
40. Xiaohui Song; Yu Zhang; Song Zhang; Shizhan Song; Jing Ma; Weibo Zhang "Active distribution network protection mode based on coordination of distributed and centralized protection" publish in 2017 China International Electrical and Energy Conference (CIEEC).

Current Differential Protection

5

Contents

5.1 Introduction 102
5.2 Fundamental Principles 103
5.3 Summation Arrangements 108
5.4 Phase Comparison Carrier Protection 110
5.5 Communication Media in Differential Protection 113
5.5.1 Pilot Wires 113
5.5.2 Power Line Carrier Circuits 114
5.5.3 Microwave Radio Links 115
5.5.4 Fiber Optical Links 115
5.6 Differential Protection of Teed Circuits 115
Self-Assessment Questions 118
Bibliography 119

Abstract

When you have completed this chapter, you should be able to define the basic principle of current differential protection. This includes a clear understanding of the concept of bias, which is a crucial principle in current differential protection. Identifying the need for bias and being able to derive both biases and operate quantities is a key skill that will make you feel more skilled and proficient. You will also be able to assess the advantages and disadvantages of different types of current differential protection schemes and understand the requirement for a sound, reliable communications system in differential protection.

Supplementary Information The online version contains supplementary material available at https://doi.org/10.1007/978-3-031-94312-6_5.

T. Masood et al., *Protection of Electrical Power Transmission Systems*, Power Systems, https://doi.org/10.1007/978-3-031-94312-6_5

Objectives
When you have completed this chapter, you should be able to

- Define the basic principle of current differential protection.
- Identify the need for bias and be able to derive both bias and operating quantities.
- Assess the advantages and disadvantages of different types of current differential protection schemes used in practice.
- Identify the requirement for a good, reliable communications system in current differential protection and consider the different types of systems, such a pilot wire, power line carrier, microwave, and fiber optics, used in practice.
- Consider the application of current differential protection for protecting complex circuits such as *teed feeders*.

5.1 Introduction

Current differential protection is an example of a unit scheme that only responds to faults inside a clearly defined zone. It utilizes only information derived from two (or occasionally more) points, which are at the boundaries of the protected zone in a system, and compares the derived quantities. Unlike *non-unit* schemes (which suffer from an element of uncertainty and therefore restriction near the boundaries of the protected zone), the big advantage of current differential protection lies in its ability to identify positively whether a fault is internal or external to the protected zone and is capable of providing high-speed protection for the whole feeder length. The other main advantages are as follows:

- No VTs required.
- Suitable for complex circuits such as *teed feeders*.
- Immunity to power swings
- Simple to set with no relay coordination problems

Current differential protection does have the disadvantage that it requires a secure and dependable communications link for data transfer between ends of the protected zone, and also, unlike non-unit schemes, it does not provide backup protection to adjacent feeders. This chapter deals with some of the basic principles of current differential protection as applied to feeders and considers the different types of communication media used in such a technique. It also looks briefly at the application of current differential protection to complex circuits such as a teed circuit [1].

5.2 Fundamental Principles

All current differential protection schemes are derived from the simple *Merz and Price* circulating system. In one arrangement, similar current transformers (CTs) at each end of the protected zone are interconnected by an auxiliary pilot circuit, as shown in Fig. 5.1. Current transmitted through the zone causes secondary current to circulate around the pilot circuit without producing any current in the relay.

A fault within the protected zone will cause secondary currents of opposite relative phase compared with the through-fault (or external fault) condition, and these will then flow in the relay. An alternative but similar technique hinges upon connecting the CTs in opposition, so that under external fault conditions, no current flows in the series-connected relays. This technique, which is often referred to as the *balanced voltage* principle, is shown in Fig. 5.2 [2].

Balanced current schemes of the type shown in Fig. 5.1 are rather limited in application because the voltage developed across the CT secondaries increases in proportion to the length of the feeder. Prohibitively large CTs are required when applying this principle to feeders of longer than about 1 km and the balanced voltage principles finds more widespread use.

In the balanced voltage system, with primary through current, the CT secondaries are opposed and produce no current in the interconnecting pilot leads or the series-connected relays. An in-zone fault leads to a circulating current condition and hence to relay operation. An immediate consequence of the arrangement is that the CTs are in effect open-circuited as no secondary current flows for any primary

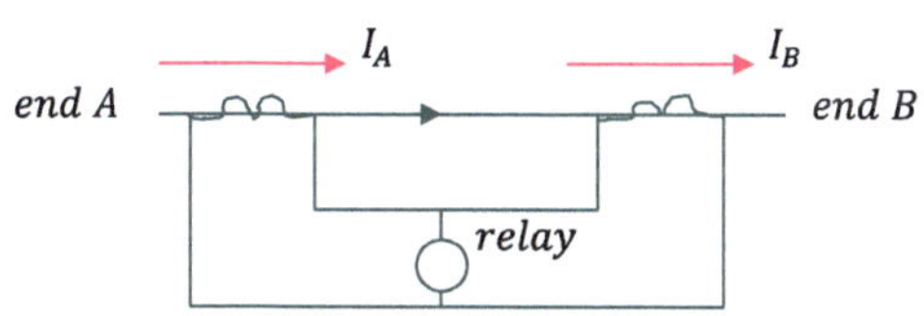

Fig. 5.1 Circulating current system. Diagram illustrating a circuit with two ends labeled "end A" and "end B." Current flows from end A to end B, indicated by arrows. A relay is connected in the circuit, depicted as a circle with a line

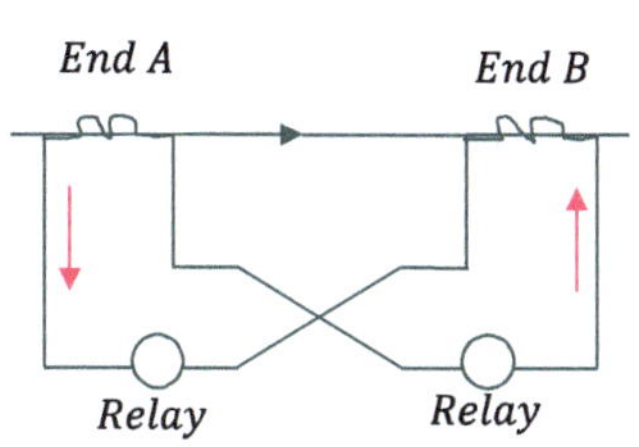

Fig. 5.2 Balanced voltage system. Diagram illustrating a relay circuit with two ends labeled "End A" and "End B." The circuit features two relays, each connected by lines with arrows indicating current flow. The left side shows a downward arrow, and the right side shows an upward arrow, representing the direction of current

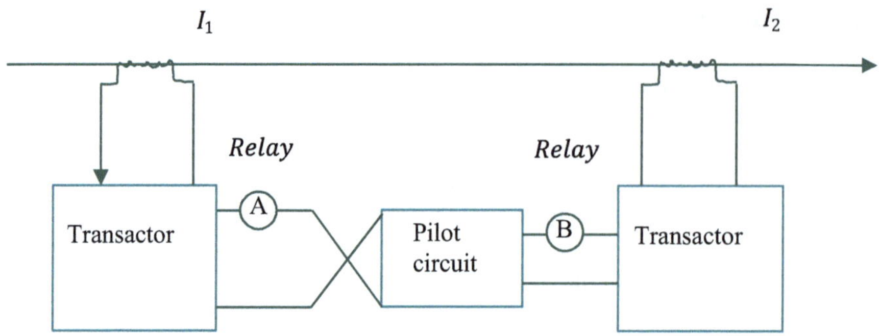

Fig. 5.3 A modified balanced voltage system. Diagram of a pilot protection system for a transmission line. It shows two transactors connected by a pilot circuit. Current flows from I_1 to I_2 through the line. Relays labeled A and B are positioned between the transactors and the pilot circuit

through-current conditions. In order to avoid severe distortion of the secondary voltage due to saturation, a voltage proportional to the current is produced by means of a transactor. The latter is an iron-cored transformer in which the core is provided with nonmagnetic gaps sufficient to support the flux at the maximum current level. Figure 5.3 shows a schematic of the principle involved.

The series-connected relays are of relatively high impedance, and the system may be satisfactorily applied in principle for the protection of feeders up to about 50 km. Some modification of the basic scheme is, however, required for two main reasons:

- Under heavy through-fault conditions, the secondary CT current is not exactly equal due to small differences in CT ratios at the two ends.
- The pilot wire circuit has shunt capacitance so that even if the voltages at either end are identical, spill current will flow in the relays. This tends to cause tripping under healthy conditions.

Because of the wide range of fault current magnitude for which operation may be required, permanent desensitization of the relays to prevent unwanted operations from spill current may not be satisfactory. The most common method used for overcoming this problem is the use of bias, the principle of which is now very widely used in different forms. Biasing of relays consists of restraining the operation of a relay of the local currents exceeding the remote current by an amount dependent on the magnitude of the currents at either end of the feeder [3]. When bias is applied to feeder protection, each relay is biased by the local fault infeed. Figure 5.4 typifies the bias characterization in terms of the zone boundary current I_1 and I_2 from which it can be clearly seen that some spill current can be tolerated, and at high fault currents, the margin between the ideal stability condition $|I_1| = |I_2|$ increases.

Remoteness of the interconnection of feeder equipment, i.e., increase in feeder length, can lead to phase displacement of currents. It therefore becomes necessary

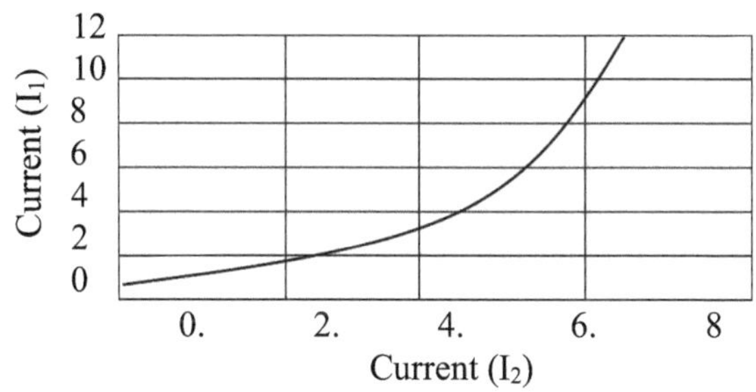

Fig. 5.4 Typical bias curve for feeder protection. A line graph showing the relationship between current (I_2) on the x-axis and current (I_1) on the y-axis. The curve starts near the origin and rises steeply, indicating an exponential increase in I_1 as I_2 increases

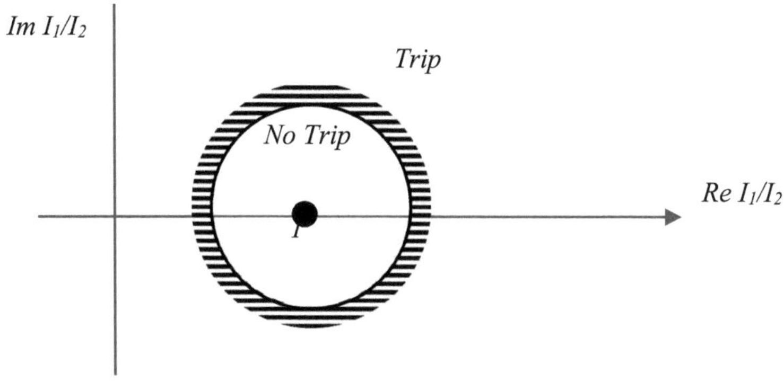

Fig. 5.5 Typical polar characteristic of a current differential feeder protection system. Diagram illustrating a trip and no-trip zone on a complex plane. The horizontal axis, a circle is centered at point 1 on the real axis. Inside the circle is labeled "No Trip" and the area outside is labeled "Trip." The circle is surrounded by striped lines indicating the boundary between the two zones

to examine relay performance with relative phase displacement as well as magnitude difference.

Exercise 5.1
What is the cause of phase displacement between currents and why is it detrimental to protection relay performance?

Phase displacement is caused by the charging current through the line shunt capacitance, which becomes more significant as the line length increases. The net effect is that, under through-fault conditions, the difference in the currents at the two ends can be so large (even with bias) as to cause the protection relay to malfunction. Figure 5.5 shows a locus of the bias characteristic (in a polar form) when both phase and magnitude of the currents are compared [4, 5].

It can be seen that the ideal balanced condition: $\frac{I_1}{I_2} = \angle 0°$ is surrounded by a circle to permit stability due to the phase and magnitude difference. In practice, it is found that a stability characteristic based on a circle of radius 0.5 is suitable for a majority of faults. The tripping criterion for the local relay is then given by:

$$\left| \frac{I_1}{I_2} - 1 \right| \geq 0.5 \quad (5.1)$$

Although the tripping criterion for the remote-end relay can also be arranged such that:

$$\left| \frac{I_2}{I_1} - 1 \right| \geq 0.5 \quad (5.2)$$

Under certain fault conditions, tripping at the two ends may not, however, be simultaneous.

Exercise 5.2
Why is it desirable to have near-simultaneous tripping of the circuit breakers at the two ends of a feeder?

In practice, over 80% of faults on overhead lines are caused by lightning strikes and are transient in nature. The arcing fault is cleared, and the system is restored to normal through the employment of an auto-reclosure technique. In such a technique, once the fault has been detected by the protection relays, the circuit breakers at the two ends of the line are tripped and, after a preset time, are automatically reclosed, restoring the system to normal [5, 6]. However, the non-simultaneous opening of the circuit breakers at both ends of the faulted line can result in insufficient time during the auto-reclosure cycle for the fault arc to be extinguished. This results in the possibility that a transient fault can cause permanent lockout of the circuit breaker at each end of the line.

Exercise 5.3
Current differential relays at the two ends of a feeder have tripping criteria given by

$$\left| \frac{I_1}{I_2} - 1 \right| \geq 0.5$$

(local relay)

$$\left|\frac{I_2}{I_1}-1\right| \geq 0.5$$

(remote-end relay)

Show that if the stability characteristics for the two relays are on the same axes:

$$\left(\textbf{i.e.} \Im\frac{I_1}{I_2} \; vs \; \Re\frac{I_1}{I_2}\right).$$

Then the tripping criterion for the remote end relay is equivalent to:

$$\left|\frac{I_1}{I_2}-\frac{4}{3}\right| \geq \frac{2}{3}$$

What is the implication of this result and what is the remedy?

$$let\,\frac{I_1}{I_2} = a + jb; then\,\frac{I_2}{I_1} = \frac{1}{a+jb}$$

For the local relay

$$\left|\frac{I_1}{I_2}-1\right| \geq 0.5$$

or

$$(a-1)^2 + b^2 \geq 0.25.$$

This is the equation of a circle with center 1.0 and radius 0.5

For remote-end relay:

$$\left|\frac{I_2}{I_1}-1\right| \geq 0.5$$

or

$$\left|\frac{I_2}{a+jb}-1\right| \geq 0.5$$

$$\left|\frac{a-jb}{a^2+b^2}-1\right| \geq 0.5$$

$$Let\ a^2 + b^2 = k$$

$$\left|(a-k) - jb\right| \geq \frac{k}{2}$$

or

$$\left(a^2 + k^2 + 2ak + b^2\right) \geq \frac{k^2}{4}$$

This can be rearranged to give $k \geq 4\frac{2a-1}{3}$ Eq. 2.5

Substituting for K into Eq. 2.5 gives

$$\left(a - \frac{4}{3}\right)^2 + b^2 \geq \frac{4}{9}$$

$$= \left|\left(a - \frac{4}{3}\right) + jb\right| \geq \frac{2}{3}$$

$$= \left|\frac{I_1}{I_2} - \frac{4}{3}\right| \geq \frac{2}{3}$$

Thus, for the remote-end relay, the stability characteristic is a circle with center:

$$\frac{4}{3}, 0\ and\ \frac{2}{3}$$

Clearly, the two characteristics are not coincidental, and this means that, for certain fault conditions, the tripping of the breakers at the two ends will be non-simultaneous. The problem can be overcome by making the tripping criterion for the remote-end relay:

$$= \left|\frac{I_2}{I_1} - \frac{4}{3}\right| \geq \frac{2}{3}$$

$$\left(= \left|\frac{I_1}{I_2} - 1\right| \geq 0.5\right)$$

5.3 Summation Arrangements

Schemes have so far been discussed as though they were applied to single-phase systems. A three-phase system could be provided with independent protection for each phase, but this would involve a corresponding number of pilot channels, thus making it prohibitively expensive. The alternative is to combine the separate phase

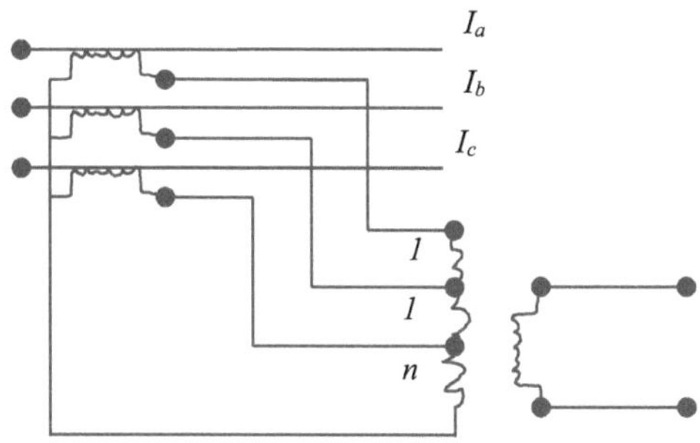

Fig. 5.6 Summation transformer. Diagram of a summation transformer circuit. Three input lines connect to coils, converging into a single coil labeled \(n\). The output is labeled "Summated output," indicating the combined electrical output from the inputs

currents into a single quantity for comparison over the pilot channel. Direct summation of the currents in a three-phase system would produce an output proportional only to the zero-sequence component, with the consequence that the relay will only operate for earth faults. Other arrangements must be used to obtain an output under all practical fault conditions [7, 8]. In one such arrangement (Fig. 5.6), a summation transformer is employed in which the interphase chapters of the winding, a–b and b–c are given an equal number of turns, whereas the neutral end of the winding c–N has a greater number of turns, giving a turns ratio of 1:1:n. The output of the transformer is proportional to $[(n + 2)I_a + (n + 1)I_b + nI_c]$, n being typically 3. However, with this technique, the differential protection does suffer from blind spot problems, i.e., it fails to detect certain internal faults.

Exercise 5.4

If an internal fault on a feeder is such that $I_{c1} = I_{c2}$ and

$$\frac{I_{c1}}{I_{c0}} = 4,$$

Show that a summation transformer with turns ratios 1:1:3 will give zero output.

The output I_s is proportional to:

$$5I_a + 4I_b + 3I_c = 5\left(I_{CO} + a^2 I_{c1} + aI_{c2}\right) + 4\left(I_{CO} + aI_{c1} + a^2 I_{c2}\right) + 3\left(I_{CO} + I_{c1} + I_{c2}\right)$$

Taking the "c" phase as a reference, where

$$a^2 = \left(-\frac{1}{2} - j\frac{\sqrt{3}}{2}\right)$$

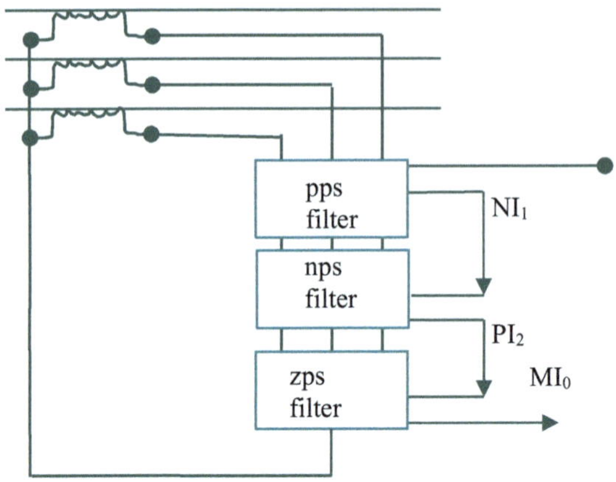

Fig. 5.7 Phase sequence network. Diagram showing an electrical circuit with three filters labeled "pps filter," "nps filter," and "zps filter." Each filter receives input from a series of coils and outputs to the right. The outputs are leading to a "Summated output"

$$a = \left(-\frac{1}{2} - j\frac{\sqrt{3}}{2} \right)$$

This then gives $I_s \propto 12I_{co} - 3I_{c1} = 0$.

An alternative and better approach is to use a combination of the phase-sequence components of the current at each end. Figure 5.7 shows the arrangement.

A detailed study has shown that a satisfactory performance for all types of practically encountered faults is obtained with a relaying quantity I_s proportional to:

$$I_s \propto (6I_2 - I_1) \quad (5.3)$$

5.4 Phase Comparison Carrier Protection

Phase comparison carrier protection, involving a comparison of only the relative phases of the currents at the two ends of the line, is another example of current differential protection widely used on extra high-voltage transmission lines. In such a scheme, the medium of communication between line ends is commonly the power line conductor itself (this is discussed in some detail later on). Figure 5.8 shows the principle of a phase comparison carrier protection, and a schematic of the arrangement is shown in Fig. 5.9 [9, 10].

Under normal healthy conditions, currents I_A *and* I_B *are approximately in phase. The derived relaying quantities (summed outputs) at each end of the line are used to modulate the carrier signal (high-frequency output from an oscillator in the range 70–700 kHz) in such a manner that a series of half-cycle carrier blocks are produced. The polarities of the CTs are so arranged that the blocks produced at each end of the line differ in phase by* 180 ° , i.e., $I_A = - I_B$. The blocks are

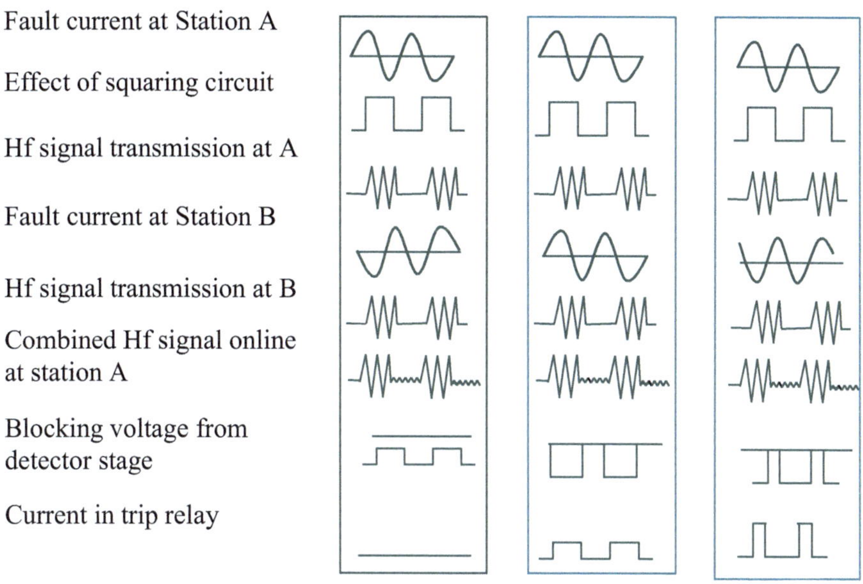

Fig. 5.8 Principle of phase comparison carrier protection. Diagram illustrating electrical signals at different stages in a transmission system. The left column shows waveforms for fault current at Station A, effect of squaring circuit, HF signal transmission at A, fault current at Station B, HF signal transmission at B, combined HF signal online at Station A, blocking voltage from detector stage, and current in trip relay. The middle and right columns display similar waveforms, indicating signal changes and interactions across the system

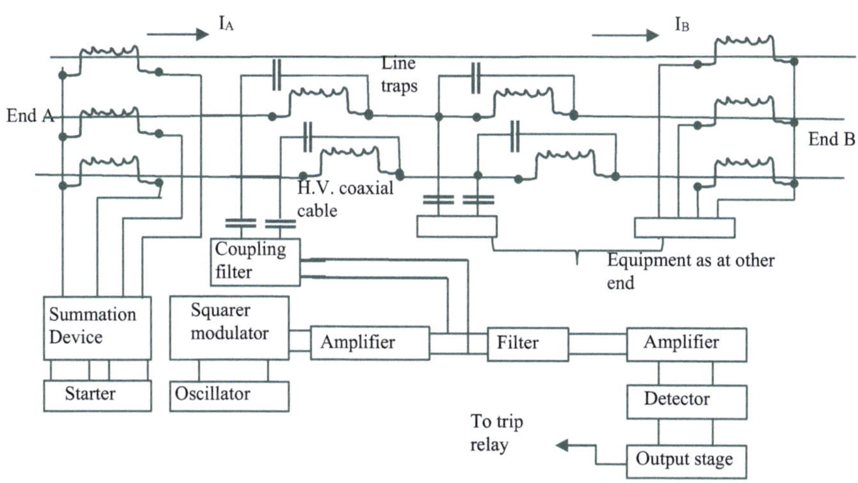

Fig. 5.9 A schematic of the arrangement. Diagram of a power line carrier communication system showing signal flow from End A to End B. Key components include line traps, H.V. coaxial cable, coupling filter, summation device, starter, squarer modulator, oscillator, amplifiers, filters, detector, and output stage. The system is designed to transmit signals to a trip relay, with equipment mirrored at both ends

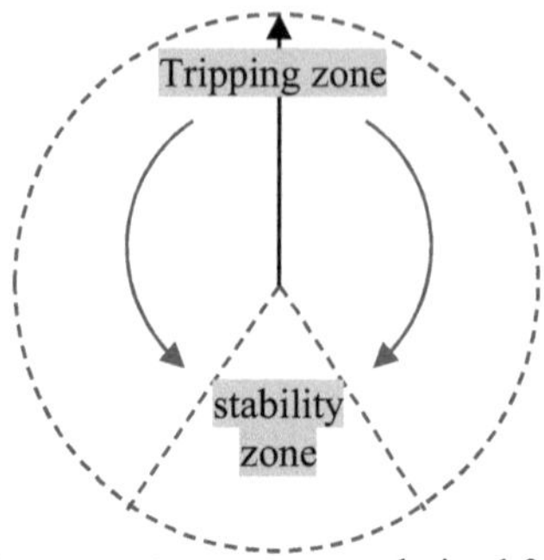

I_A, I_B output currents to comparator derived from summated phase currents

Fig. 5.10 Tripping characteristic of phase comparison carrier protection. Diagram illustrating a tripping and stability zone within a dashed circle. The tripping zone is at the top with an upward arrow, while the stability zone is at the bottom with downward arrows. Output currents to the comparator are derived from summated phase currents

amplified and injected via line couplers at each end of the line. Under normal healthy conditions (or an external fault), the modulated carrier is continuous at both ends, and tripping is blocked. Under internal fault conditions, gaps in the carrier are obtained, and tripping is initiated as shown in Fig. 5.8. Most phase comparison carrier schemes have a ±30° stability zone, as shown in Fig. 5.10. On lines longer than 200 km, it is often necessary to have a large stability angle because, even under healthy or out-of-zone faults, the line capacitance charging current can cause a significant phase shift between the compared currents. Increasing the stability margin does, however, increase the likelihood of blind spots for internal faults, and the limit of applicability of phase comparison carrier protection is reached at around 500 km [11].

Exercise 5.5
In phase comparison carrier protection, what will happen if carrier transmission from one end is interrupted for some reason?

Phase comparison carrier protection is really a blocking scheme, and this effectively means that if the carrier signal is not received from one end, then gaps in the signal at either end would still occur, resulting in tripping. Thus, although the scheme is fail-safe for internal faults, there will nonetheless be false tripping for external faults.

5.5 Communication Media in Differential Protection

As mentioned before, a secure and reliable communications system for the transfer of signals between line ends is vitally important for maintaining the integrity and dependability of unit protection, such as current differential protection. To achieve this, we need to have a transmitter, a receiver, and a communication medium between line ends [12, 13]. Media currently in use for signaling for protection purposes are as follows:

- Pilot wires.
- Power line carrier circuits.
- Microwave radio links.
- Fiber optical links.

5.5.1 Pilot Wires

These can be either rented wires (from the local telephone company) or private (owned by the electricity supply authorities).

In the case of a rented pilot, the routing is through the national telecommunications network. One of the major drawbacks of such a system is the very considerable electrical and human interference to which the protection signal systems can be subjected. For example, some degree of security and protection against induced voltages must be built into signaling systems. A rise in the earth potential of the signaling system terminal stations can be significant, and isolation must be provided to protect both the personnel and equipment of the telephone company. Communication between relaying points may be over a two-wire or four-wire link. Consequently, the effect of a particular human action, for example, an incorrect disconnection, may disrupt communication in one direction of both. It is not only hazards that place limitations on signaling systems. The telephone company itself imposes stringent requirements upon privately owned equipment connected to its network. The signaling transmitted must be limited in both level and bandwidth to avoid interference with other signaling systems. In the United Kingdom, the maximum 1-minute mean power permitted is −13 dBm for frequencies in the band 200–3200.

Private pilot wires are owned by the power authority and may be laid in a trench with high voltage cables, laid by a separate route, or strung as an open wire on a separate wood pole route. Voltages due to rises in the station earth potential at the terminal station can be very high, so terminal isolation transformers are used to reduce the risk if injury to people and damage to equipment on the pilot side. Voltage impulses due to lightning and transients due to power circuit isolator switching are also hazards that the signaling equipment must withstand without damage. Signaling can be carried out at a relatively high level (+10 dBm) on private pilot links, as there is normally no limit placed by telephone companies. Private pilot wires can be attractive to a power authority running a very dense power system with short distances between stations, particularly if it is not possible to rent channels from the telephone company [14–16].

Exercise 5.6
What do you understand by signaling power in dBm and what are −13 dBm and +10 dBm power equivalent to in terms of watts? Why is it desirable to transmit signals at high levels?

Power in dBm is a measure of power normalized to an input power of 1 mW, i.e., 0 dBm = 1 mW. Thus, the power of −13 dBm is equivalent to:

$$-13 = 10\log_{10}\frac{P}{1}$$

$$P = 0.05\,mW$$

Likewise

$$10 = 10\log_{10}\frac{P}{1}$$

$$P = 10\,mW$$

In practice, the transmitted signals are attenuated by the losses in the pilot wire, isolating transformers, and so on, and this therefore imposes a limit on the length of the wire over which signals can be transmitted. It is thus apparent that higher-level signals would permit applications of protection to longer feeders.

5.5.2 Power Line Carrier Circuits

For the protection of long electrical transmission lines, or if the route involves installation difficulties, it may be too expensive to provide a pilot link for protection signaling. In this respect, high-frequency signal transmission along the actual overhead power line offers advantages over the use of pilot links. It is robust and therefore reliable, constituting a low-loss transmission path that is under the full control of the power authority. Figure 5.9 shows a complete schematic of the PLC system.

High-voltage capacitors are used, along with drainage coils, for the purpose of injecting the signal onto and extracting it from the line. Injection can be carried out by impressing the carrier signaling voltage (in the frequency range 70–700 kHz) between one conductor and earth or between any two-phase conductors.

To raise the efficiency of transmission, we need to minimize the loss of signal into other parts of the power system. This is done with a line trap, which is essentially a parallel circuit tuned to present a very high impedance to the signal frequency, and, as shown in Fig. 5.9, it is connected in the phase conductor on the station side of the injection equipment [17, 18]. The drawbacks of this communication medium are that it suffers from restricted capacity due to the overhead line characteristics and the licensing terms imposed by the national communication authorities to restrict the PLC transmitter power outputs; this is done in order to avoid undue interference caused by stray radiation from power lines with radio navigational systems and domestic broadcasting bands. Furthermore, the interfacing equipment(s) are expensive [19, 20].

5.5.3 Microwave Radio Links

The microwave band is that part of the frequency spectrum above 1000 MHz. Usually, microwave radio schemes provide a highly reliable service, but difficulties may be encountered in obtaining frequency allocations. Furthermore, in order to minimize *fading* of signals (due, for example, to ground reflected waves), a very careful choice of position, direction, and shape of the transmitting and receiving aerials is required. Aerial height is another very important factor in reducing interference. Sufficient clearance must be provided from the ground and other obstacles, and this can be a particular problem for protection signaling where the presence of transmission towers, aerial wires, and cooling towers adds to the difficulty of finding a suitable transmission path [21, 22].

5.5.4 Fiber Optical Links

Optical fibers are fine strands of glass, which behave as waveguides for light. This ability to transmit light over considerable distances can be used to provide optical communication links with enormous information-carrying capacity and an inherent immunity to electromagnetic interference.

To communicate information, a beam of light is modulated in accordance with the signal to be transmitted. This modulated beam travels along the optical fiber and is subsequently decoded at the remote terminal into the received signal. On/off modulation (i.e., binary) of the light source is normally preferred to linear modulation since the distortion caused by non-linearities in the light source detector, as well as variations in received light power, is largely avoided. Optical fiber communications are rapidly becoming established in the electrical supply industry, and many authorities are installing their own links. One common practice is to produce composite cables comprising optical fibers embedded within the power line conductors, either earth or phase. Optical fiber communication is expected to play an important role in future protection signaling systems.

5.6 Differential Protection of Teed Circuits

Teed feeders (or three-terminal lines) are an attractive alternative to two-terminal lines, both environmentally and economically. However, from a protection point of view, they have always presented difficult problems. Of the alternatives, differential protection is generally regarded as the method that is best suited to teed circuits.

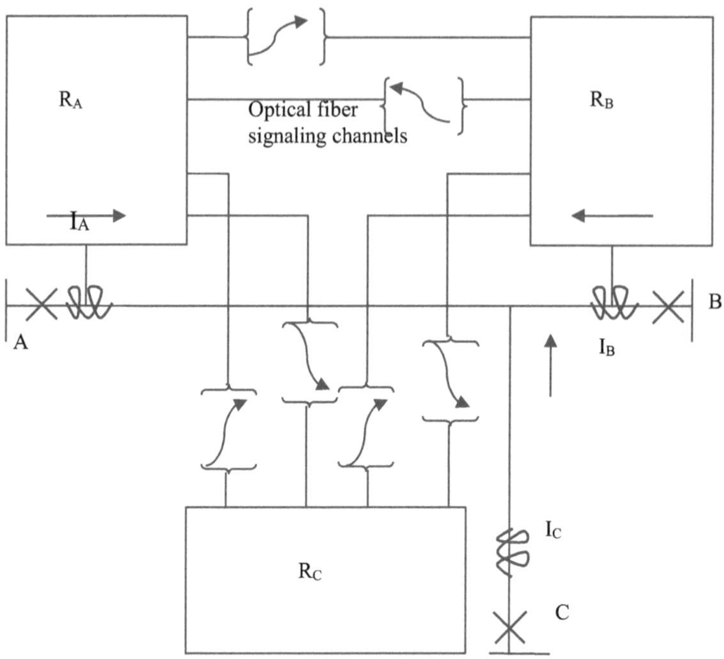

Fig. 5.11 Current differential protection for teed feeders using optical fiber signaling. Diagram illustrating an optical fiber signaling system with three main components. The components are interconnected by optical fiber channels, with directional arrows indicating signal flow. Points A, B, and C are marked with crossed lines and coiled symbols, representing connections or interfaces. If I_A, I_B, and I_C are the current vector signals at line ends A, B, C, then for an unfaulted circuit

Current differential relays can provide unit protection for teed circuits without the restrictions associated with other forms of protection. In this respect, the advent of fiber optical links to provide continuous signaling between the three ends of the circuit has resulted in the development of a number of different current differential schemes, based on numeric relaying techniques. Numeric relays, relays based on microprocessors, are further described in Chap. 7. One such three-ended system is shown in Fig. 5.11, where numeric current differential relays are interconnected by optical fiber links so that each can send information to the others [23, 24].

$$I_A + I_B + I_C = 0 \tag{5.4}$$

The basic principles of operation of the system are that each relay measures its local three-phase currents and sends its values to the other relays. Each relay then calculates, for each phase, a restaurant differential current and also a bias current, which is used to restrain the relay in the manner conventional for biased differential unit protection. The bias feature is necessary in this scheme because it is designed to operate from conventional current transformers, which are subject to transient transformation errors [24, 25]. The quantities are as follows:

$$\left|I_{bias}\right| = \left|I_A + I_B + I_C\right| \tag{5.5}$$

$$\left|I_{bias}\right| = \frac{1}{2}\left|I_A + I_B + I_C\right| \tag{5.6}$$

Figure 5.12 shows the percentage biased differential characteristics used, the tripping criteria being:

$$\left|I_{diff}\right| > K\left|I_{bias}\right|$$

$$and\left|I_{diff}\right| > I_S \tag{5.7}$$

where K = percentage bias setting; I_s = minimum differential current setting.

Operations

The three-phase CT secondary currents at each line end are fed to the respective relays, where they are sampled at a fixed frequency and converted to digital form by an analog-to-digital converter. An associated microcomputer carries out signal processing algorithms to filter out unwanted d.c. and harmonic components. This process yields the power frequency at each relay that is transmitted to the others in a binary-coded form, at the data rate of 64 k bits/s, and each set of current samples transmitted is accompanied by error checking, status, and synchronization data. The message frames are received in the relays, which then carry out the calculations described above. If the magnitudes of the differential currents indicate that a fault has occurred, the relays trip their local circuit breaker. The relays also continuously monitor the communication channel performance and carry out self-testing and diagnostic operations [25, 26].

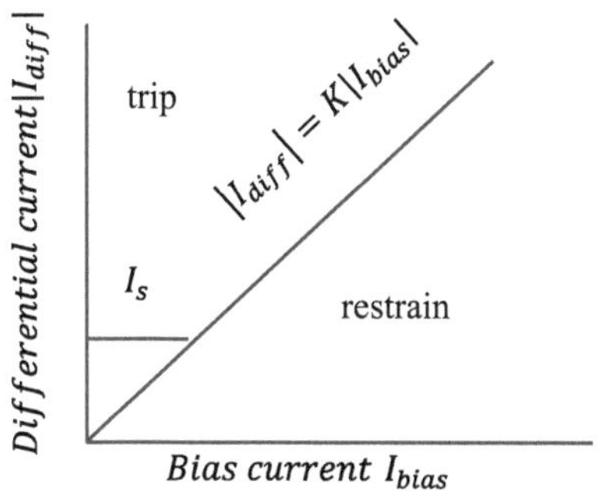

Fig. 5.12 Percentage bias differential protection characteristic. Graph illustrating the relationship between differential current and bias current. A diagonal line represents the equation $|I_{diff}| = K|I_{bias}|$, dividing the graph into two regions: "trip" above the line and "restrain." A horizontal line labeled "I_s" intersects the y-axis, indicating a threshold level

Self-Assessment Questions

? Self-assessment question 5.1a

Explain why it is necessary to condense information into single-phase quantities at each end of a multiphase feeder employing a unit protection scheme of the current differential type. Outline the principles of the two main methods, i.e., the summation transformer and the phase sequence segregating network, used in practice, and briefly discuss the reasons why the former can sometimes lead to "blind spots" under certain fault conditions.

? Self-assessment question 5.1b

An EHV feeder employing a phase-comparison carrier protection scheme has a "c" phase–earth fault at point F as shown in Fig. 5.13 when the source voltages are cophasal. The total system impedance in per unit (p.u) on either side of the fault is as shown.

It illustrates the following points:

(i) A 1:1:3 ratio summation device will be unsatisfactory.
(ii) Summation based upon phase segregating networks compounding the quantity $(6I_{a2} - I_{a1})$ will give satisfactory fault clearance (phase sequence is a, b, and c).

? Self-assessment question 5.2a

Outline the principles of a balanced voltage differential feeder protection and give reasons as to why, in a double-end-fed system, it is desirable to have near simultaneous tripping of the two circuit breakers under internal fault conditions.

? Self-assessment question 5.2b

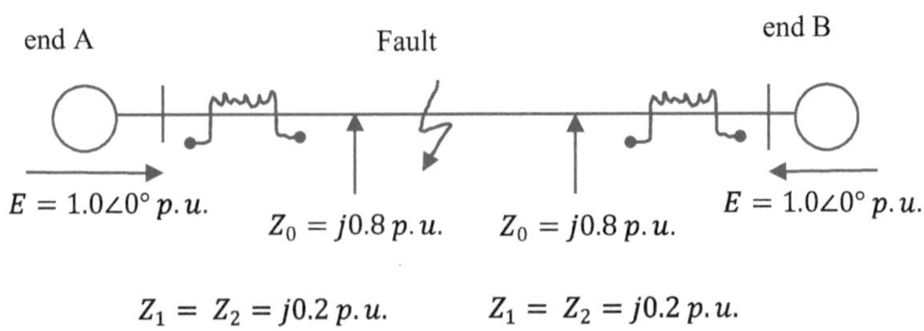

Fig. 5.13 Power flow schematic. Diagram of an electrical transmission line with a fault in the center. The fault is indicated by a lightning bolt symbol between the two sources

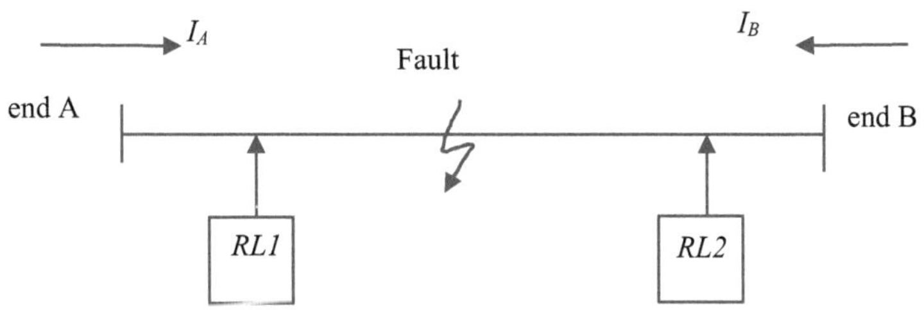

Fig. 5.14 Plain feeder. Diagram illustrating a power line with a fault in the center. Indicate current flow toward the fault from ends A and B, respectively. Two relays, RL1 and RL2, are positioned below the line near each end, connected by arrows pointing upward toward the line

A plain feeder is protected by two relays, RL1 and RL2, as shown in Fig. 5.14. The tripping criterion for the two relays is such that:

$$\left|\frac{I_A}{I_B} - 1\right| \geq 0.5 \, for \, RL1$$

$$\left|\frac{I_B}{I_A} - 1\right| \geq 0.5 \, for \, RL2$$

By drawing the characteristics of the two relays on the same axes, show that when an internal fault occurs on the line, such that $I_A/I_B = (1.5 + j0.5)$ RL2 will fail to detect the fault at the same time as RL1. Suggest an alternative tripping criterion for RL2 that would overcome this problem.

Bibliography

1. Bin Li; Chao Li; Fengrui Guo; Ying Xin "Overcurrent Protection Coordination in a Power Distribution Network With the Active Superconductive Fault Current Limiter" Published in IEEE Transactions on Applied Superconductivity
2. Chong Tong; Yunfeng Cai; Ziyang Zhang; "Artificial intelligence-based lightning protection of smart grid distribution system" published in International Symposium on Lightning Protection (XIV SIPDA) in 2017
3. Harsha Ravindra; M. Omar Faruque; "Impact of PV on distribution protection system" Published in North American Power Symposium (NAPS) in 2012
4. Hesam-Edin Hayati Soloot; Edris Agheb; Amir Hayati Soloot; Shahab Moghadam "A SWOT Analysis of Two Protection Strategies Due to the Expansion of Renewable Distributed Generation Distribution Network" in published in 15th International Conference on Protection and Automation of Power Systems (IPAPS) in 2020
5. Da Lei; Yunlei Yang; Mingjian Hu; "Solutions for the Distributed Photovoltaic Access Distribution Network Relay Protection" Published in 5th Asia Conference on Power and Electrical Engineering (ACPEE) in 2020.
6. Ping Zeng; Zhaoxin Du; Desheng Zhou; Xiaofeng Zou; Qian Feng; "A Study on Adaptive Protection System of Distribution Network Based on Passive Protection Terminal of Ring Network Cabinet" Published in IEEE 6th Conference on Energy Internet and Energy System Integration (EI2) in 2022

7. Kun Yang; Zhen Xiang; Ruixiong Yang; "Decentralized Protection Coordination Based on multi-agent system for Active Distribution Networks Published in 7th Asia Conference on Power and Electrical Engineering (ACPEE) in 2022
8. Renan H. Furlan; Carlos H. Beuter; Rodrigo P. Bataglioli; "Improvement of overcurrent protection considering distribution systems with distributed generation"published in 18th International Conference on Harmonics and Quality of Power (ICHQP) in 2018
9. Bruno Fazio; Sanja Bajic; Serge St-Antoine "Relay Protection Coordination Analysis using Fault Current Measurements:" published in 2022 IEEE Power & Energy Society General Meeting (PESGM).
10. Dong Chen; Bing Hu; Mingjun Xue; Yinian Qi; Chang Song "Research and Application of the System of Relay Protection and Self-Healing for Distribution Network with Mixed Use of Circuit Breaker and Load Switch" published in 2023 International Conference on Power System Technology (PowerCon)
11. Hana A. Halim; Melaty Amirruddin; Nur S. Noorpi; Nurhakimah M. Mukhtar "An Improved Protection Scheme for Smart Distribution Grid published in 2013 1st International Conference on Artificial Intelligence, Modelling and Simulation.
12. Jamil Abdo & Tariq Masood "New Technique to develop Intelligent, highly reliable & Efficient Unmanned Power Generation Modular" 22nd International Conference on Renewable Energies and Power Quality (ICREPQ'24) Bilbao (Spain), 26th to 28th June 2024. https://www.icrepq.com/icrepq24/392-24-abdo.pdf
13. Jamil Abdo & Tariq Masood "Operational Reliability- Production Best Practice Pragmatic Case Studies & Improvement" published in Proceedings of the 9th North American Conference on Industrial Engineering and Operations Management Washington D.C., USA, June 4–6, 2020
14. Jamil Abdo & Tariq Masood "Techno Economics Analysis to Reshape Global Clean and Green Energy" 22nd International Conference on Renewable Energies and Power Quality (ICREPQ'24) Bilbao (Spain), 26th to 28th June 2024. https://www.icrepq.com/icrepq24/415-24-abdo.pdf
15. Jane Courtney; Aaron McDonnell "Impact on Distribution System Protection with the Integration of EG on the Distribution Network" published in 2019 54th International Universities Power Engineering Conference (UPEC).
16. Kun Yang; Zhen Xiang; Ruixiong Yang; Lin Jiang "Decentralized Protection Coordination Based on multi-agent system for Active Distribution Networks" published in 2022 7th Asia Conference on Power and Electrical Engineering (ACPEE).
17. M.Saeed Ullah, Tariq Masood, Umair Zahid, H. Abuzar Ahsan "Electrochemical Oxidation of Dissolved organic Dye in Aqueous Solution' Published in IMASCON-2019
18. Nasser Alemadi, Tariq Masood, R Aggarwal, Suhail A Qureshi, D Kothari "Novel selection approach of STATCOM device tailored and optimized engineering configuration to implement in between Qatar and Bahrain" published in Int. J. Autom. Power Eng. https://doi.org/10.14355/ijape.2014.0302.01
19. Qin Lijun; Wang Ying; Zheng Shenrui; Guo Qing " Situation of study on distribution network protection containing distributed generation and application scheme of intelligent protection" 2011 4th International Conference on Electric Utility Deregulation and Restructuring and Power Technologies (DRPT)
20. Qingfa Chai; Jiayin Jiang; Yi Tang; Anning Wang; Congcong Li " High Reliability Relay Protection Setting Scheme of Distribution Network" published in 2023 International Conference on Power System Technology (PowerCon)
21. Rani Kumari; Bhukya K. Naick "Protection Coordination Impact on Reliability of Distribution Systems With Distributed Energy Resources" 2024 Third International Conference on Power, Control and Computing Technologies (ICPC2T).
22. V. S. Sazanov; A. I. Kovalenko; A. A. Voloshin "The Development of System for Automatic Adaptive Change of Relay Protection Settings in Distribution Networks" published 2022 5th International Youth Scientific and Technical Conference on Relay Protection and Automation (RPA)

23. Vassilis C. Nikolaidis; Evangelos Papanikolaou; Anastasia S. Safigianni "A Communication-Assisted Overcurrent Protection Scheme for Radial Distribution Systems With Distributed Generation" IEEE Transactions on Smart Grid
24. Xiaohui Song; Yu Zhang; Song Zhang; Shizhan Song; Jing Ma; Weibo Zhang "Active distribution network protection mode based on coordination of distributed and centralized protection" publish in 2017 China International Electrical and Energy Conference (CIEEC).
25. Yu Zou; Minchuan Liao; Ruiqing Lian "Discussion on lightning protection of distribution network and its differentiation" published in 2020 International Conference on Intelligent Computing, Automation and Systems (ICICAS)
26. Yuwei Xiong; Wei Chen "Positive sequence current differential protection based on least square method for distribution network with distribution generation" 2021 IEEE 4th International Electrical and Energy Conference (CIEEC).

Auto-Reclose for Transmission Systems

6

Contents

6.1 Introduction 124
6.2 The Main Causes of Faults on Transmission Lines 124
6.3 Transient and Permanent Faults 125
6.4 Fault Clearance 125
6.5 Circuit Breaker Operation 126
6.6 Auto-Reclosing Relays 127
6.6.1 The Auto-Reclosing Relay Sequence 127
6.6.2 Multiple End-Fed Transmission Lines 128
6.6.3 Delaying Auto-Reclosing Relay Schemes 128
6.6.4 High-Speed Auto-Reclosing Relay Schemes 128
6.6.5 Check Synchronism Relays 130
6.6.6 Live Line/Dead Bus, Live Bus/Deadline Check Relays 131
Self-Assessment Questions 131
Bibliography 132

Abstract

Upon completing this chapter, you will be able to list the leading causes of faults in transmission systems and understand the practical implications of this knowledge. This understanding will enable you to describe the likelihood of faults remaining permanently and the mechanism by which an arc may be re-struck. You will also be able to explain the delay settings on auto-reclose relays and their role in preventing system failures in real-world scenarios.

Supplementary Information The online version contains supplementary material available at https://doi.org/10.1007/978-3-031-94312-6_6.

T. Masood et al., *Protection of Electrical Power Transmission Systems*, Power Systems, https://doi.org/10.1007/978-3-031-94312-6_6

Objectives

When you have completed this chapter, you will be able to:

- List the main causes of faults on transmission systems.
- Describe the likelihood of faults remaining permanently.
- Show the mechanism by which an arc may be restruck.
- Explain the delay settings on auto-reclose relays.

6.1 Introduction

The objective of auto-reclose relaying is to return a power system to its normal operating condition automatically as quickly as possible following a break in service caused by a short-circuit fault on that system. The fault will be detected by the relevant protection relay, and either its operation or the opening of the associated circuit breaker will be used to initiate the operation of the auto-reclose scheme. This will reclose the circuit breaker after a preset time delay and, in the majority of cases, restore the power system to its healthy prefault configuration. If the transmission line is part of a heavy interconnected power system, returning the line to service improves the power transfer capabilities of the system and enhances power system stability. In a loosely connected system, where the transmission line provides an intertie between two centers of generation, it is important that synchronization between the two centers is maintained.

6.2 The Main Causes of Faults on Transmission Lines

The most common cause of short-circuit faults on overhead transmission lines is a lightning strike close to the overhead conductors. This type of fault accounts for more than 80% of all faults affecting transmission lines. Indirect lightning strikes induce high voltage into the conductors, which, in turn, leads to insulation failures. If the fault is discounted quickly, there is a high probability that there has been no permanent damage to the power system plant, and the line can be returned to service. This type of fault is referred to as a transient fault.

On long spans of overhead transmission lines, particularly over exposed country, high winds can cause clashing between adjacent conductors, resulting in short-circuit faults. Again, the resultant faults are generally transient faults, and if the line is quickly disconnected from the source of supply, serious damage will not occur.

Direct lightning strikes, where the lightning hits the conductors, contain sufficient concentrated energy to cause permanent damage, resulting in a physical break to the conductors. Clearly, for such a fault, the damage must be repaired before the line can be returned to service. The fault is therefore defined as a permanent fault.

Other causes of permanent faults include contamination buildup on insulators, which eventually leads to insulation breakdown, and foreign objects coming into contact with the conductors. The latter can be caused by accident, vandalism, or vegetation providing the fault path. With underground cables, most faults are permanent and are caused by insulation degradation resulting from aging overvoltage, overheating, or a combination of these [1, 2].

6.3 Transient and Permanent Faults

The majority of faults, between 80% and 90%, on overhead transmission lines are transient in nature and can be cleared by disconnecting the fault from the source of power. As soon as the circuit breaker or breaker has isolated the faulted plant from the source of power, the arc of a transient fault is extinguished, and when the dielectric strength of the insulation has been restored, the plant can be reconnected to the power system [3, 4].

The remaining 10–20% of faults are classified as permanent faults, and any attempt to reconnect them to the power system will result in fault current and the re-tripping of the protection system. The damage caused by permanent faults must be repaired before the plant involved can be restored to normal service. Naturally, any transient fault that is not disconnected from the source of supply will eventually cause permanent damage to the power system plant and hence develop into a permanent fault. The shorter the duration of the fault, the less likely that the fault will cause long-lasting damage, and it is therefore beneficial to reduce fault clearance times [5, 6].

6.4 Fault Clearance

Tripping the circuit breaker when connecting the faulted plant to the source or sources of supply does not instantaneously remove the fault from the power system. The circuit breaker requires time to separate its contacts mechanically, and then the arc formed by the opening contacts must be ruptured, finally breaking the flow of current into the fault. The process involved in rupturing the circuit breaker arc also leads to the rupturing of the fault arc. These two arcs are in series. This may require only a few half cycles but can last for longer depending on the type of switch gear and the characteristics of the circuit involved [7, 8]. Before the circuit breaker can be reclosed, thereby re-energizing the previously faulted plant, the dielectric strengths of the insulation must have covered sufficiently to withstand the system voltages involved. If the dielectric strength has not recovered, the re-energization voltage will restrike the arc, and the fault will be re-established [9, 10].

The time required for gases involved in the fault arc to recover their dielectric strengths for three-pole tripping has been analyzed over several decades and has led to the empirical curves from separate work undertaken in the United Kingdom and the United States. The differences in the findings illustrate the complexity of the subject. Both curves show that for a 275-kV transmission line, a delay of

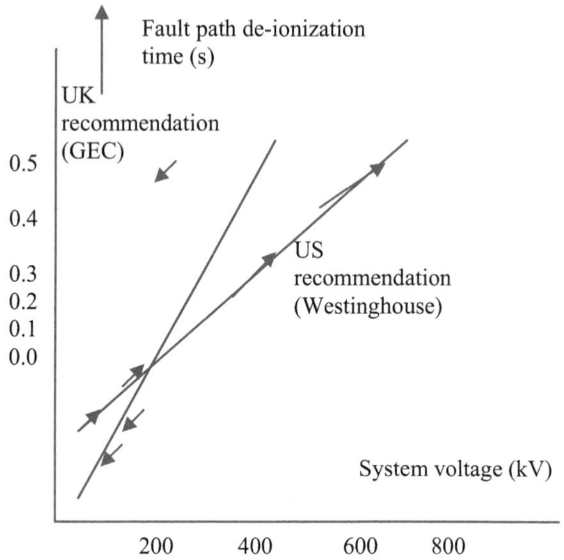

Fig. 6.1 (Fault de-ionization time for three pole tripping) Graph showing fault path de-ionization time in seconds versus system voltage in kilovolts. Two lines represent recommendations: the UK recommendation (GEC) with a steeper slope, and the US recommendation (Westinghouse) with a more gradual slope. The UK line starts at 0.1 seconds for 200 kV, while the US line starts slightly higher, indicating different de-ionization times for the same voltages

approximately 0.3 s is required to allow the ionization caused by the fault to disperse (Fig. 6.1).

If the fault is a single-phase-to-earth fault and only one pole of the circuit breaker is tripping, longer times are required for the de-ionization of the fault path. This is because of voltages induced onto the tripped phase conductor from the untripped phases.

6.5 Circuit Breaker Operation

In addition to considering the state of the power system in assessing the criteria for successful auto-reclose, it is important that the switchgear is also suitable for this duty. The auto-reclose cycle imposes severe stresses on the circuit breaker, and therefore, before reclosure can be contemplated, the circuit breaker must be ready to close its contacts and also trip, without delay, should the fault persist. In most cases, the times required to allow the circuit breaker to reset are less than those required for the de-ionization of the fault path [11, 12]. The resetting time of a circuit breaker depends on its type and design. For EHV breakers, rapid single-shot reclosures present few problems; however, multiple operations can be impeded by the need to recharge the closing mechanism.

Exercise 6.1
Explain why the operation of the circuit breaker reclosure contact should be delayed after the circuit breaker tripping contact has operated.

Before the circuit can be reclosed, time must be allowed for the circuit breaker to complete to operation and rest following a trip. The circuit breaker has to be ready not only to reclose but also to trip again should the fault not be cleared. Time must also be allowed for the ionization caused by the fault to disperse before the transmission line can be re-energized so that the dielectric strength of the insulation can recover. In general, the time required for de-ionization of the fault path is greater than that required for the circuit breaker to reset [13, 14].

6.6 Auto-Reclosing Relays

Although included in the general definition of protection relays, auto-reclosing relays do not detect fault conditions, do not cause circuit breaker tripping, are instead involved in circuit breaker tripping, and are instead involved in circuit breaker reclosing. Auto-reclosing relays are switching sequence controllers and can be implemented using electromechanical, static, programmable logic controller and microprocess technology. They do, however, have to operate in a scheme including and alongside protection relays, and they share the same harsh electrical environment as the protection relays.

6.6.1 The Auto-Reclosing Relay Sequence

The auto-reclosing relay's switching sequence can be triggered either from the operation of the protection trip relay's output or from an auxiliary contact on the circuit breaker. In general, using an auxiliary contact on the circuit breaker has some advantages over using the protection trip output. The circuit breaker contact operates when the breaker's main contact opens and not when the instruction to trip is given [14, 15]. Assuming that the sequence is triggered by the protection relay, the switching sequence begins with a time delay to allow the fault current to be broken and for de-ionization of the fault path. This time delay between the auto-reclose scheme being initiated and the operation of the relay to close the circuit breaker is defined as the *dead time* of the auto-reclose relay. After the dead time, the auto-reclose relay-initiated circuit breaker closure, and after the time required for the contacts to close fully, power is restored to the circuit. For a transient fault, the fault will have been cleared by the first trip, and the power will be successfully restored. If, though, the fault was a permanent fault, the reclosing may provide a *single-shot reclosure or multiple shots*. With the single-shot scheme, the second trip would lead to a lock-out condition where subsequent reclosure of the circuit breaker would demand operator intervention, and with multiple-shot schemes, multiple operator intervention. With multiple shot schemes, multiple dead time and reclosures would be introduced before lock-out [16].

The auto-reclose scheme output contact, which initiates circuit breaker reclosure, also initiates the scheme's resetting timer, the reclaim timer. If the reclosure is successful and the protection does not trip during the reclaim time, the auto-reclose scheme is reset, awaiting the next fault condition. The reclaim time is chosen to be sufficiently long to ensure that the system has been safely restored but short enough to ensure that an independent transient fault is not mistaken for a continuous part of the original fault. This is particularly important in areas that are prone to lightning strikes and where storms could produce a series of independent faults. These switching operations are illustrated in Fig. 6.2, which shows the timing sequence for a single-shot auto-reclose in response to a transient and a permanent fault [17]. A typical schematic for a single-shot auto-reclose scheme with lock-out is shown in Fig. 6.3.

6.6.2 Multiple End-Fed Transmission Lines

Most transmission lines are fed from two or more ends, and hence consideration must be given to the operation of the protection and auto-reclose schemes at all of the feeders' terminals when designing the auto-reclose scheme. It is generally assumed that the relaying schemes at the different terminals are not synchronized and that communication systems do not exist for this duty.

The fault path will be maintained until all sources of power are disconnected from the faulted feeder. The delay between the tripping of the protection and the different ends of the feeder therefore requires increased dead times of the auto-reclose relays to ensure that sufficient time is allowed for the fault path to de-ionize [18].

6.6.3 Delaying Auto-Reclosing Relay Schemes

With a heavily interconnected power system, the tripping of one transmission line will not break the link between generation units, and therefore, in the short term, the generators will not significantly drift apart and lose synchronism. The preferred approach to reconnecting the transmission line is to wait for any oscillation resulting from the tripping to die down and then reclose the circuit breaker or breakers, returning the line to service. Providing the system with time to wait is achieved by delaying the auto-reclosure, i.e. using a delayed or low-speed auto-reclose scheme.

Technically, delayed auto-reclose schemes are defined as those where the time delay before reclosing the breaker, the dead time, is greater than 1 s.

6.6.4 High-Speed Auto-Reclosing Relay Schemes

High-speed auto reclosure schemes are those where the dead time is less than 1 s. For loosely interconnected power systems where the transmission line provides the intertie between major sources of generation, they offer the possibility of reconnecting these sources before the systems have lost synchronism or before the system as a whole loses stability. For both synchronization and stability, the fault duration

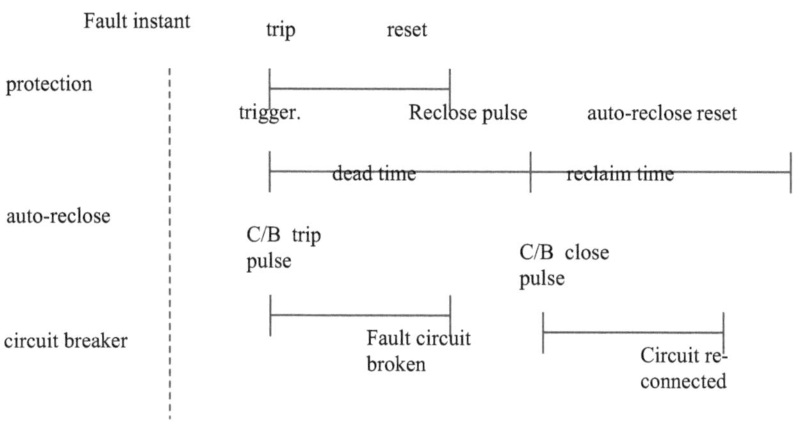

(a) **reclosure for a transient fault -** Diagram illustrating the sequence of events in a circuit breaker autoreclose system. It shows the timeline from a fault instant, including the protection trigger, reclose pulse, and autoreclose reset. Key phases are marked: "trip" with "trigger," "reset" with "Reclose pulse," "dead time," and "reclaim time." The circuit breaker (C/B) processes include "C/B trip pulse," "Fault circuit broken," "C/B close pulse," and "Circuit re-connected."

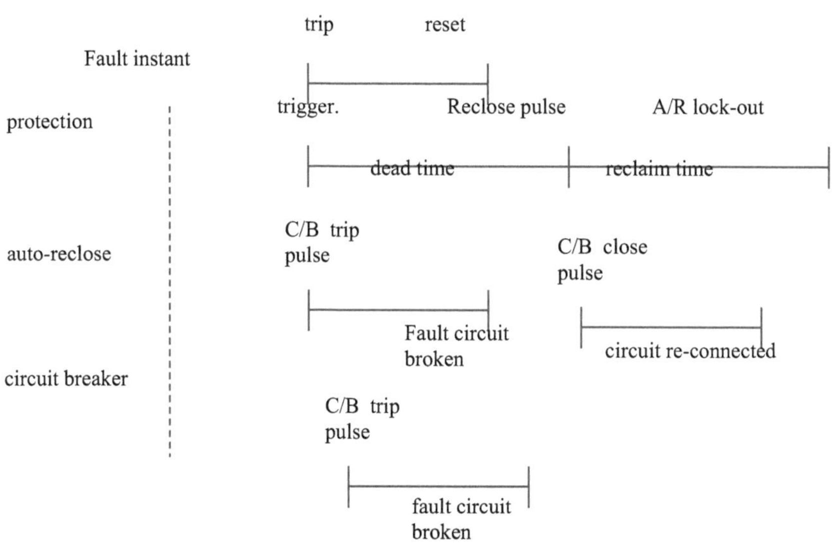

(a) reclosure for a permanent fault

Fig. 6.2 (Auto-reclose switching sequence for transient and permanent faults) Diagram illustrating the sequence of events in an auto-reclose protection system for circuit breakers. It shows the timeline from fault instant to circuit re-connection. Key phases include the trip trigger, dead time, reclose pulse, reclaim time, and A/R lock-out. Labels indicate "Fault circuit broken" and "circuit re-connected" with corresponding C/B trip and close pulses

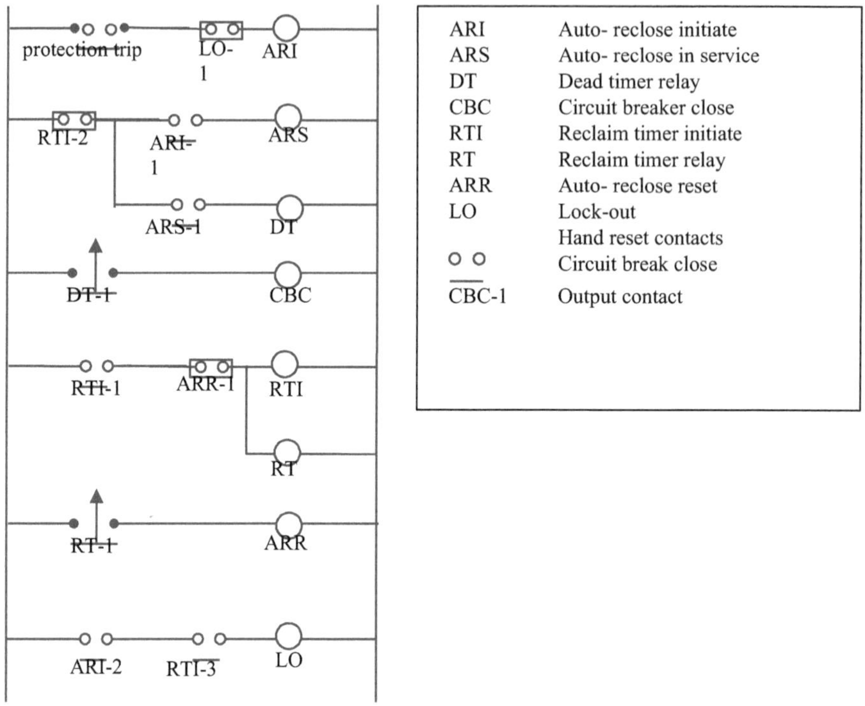

Fig. 6.3 (Typical schematic for single-shot auto-reclose scheme) Diagram of an electrical circuit with labeled components and a key. The circuit includes elements like protection trip, auto-reclose initiate (ARI), auto-reclose in service (ARS), dead timer relay (DT), and others. The key explains abbreviations: ARI as auto-reclose initiate, ARS as auto-reclose in service, and DT as dead timer relay. The diagram shows connections between these components, illustrating their interactions in the circuit

must be kept to a minimum, and therefore, high-speed protection is required. The auto-reclose scheme's dead time should be short such that the generation systems are separated for as short a time as possible. Also, the circuit breakers at the line's terminals would be tripping simultaneously, and special schemes are available to provide this feature [19].

6.6.5 Check Synchronism Relays

Check synchronism relays are used to provide an automatic check on phase angle and voltage difference between the two power systems, which are to be connected by the circuit breaker. They operate when these differences are within preset limits and are restrained whenever these differences are outside these limits. The relay's output is used to permit the closing operation of the circuit breaker when the reclosure will not cause a major disturbance to the system.

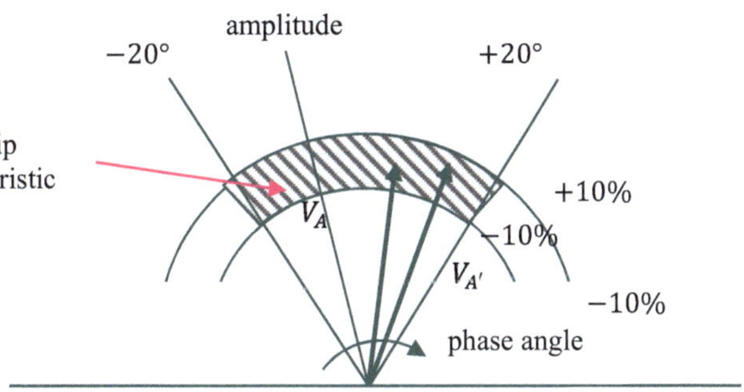

Fig. 6.4 (Typical operating characteristic for a check synchronization relay) Diagram illustrating a relay trip characteristic with a shaded arc representing the operational range. The arc spans from −20° to +20° in phase angle and from −10% to +10% in amplitude. A pink arrow points to the shaded area, labeled "Relay trip characteristic," indicating the conditions under which the relay will trip

Typical operating characteristic for a check synchronizing relay is shown in Fig. 6.4. The relay inputs, VA and VA', are obtained from voltage transformers on either side of the circuit breaker. This figure shows the characteristic where the phase angle setting is ±20° and its voltage unbalance setting is ±10% by including a time setting to operate with the phase angle setting. This enables the relay to be set to restraint whenever the slip frequency between the two system is above a set time.

6.6.6 Live Line/Dead Bus, Live Bus/Deadline Check Relays

Whenever the system connected to one side of an open circuit breaker is de-energized, i.e., dead, the closing of the circuit breaker can be executed at any time without the problem of causing a system disturbance. A live line/dead bus or live bus deadline check relay will operate whenever the system on one side of the circuit breaker is de-energized allowing it to be closed safely. This type of relay uses voltage relays in a scheme similar to that shown in Fig. 6.5. The sensitivity of these relays is chosen such that any induced voltage from parallel lines sharing the same towers will not cause malfunctioning [20].

Self-Assessment Questions

? Self-assessment question 6.1

Starting with the auto-reclose scheme shown in Fig. 6.3, construct a scheme providing single-shot auto-reclose for single-phase faults and immediate lock-out for phase-to-phase and three-phase faults. Include check synchronization and live line/

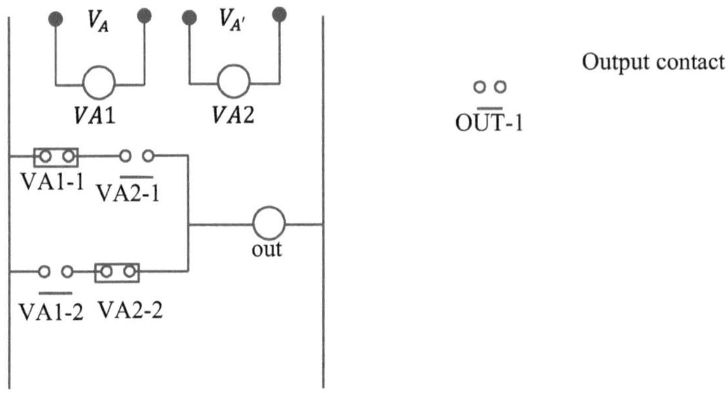

Fig. 6.5 (Live line/ dead bus, line bus/deadline check supervision relay) Diagram of an electrical circuit with two vertical lines. The left line has two switches labeled VA1 and VA2, connected to VA1-1, VA2-1, VAT-2, and VA2-2. The right line is labeled "OUT-1" and connects to an output labeled "out"

dead bus and deadline live bus supervision into the scheme as an automatic check on the circuit breaker closure.

Bibliography

1. Vassilis C. Nikolaidis; Evangelos Papanikolaou; Anastasia S. Safigianni "A Communication-Assisted Overcurrent Protection Scheme forRadial Distribution Systems With Distributed Generation" IEEE Transactions on Smart Grid
2. Da Lei; Yunlei Yang; Mingjian Hu; "Solutions for the Distributed Photovoltaic Access Distribution Network Relay Protection" Published in 5th Asia Conference on Power and Electrical Engineering (ACPEE) in 2020.
3. L. Ndahepele; S. Chowdhurym "Impact of Distributed Generation on Traditional Protection in Distribution and Transmission Systems: A Review" Published in IEEE PES/IAS PowerAfrica in 2020
4. Cleberton Reiz; Jonatas Boas Leit "Impact Analysis of Distributed Generation on Protection Devices Coordination in Power Distribution Systems" 2021 IEEE PES Innovative Smart Grid Technologies Conference - Latin America (ISGT Latin America) in 2021
5. Kun Yang; Zhen Xiang; Ruixiong Yang; "Decentralized Protection Coordination Based on multi-agent system for Active Distribution Networks Published in 7th Asia Conference on Power and Electrical Engineering (ACPEE) in 2022
6. Renan H. Furlan; Carlos H. Beuter; Rodrigo P. Bataglioli; "Improvement of overcurrent protection considering distribution systems with distributed generation" published in 18th International Conference on Harmonics and Quality of Power (ICHQP) in 2018
7. Ping Zeng; Zhaoxin Du; Desheng Zhou; Xiaofeng Zou; Qian Feng; "A Study on Adaptive Protection System of Distribution Network Based on Passive Protection Terminal of Ring Network Cabinet" Published in IEEE 6th Conference on Energy Internet and Energy System Integration (EI2) in 2022
8. Xiaohui Song; Yu Zhang; Song Zhang; Shizhan Song; Jing Ma; Weibo Zhang "Active distribution network protection mode based on coordination of distributed and centralized protection" publish in 2017 China International Electrical and Energy Conference (CIEEC).

9. Yuwei Xiong; Wei Chen "Positive sequence current differential protection based on least square method for distribution network with distribution generation" 2021 IEEE 4th International Electrical and Energy Conference (CIEEC).
10. Zong Guo; Jiarui Li; Hongbo Lian; Shihua Zhao; Yang Chen; Renjie Dai; Xiuneng Xu; Yufeng Gu; Yangyang Liu "Optimization Method for Protection of Distribution Network with MW-PV Based on Multi-level Cooperative Protection Strategy" 2019 IEEE Sustainable Power and Energy Conference (iSPEC).
11. Jamil Abdo & Tariq Masood "New Technique to develop Intelligent, highly reliable & Efficient Unmanned Power Generation Modular" 22nd International Conference on Renewable Energies and Power Quality (ICREPQ'24) Bilbao (Spain), 26th to 28th June 2024. https://www.icrepq.com/icrepq24/392-24-abdo.pdf
12. Jamil Abdo & Tariq Masood "Operational Reliability- Production Best Practice Pragmatic Case Studies & Improvement" published in Proceedings of the 9th North American Conference on Industrial Engineering and Operations Management Washington D.C., USA, June 4-6, 2020
13. Jamil Abdo & Tariq Masood "Techno Economics Analysis to Reshape Global Clean and Green Energy" 22nd International Conference on Renewable Energies and Power Quality (ICREPQ'24) Bilbao (Spain), 26th to 28th June 2024. https://www.icrepq.com/icrepq24/415-24-abdo.pdf
14. Jane Courtney; Aaron McDonnell "Impact on Distribution System Protection with the Integration of EG on the Distribution Network" published in 2019 54th International Universities Power Engineering Conference (UPEC).
15. M. Saeed Ullah, Tariq Masood, Umair Zahid, H. Abuzar Ahsan "Electrochemical Oxidation of Dissolved organic Dye in Aqueous Solution' Published in IMASCON-2019
16. Nasser Alemadi, Tariq Masood, R Aggarwal, Suhail A Qureshi, D Kothari "Novel selection approach of STATCOM device tailored and optimized engineering configuration to implement in between Qatar and Bahrain" published in Int. J. Autom. Power Eng. https://doi.org/10.14355/ijape.2014.0302.01
17. Qin Lijun; Wang Ying; Zheng Shenrui; Guo Qing "Situation of study on distribution network protection containing distributed generation and application scheme of intelligent protection" 2011 4th International Conference on Electric Utility Deregulation and Restructuring and Power Technologies (DRPT)
18. Rani Kumari; Bhukya K. Naick "Protection Coordination Impact on Reliability of Distribution Systems With Distributed Energy Resources" 2024 Third International Conference on Power, Control and Computing Technologies (ICPC2T).
19. Tariq Masood & Jamil Abdo "New Technique to Enhance Hybrid Renewable Energy Resources' Operational Flexibility and Reliability" 22nd International Conference on Renewable Energies and Power Quality (ICREPQ'24) Bilbao (Spain), 26th to 28th June 2024. https://www.icrepq.com/icrepq24/396-24-masood.pdf
20. Tariq Masood and Jamil Abdo "Review: Uninterrupted Liquid Hydrogen (E-Methanol) Production framework to deliver Clean & Green Energy" Published 22nd International Conference on Renewable Energies and Power Quality (ICREPQ'24) Bilbao (Spain), 26th to 28th June 2024. https://www.icrepq.com/icrepq24/414-24-masood.pdf

Numeric Protection

7

Contents

7.1 Introduction.......... 136
7.2 Microprocessors.......... 137
7.2.1 General Description.......... 137
7.2.2 Communication between a Microprocessor and its Peripherals.......... 139
7.2.3 Number Representation.......... 141
7.3 Relay Hardware.......... 142
7.3.1 Introduction.......... 142
7.3.2 Data Acquisition.......... 142
7.3.3 Galvanic Isolation Transformers.......... 145
7.3.4 Memory.......... 146
7.3.5 Timer.......... 147
7.3.6 Microprocessor.......... 147
7.3.7 Serial Communications.......... 148
7.3.8 Digital I/O.......... 148
7.3.9 Keypad and Display.......... 149
7.3.10 Electromagnetic Compatibility.......... 149
7.4 Digital Signal Processing.......... 149
7.4.1 Sampling.......... 149
7.4.2 Sampling Theorem.......... 149
7.4.3 Preventing Aliasing in a Protection Relay.......... 151
7.4.4 Digital Filtering.......... 152
7.4.5 Fourier Transforms.......... 155
7.4.6 The Fast Fourier Transforms.......... 156
7.5 Distance Protection.......... 157
7.5.1 Introduction.......... 157
7.5.2 Filtering.......... 157

Supplementary Information The online version contains supplementary material available at https://doi.org/10.1007/978-3-031-94312-6_7.

T. Masood et al., *Protection of Electrical Power Transmission Systems*, Power Systems, https://doi.org/10.1007/978-3-031-94312-6_7

7.5.3 Impedance Calculation........ 158
7.5.4 Fault Evaluation........ 159
7.5.5 Relay Elements........ 161
7.6 Differential Protection........ 162
7.6.1 Introduction........ 162
7.6.2 Hardware........ 162
7.6.3 Current Vector Calculation........ 163
7.6.4 Source of Error in a Digital Differential Relay........ 163
7.6.5 Propagation Delay Compensation........ 164
7.6.6 Differential Characteristics........ 166
Self-Assessment Questions........ 167
Bibliography........ 170

Abstract

When you have completed this chapter, you will be able to: Describe how a microprocessor communicates with its peripherals. List the primary components found within the hardware of a numeric relay. Explain how analog busbar signals are converted into numeric values. Explain the cause of the aliasing of sampled signals. Describe how a digital filter works and why analog and digital filters are used in a numeric relay. Describe how values may be converted into phasor form using the discrete Fourier transform. In addition, you will have a comprehensive understanding of the operation of a numeric distance relay and a numeric current differential relay, which will make you feel confident and competent in dealing with these critical components in protection systems.

Objectives

When you have completed this chapter, you will be able to:

- Describe how a microprocessor communicates with its peripherals.
- List the main components found within the hardware of a numeric relay.
- Explain how analog busbar signals are converted into numeric values.
- Explain the cause of aliasing of sampled signals.
- Describe how a digital filter works and why both analog and digital filters are used in a numeric relay.
- Describe how values may be converted into phasor form using the discrete Fourier transform (FT).
- Explain the operation of a numeric distance relay.
- Explain the operation of a numeric current differential relay.

7.1 Introduction

This chapter is an introduction to the field of numeric protection, that is, relay, which forms an operating decision based upon processing of the input signals by a computer or, more pertinently, a microprocessor. The term numeric is adopted to avoid confusion with static (i.e., electronic) relay, which also uses microprocessors for some aspects of the overall relay function, e.g., scheme logic; such relays can be classed as digital, but not numeric.

Numeric protection is an important and fast-growing area; numeric relays are likely to be more common in the future due to their low cost, high reliability, and ability to provide better and more versatile protection functions than have been possible with conventional static relays. Since the mode of operation of numeric relays differs so much from that of conventional relays, this chapter is provided to allow you to gain an understanding of numeric relay fundamentals.

To understand numeric relays requires an understanding of microprocessor-based systems. First of all, you are given a brief introduction to the operation of a microprocessor, which then leads on to a description of typical numeric relay hardware. In order to understand how analog signals are brought into the digital environment, a description of the *data acquisition* circuitry is provided. Having digitized such signals, some theory is required relating to how the signals may be subsequently processed, and this is given in the chapter on *digital signal processing*.

Upon reaching this point, you should have the necessary background knowledge to understand the operation of a numeric relay. Although many types of numeric relays are available, only the two main types used in transmission protection will be described. These are distance and current differential relays with which you will already be familiar since they have been described in earlier chapters of this book. This conveniently allows you to draw a technical comparison between the differing technologies [1].

7.2 Microprocessors

7.2.1 General Description

All digital relays are based upon microprocessors, so a logical starting point for these notes is a brief introduction to microprocessors. No detailed explanation of the internal machinations of the microprocessor will be given here, nor would it be appropriate; however, readers wishing to understand more can find a readily available, if not bewildering, selection of textbooks dealing with this area.

Computers are essentially sequential, binary devices to which the logic can be arbitrarily specified or programmed. Binary means that the signals in a microprocessor can only assume that it is common to use the terminology "0" and "1" to denote the state of a signal. Prior to the development of the integration technique that allows, currently, millions of transistors to be fabricated onto the same piece of the silicon substrate, computers were composed of separate circuits suitably interconnected. When it became possible to integrate most of these circuits onto the same piece of silicon, computers became known as microprocessors.

The basis of the microprocessor system is shown in Fig. 7.1. Devices that are directly connected to a microprocessor are usually called peripherals. The most important peripheral of Fig. 7.1 is marked as I/O, which stands for input and output, in essence, the connection or interface to the outside world; a microprocessor system that is not connected to the outside world; a microprocessor system that is not connected to the outside world is useless. In the context of protection relays, the I/O

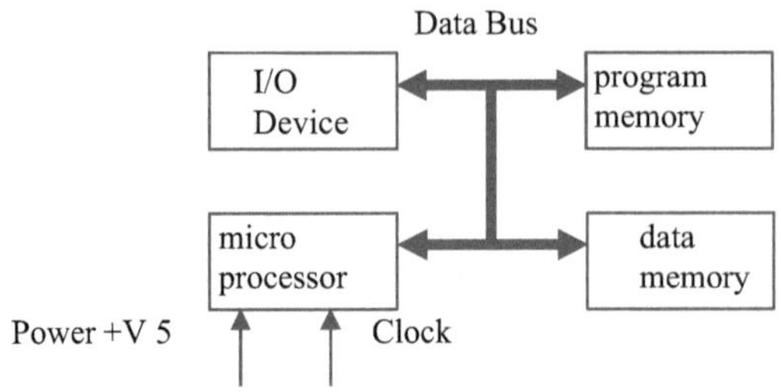

Fig. 7.1 Microprocessor and peripherals. Diagram of a microprocessor system architecture. The central component is a data bus connecting four elements: I/O device, program memory, microprocessor, and data memory. The microprocessor receives power labeled as "Power +V 5" and a "Clock" signal. Arrows indicate data flow directions between components

Table 7.1 Input and output forms of information

Input	Application	Output
Optical pulses	Compact disk player	Analog signal
Keystrokes	Word processor	VDU, magnetic media, this workbook
Analog voltage and current signals	Protection relay	Binary output (trip or no trip)

would be represented by the connection to busbar voltage transformers and current transformers (input) and circuit breaker trip contacts and alarms (output). In addition, the microprocessor is connected to two types of memory, program and data, a power supply input typically +5 V, and a timing signal input, or clock, which will be derived from a quartz crystal [2]. It is the job of the I/O devices to convert between the different representations of information used in the outside world and inside the microprocessor. Inside a microprocessor, the only representation used is in the form of numbers.

Exercise 7.1
In what form are the input and output signals to a digital protection relay?

The information exchanged with a digital protection relay is shown in Table 7.1. Two other examples are also shown in this table.

In Fig. 7.1, the microprocessor is connected to different types of memory. Program memory contains a list of instructions to the microprocessor that makes it behave in a predetermined fashion. Generally, a microprocessor is unable to change its program memory. Data memory is used for the transient store of information necessary for the microprocessor to perform its designed task.

Exercise 7.2
Can you deduce the uses of program and data memory in a digital protection relay?

Program memory is used to contain the specific list of instructions that will allow the microprocessor to perform the desired relaying task, i.e., behave as a distance relay as opposed to a current differential relay. Data memory is used for recording numbers representing VT or CT signals and the results of intermediate calculations; after the microprocessor has processed such numbers, they are of no further use, and the relevant part of the memory can be reused. Hence, a microprocessor is able to change the contents of the data memory [3].

If you have understood the basic devices (memory and I/O) to which a microprocessor is connected, you are now ready to understand, in simple terms, how a microprocessor operates. A microprocessor can perform four different types of tasks, or operations. These are:

- The ability to move numbers around, e.g., from an I/O device to memory.
- The ability to perform arithmetic operations on number, e.g., add two numbers together.
- The ability to perform logic operations on numbers, e.g., "and" two numbers together.
- The ability to perform operations conditional on the outcome of some previous operation, e.g., add two numbers together *if* the result of the previous addition was zero.

Of course, within each of these categories, there are many different types of operations. All complex applications of microprocessors are based upon the repeated use of the above operations. To specify in advance which particular type of operation a microprocessor is to perform, the microprocessor manufacturer provides a list of numbers that correspond to instructions to the microprocessor. If the correct numbers are placed in the program memory, then the microprocessor will perform the desired task. This process is referred to as *programming* [4]. Generally, microprocessors are programmed through a *programming language* that converts the programming instructions written by a human, and understood by a human, into instructions understood by the microprocessor. In general, programming requires some form of algorithm—a procedure, usually mathematically based, which describes how a particular task may be performed on a computer or microprocessor. Algorithms are subsequently converted into computer programs. Another generic term describing computer programs is *software.*

7.2.2 Communication between a Microprocessor and its Peripherals

Figure 7.1 shows that the microprocessor is connected to its memory and I/O through a data bus. The data bus allows binary data, i.e., numbers, to be transferred to and from the microprocessor; the data bus is hence described as being *bidirectional.* A data bus is composed of a number of separate signal lines, or bits, each of which can convey a binary value, i.e., 1 or 0. It is usual to classify a microprocessor according to the number of bits or 32-bit microprocessor. This assumes that the number of bits in the data bus is the same as the number of bits that the microprocessor can handle internally—this is usually the case, although there are exceptions. The number of bits in the microprocessors with large word lengths can handle greater precision of calculation more easily than microprocessors with smaller word lengths.

In addition to the data bus, other signals are required to effect data transfer between the microprocessor and its peripherals. These signals are not shown in Fig. 7.1. A memory device, for example, can typically hold over 16,000 separate 8-bit numbers. The exact position in the memory where a number is stored is referred to as the *location*. In order for the microprocessor to specify which particular location in the memory it wishes to access, another bus is used—the *address bus.* The address bus is not bidirectional since the addressing information only travels from the microprocessor to the memory or peripheral device. The size of the address bus dictates the total amount of memory to which the microprocessor may be connected—address buses are typically 20 bits in length [5].

Figure 7.2 shows the arrangement of the data and address buses for a simple microprocessor/memory system. If the microprocessor sends a number to the memory, this is referred to as a *memory write*. Conversely, if the microprocessor receives a number from the memory, this is known as a memory read. To allow the memory to differentiate between reads and writes, an extra signal is provided from the microprocessor, shown as $R/\overline{W}$ in Fig. 7.2. The $R/\overline{W}$ is set to a 1 if the ensuing operation is a read, and to a 0 if the operation is to be a write.

Exercise 7.3

Explain how the data bus, address bus, and $R/\overline{W}$ signal are used when the contents of a location in memory are transferred into the microprocessor.

This transfer is a memory read operation, and hence the $R/\overline{W}$ signal is set to 1 to inform the memory to place the contents of the required location, indicated by the

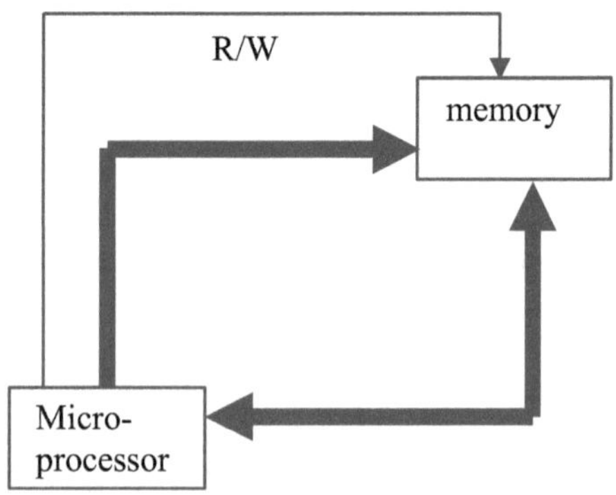

Fig. 7.2 Address and data buses. Diagram illustrating data flow between a microprocessor and memory. The microprocessor sends data to memory, indicated by an arrow pointing from the microprocessor to memory. Another arrow shows data returning from memory to the microprocessor. A separate line labeled "R/W" connects the two, indicating read and write operations

address bus, onto the data bus, which will be configured to convey the data from the memory to the microprocessor [6].

7.2.3 Number Representation

Humans count according to the number base 10 since we all have 10 fingers. Since a microprocessor works in binary states, i.e., 0 or 1, it counts according to the number base 2. Considering an 8-bit data bus, the arrangement of the number representation is shown below.

2^7	2^6	2^5	2^4	2^3	2^2	2^1	2^0

The first data bit, on the right-hand side, represents whether 2°, i.e., a unit, is contained in the whole number; this is also referred to as the *least significant* bit (LSB). Conversely, the last data bit, on the left-hand side, represents the value of 2^7, i. e. , 128. This is the highest-ranked data bit and is also referred to as the most significant bit (MSB). This form of representation is called an unsigned integer and can represent numbers from 0 *to* $2^8 - 1$, i.e., 255 decimal or 111,111 11 in binary. In order to represent a signed number, microprocessors use a convention called *two's complement.*

-2^7	2^6	2^5	2^4	2^3	2^2	2^1	2^0

In the two's complement notation, the smallest number is -2^7, *i. e.* , – 128 in decimal or 1000,000,0 in binary and the largest number $2^7 - 1$, *i. e.*, 127 in decimal or 0111,111,1 in binary. Finally, a fractional representation is also possible:

-2^0	2^{-1}	2^{-2}	2^{-3}	2^{-4}	2^{-5}	2^{-6}	2^{-7}

In all of the previously mentioned number representations, the binary equivalent of the decimal point is in a fixed position. Thus, these representations are referred to as fixed points. Another representation that is used is called floating point. Here, a number, *X*, is expressed as:

$$X = M \times 10^e \tag{7.1}$$

where *M*, the mantissa, and *e*, the exponent, are both signed integer numbers and are both contained within a data word. Floating-point numbers can represent a far wider range of values but are more difficult for the microprocessor to deal with than fixed-point numbers [7].

Exercise 7.4

Convert 10,001,001 into its decimal equivalent using: (a) unsigned integer, (b) signed integer, and (c) signed fractional representations.

(a) Decimal equivalent:$=2^7 + 2^3 + 2^0 = 128 + 8 + 1 = 137$
(b) Decimal equivalent:=. $-2^7 + 2^3 + 2^0 = -128 + 8 + 1 = -119$
(c) Decimal equivalent:=. $-2^0 + 2^{-4} + 2^{-7} = -1 + \frac{1}{16} + \frac{1}{128} = -0.9296875$

7.3 Relay Hardware

7.3.1 Introduction

By now, you should have a basic understanding of microprocessors and, more importantly, understand the meaning of some of the jargon terms. We will now examine the basic hardware structure of a numeric protection relay, a diagram of which is shown in Fig. 7.3. With our current rudimentary understanding of microprocessor systems, there are several aspects of Fig. 7.3, which will not be readily understood. However, we can see there is a microprocessor in the relay. There are also different types of memory, labeled as read-only memory (ROM), random access memory (RAM), and non-volatile random-access memory (VNRAM). With the exception of the block marked "timer," we can assume that all other connections to the microprocessor come under the category I/O. We will now consider each stage of the relay in detail. All circuitry to the left of the block marked ADC is referred to as data acquisition. We will begin by considering this chapter first [8].

7.3.2 Data Acquisition

A numeric relay must convert the analog relaying signals from the power system transducers, VTs, and CTs into numerical values that the microprocessor can subsequently manipulate. This process is referred to as *data acquisition, or sampling*. We refer to converted, numeric values of the analog relaying signals as samples. Sampling normally occurs at a regular time interval; the sampling frequency can be anything from 300 Hz to 4 kHz, depending on the type of relay.

Analog-to-Digital Converter

The most important component found in a data acquisition system is the analog-to-digital converter, shown as ADC in Fig. 7.3. This device accepts an analog voltage as an input and converts it to a numeric, binary value. For details on how ADCs work and the types available, see any good textbook on digital electronics. The ADC most frequently used in a protection relay is the *successive approximation* type. When the analog voltage to be converted is present on the input to a *successive approximation* converter (SAC), a "start conversion" signal is required from the microprocessor [9, 10]. The SAC then begins to form its digital estimate of the input voltage. This process takes a short internal time, called the conversion period, which is typically $20\mu s$. When the conversion process is completed, the converter informs the microprocessor with an "end of conversion" signal. The digital value can then

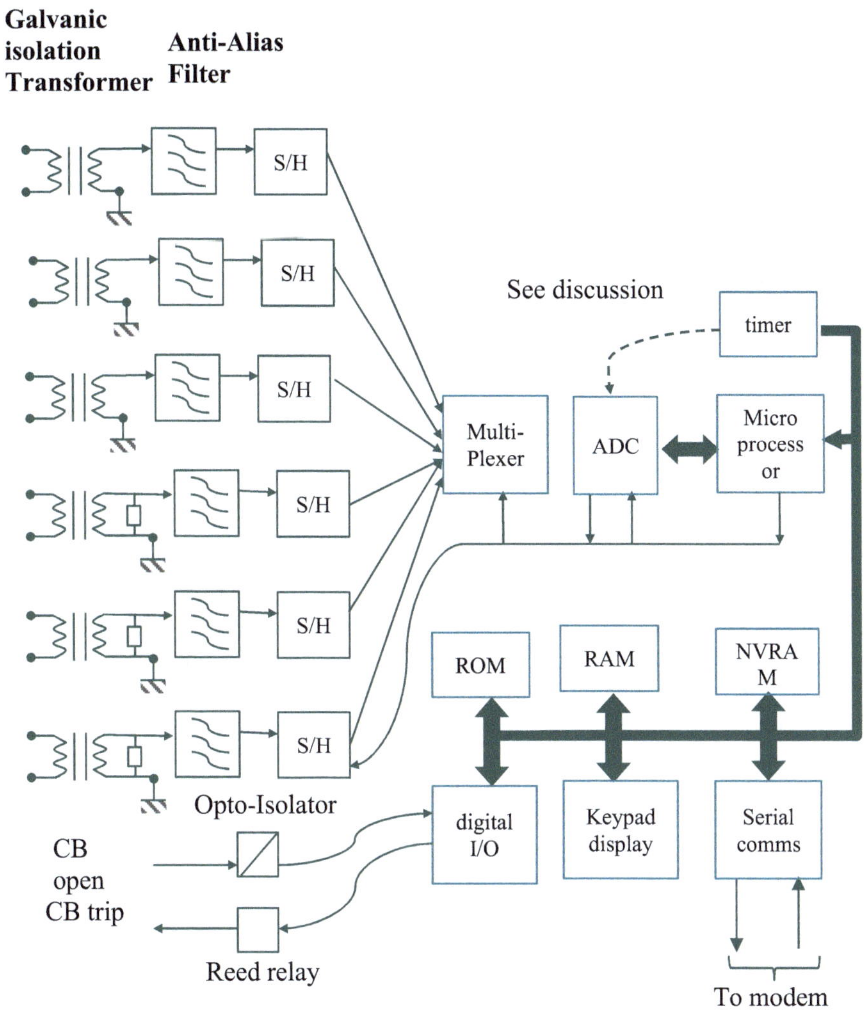

Fig. 7.3 Digital relay hardware structure. Diagram illustrating a data acquisition system. It shows multiple inputs passing through galvanic isolation transformers and anti-alias filters to sample and hold (S/H) circuits. These inputs are then routed to a multiplexer, which connects to an analog-to-digital converter (ADC). The ADC interfaces with a microprocessor, which also connects to a timer

be read into the microprocessor via the data bus. The resolution of the ADC, the number of bits to which the input signals are converted, is usually in the range of 12–16 bits.

Multiplexer

ADCs tend to be expensive devices, and thus, manufacturers do not provide an ADC for every signal coming into the relay. In Fig. 7.3, there are six input channels to the relay: three voltage and three current. Some manufacturers include as many as eight inputs to a relay: three voltage, three current, plus neutral current and mutual current. In order that all inputs can be routed through to the ADC, a multiplexer is used. The multiplexer can be thought of as a multiway switch where each of the inputs can be individually switched through to the ADC input; in Fig. 7.3, there are six possible inputs. The microprocessor controls which input reaches the ADC. This way, the ADC converts each input signal sequentially [11–13].

Sample and Hold

It was stated earlier that the ADC takes typically 20 μs to convert an analog signal. To ensure high accuracy, it is important that the input signal amplitude remains constant during the conversion period. Due to the sinusoidal nature of the power system waveforms, it is clear that the input signals cannot be guaranteed to remain constant. Furthermore, the sequential conversion arrangement of the ADC and multiplexer implies that, if no action is taken, each input signal will be converted at a slightly different point in time. This is generally undesirable from the point of view of the protection software, and so it is important that each input signal appears to be converted at the same point in time, and that the signals remain constant during the ADC conversion period. This is achieved with a relatively simple device called a *sample and hold amplifier*; a simplified diagram of which is shown in Fig. 7.4.

Figure 7.4 shows a switch, downstream from a shunt-connected capacitor, both of which are buffered by unity-gain amplifiers. The switch is under the control of the microprocessor. When it is closed, the circuit is said to sample; when it is open, the circuit holds. When sampling, the input is directly connected to the output, and hence the capacitor voltage changes in sympathy; the input buffer amplifier ensures that the capacitor does not load the input. When the microprocessor opens the switch, the capacitor will retain its voltage. Due to the presence of the output buffer amplifier, the capacitor retains its charge, and so the voltage remains constant, or holds—at least for the duration required in a protection relay, generally <1 *ms*. By using a separate sample and hold amplifier for each input relaying signal, all the signals can be converted without error [14, 15].

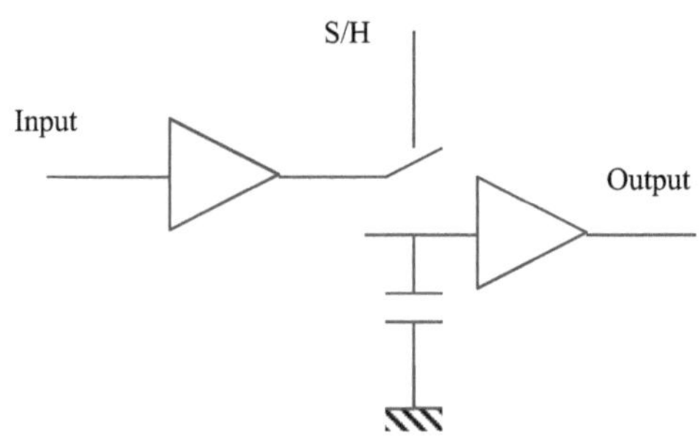

Fig. 7.4 Sample and hold circuit. Diagram of a sample and hold circuit. It shows an input signal entering a triangular amplifier symbol, labeled "Input." The signal passes through a switch labeled "S/H" and a capacitor and then enters another amplifier symbol, leading to the "Output." The capacitor is connected to ground

Anti-Aliasing Filter

The chapter on digital signal processing will explain fully why this filter is required. Its presence is to ensure that the regular samples taken from the analog relaying inputs are truly representative of the input waveforms. The anti-aliasing filter is a low-pass filter that will remove any transient components found in the input which are at a frequency greater than half of the sampling frequency. For example, if the relay is sampled at a frequency of 1 kHz, then the anti-aliasing filter would pass all frequencies up to 500 Hz. Anti-aliasing filters are usually very simple filters—on some relays, they are no more than a resistor and a capacitor! The use of complicated high-order filters is rarely found since it uses a lot of circuit board space, which increases the overall relay cost. Furthermore, the higher the order of the filter longer the filter group delay, i.e., time required for signals to propagate through the filter. Filters with long group delays lead to relays with long operating times. In general, it is far better to use digital filters to remove transients.

Exercise 7.5

Describe the actions taken by the microprocessor to enable V_a of Fig. 7.3 to be converted to a digital value and placed in the memory.

The following sequence of events is required:

1. At the required instant of sampling, the microprocessor sets the sample and hold amplifiers into the hold mode.
2. The microprocessor sets the multiplexer such that the analog signal from the V_a anti-aliasing filter is routed through to the ADC.
3. The microprocessor issues the start conversion command to the ADC.
4. The microprocessor waits while the ADC performs the conversion.
5. When the conversion is complete, the end-of-conversion signal will be issued by the ADC.
6. The microprocessor reads the digital value from the ADC, using the data bus, and then places this value into the required memory location.

7.3.3 Galvanic Isolation Transformers

In common with conventional, static types of relay, the input signals must be isolated from the substation environment in order to protect the relay electronics under adverse conditions. This is achieved through the use of wound voltage and current transformers. In the case of the current input, the CT secondary is connected to a precision resistor, thus providing a voltage proportional to the current. This is required since the data acquisition circuitry works on voltage, rather than current, signals [16, 17].

7.3.4 Memory

The requirements for program and data memory, described earlier, are met with different types of memory devices in a digital protection relay.

Read-Only Memory (ROM)
Program memory is implemented from ROM in a relay. The contents of a ROM are fixed and will be unaffected if the power to the relay is switched off. When the relay manufacturer has developed the relay program, this program is then transferred into a ROM. This may be achieved by sending the program to the ROM manufacturer may use some form of *programmable* ROM (PROM) which, with the use of special equipment, can be loaded with the relay program—but only once. You may also have heard of the term EPROM, which stands for erasable programmable ROM. This type of ROM has a small quartz window in the integrated circuit. By exposure to ultraviolet light, the ROM contents can be wiped clean and then subsequently reprogrammed. EPROMs are very useful to relay manufacturer for developing prototype relays but, due to their high cost, are unlikely to be found in production relays [18, 19].

Random Access Memory (RAM)
The term "random access" dates back to the early days of computing and has little descriptive benefit; it is more useful to think of RAM as being memory that can be both written to and read from by a microprocessor. RAM is used in a relay as data memory—used for restoring voltage and current samples, and so on. There are two different types of RAM: *Static RAM (SRAM) and dynamic RAM (DRAM).* Static RAM (SRAM) is composed of logic circuits called flip-flops, which can be set to one of two stable states. SRAM has the benefit that the time taken for the microprocessor to access its contents, the access time, is very fast; SRAM is also very expensive. Dynamic RAM (DRAM) is composed of a capacitor reflecting its binary content. Since the charge in the capacitors continually leaks away, it has to be periodically refreshed by a special circuit in the DRAM. The process of refreshing can inhibit the microprocessor from accessing the DRAM contents, and, consequently, its access time is longer compared to SRAMs. However, SRAMs are far cheaper. At the time of writing, it is possible that both types of RAM are found in a digital relay. In the event of the relay power being discounted, both types of RAM will lose their contents [20, 21].

Nonvolatile RAM (NVRAM)
In a conventional protection relay, settings are adjusted by the use of switches mounted on the relay case. In a digital relay, these switches can be eliminated, and hardware costs reduced, by allowing settings to be entered directly into the relay memory by the engineer.

Exercise 7.6
Assess the suitability of these types of ROM and RAM for storing relay settings.

ROM/PROM/EPROM: If this type of device were used, then the relay engineer would need to specify the settings before the relays were ordered so that they could be incorporated into the program memory by the relay manufacturer. The settings would be unchangeable—this would clearly be an unworkable solution.

SRAM/DRAM: The use of data memory would be feasible. However, since RAM contents are lost when the power is removed, the relay would need to be reprogrammed every time the relay is powered down. This is unlikely to be acceptable to power utilities [22].

To overcome the difficulties described above, a special type of RAM called a *nonvolatile* RAM (NVRAM) is used. The simplest form of VNRAM is an SRAM with a permanently connected lithium battery to ensure that the SRAM retains its contents even without external relay power. Although lithium batteries have a long working life, their use is not preferred and, instead, a variant of the EPROM family—the electrically alterable read-only memory (EAROM)—is used. EAROMs can be erased without UV irradiation and can be readily programmed by standard digital relay hardware; thus, they are used to store setting information.

7.3.5 Timer

A timer is a quartz crystal-controlled device that provides periodic signals to the relay—this is in addition to the microprocessor clock signal. The timer is used, for example, to generate regular timing pulses for the data acquisition circuitry. Furthermore, the timer can be used as a real-time clock to log relay operations—in effect, acting like a disturbance recorder. Note that some manufacturers link the relay sampling frequency to the power system frequency through the use of a phase-locked loop. This ensures that a fixed number of samples are taken every power system cycle, irrespective of the power system frequency.

7.3.6 Microprocessor

There are very few numeric relays that use only one microprocessor—those that do tend to be simple in nature, e.g., overcurrent relays. The exact arrangement of the microprocessors in a numeric relay varies from manufacturer to manufacturer, so it is difficult to be general. The most important features of a protection relay microprocessor are the data word length and the processing speed. Since ADCs need to be a minimum of 12 bits in length, the usual standard for a relay microprocessor is 16 bits. The processing speed is, however, more difficult to assess.

All tasks in a protection relay need to be processed quickly, yet some more quickly than others. Taking the case of a distance relay, calculation of the apparent impedance from the sampled voltage and current values and, subsequently, comparison of the impedance to a characteristic are tasks that need to be accomplished. These tasks are said to be real-time tasks, since over time a new set of voltage and current samples is taken, a new set of impedances and a comparison is be executed.

However, some tasks, such as checking relay elements, implementing zone timers, and handling transfer trip arrangements, do not require such rapid processing. Real-time tasks usually require many arithmetic calculations: addition and multiplication. Such calculations are best performed on a type of microprocessor especially designed for this purpose called a digital signal processor. Digital signal processors owe their development more to the growth in digital speech communication rather than protection relays; however, they are ideally suited to processing signals with bandwidths in the kHz region [23].

The main features of a digital signal processor are its ability to perform fast multiplication and its suitability for performing certain mathematical operations associated with digital filtering and discrete Fourier transforms (DFTs) (see pages 7–25). Presently, all relay algorithms use fixed-point arithmetic. This is not because fixed point is best suited to the task, far from it, but because floating point is best suited to the task, far from it, but because floating point digital signal processors have only recently come onto the market. Manufacturers will, in the future, undoubtedly use floating-point processing. Although digital signal processors are suited to relay algorithm processing, they are not ideal for performing more mundane tasks, and so it is common to find more conventional microprocessors implementing zone timers, handling communication, interfacing with the user, and so on.

7.3.7 Serial Communications

Remote communication with a relay is important for unit-type schemes, such as current differential, where the digital values of measured current are passed from end to end. It is also necessary to communicate with a relay so that the user can change relay settings, either on-site with the aid of a portable computer or possibly over a long distance by the use of a modem. These features are achieved by some form of serial communications for which the relays will have the necessary hardware devices. Currently, relays are also being provided with direct optical connections for communications, thus avoiding the danger of electromagnetic interference (EMI) in the substation environment affecting relay integrity [24].

7.3.8 Digital I/O

This peripheral device allows the microprocessor to change binary information to the outside world. Typical outputs include circuit breaker trip and alarm signals; typical inputs include circuit breaker open and communications channel out of service signals. These signals are, of course, binary in nature. Again, due to the requirement to isolate internal relay signals from the substation environment, typically input signals are isolated through the use of opto-isolators and output signals by the use of reed relays.

7.3.9 Keypad and Display

A common trend among manufacturers is to provide a liquid crystal display and keypad on the front of the relay panel to allow an engineer to interface directly with the relay, in order to change settings, and so on. This form of interface is usually in addition to a remote communications facility described earlier.

7.3.10 Electromagnetic Compatibility

The substation is a very harsh electromagnetic environment for the operation of microprocessor-based equipment. It is important, to ensure relay dependability, that some form of compatibility between the relay and its environment is achieved to prevent the, sometimes excessive, electric and magnetic fields found therein from affecting relay operation. Some aspects of electromagnetic compatibility (EMC) have already been touched upon, such as the isolation between any signal entering a relay and the relay electronics. Further to this, digital relays are provided with full metallic shielding to provide immunity to radiated electromagnetic interference (EMI). Ultimately, both national and international standards ensure that adequate attention to EMC is paid by relay manufacturers [25].

7.4 Digital Signal Processing

7.4.1 Sampling

The acquisition circuit of the relay converts, at regular intervals, the analog signals from the busbar transducers into numeric values. This process is shown in Fig. 7.5. The analog signal, $x(t)$, is shown as continuous solid lines, and the discrete samples $X(nT)$ taken from $x(t)$ are shown as square boxes. The instants of sampling are denoted by the vertical dotted lines, which are spaced apart by T seconds. When sampled, the numeric values are described as being *discrete time values or* just simply *samples*. Samples of $x(t)$ are given the symbol $X(nT)$, where n is an integer value referred to as the *time index*. It should be clear that $X(nT)$ is only defined at times which are multiples of T. The rate of sampling or the *sampling frequency* f_s is given by $f_s = 1/T$. In addition to the discrete time nature of $X(nT)$, it is also a discrete value since if, for example, a 14-bit ADC was used to perform the conversion, then $X(nT)$ would assume one of $2^{14} = 16{,}384$ possible values. The original analog signals $x(t)$, however, are continuous and so are represented by an infinite range of values [26].

7.4.2 Sampling Theorem

Figure 7.5 shows a signal $x(t)$ being sampled. $x(t)$ has a frequency, relative to the sampling frequency, of

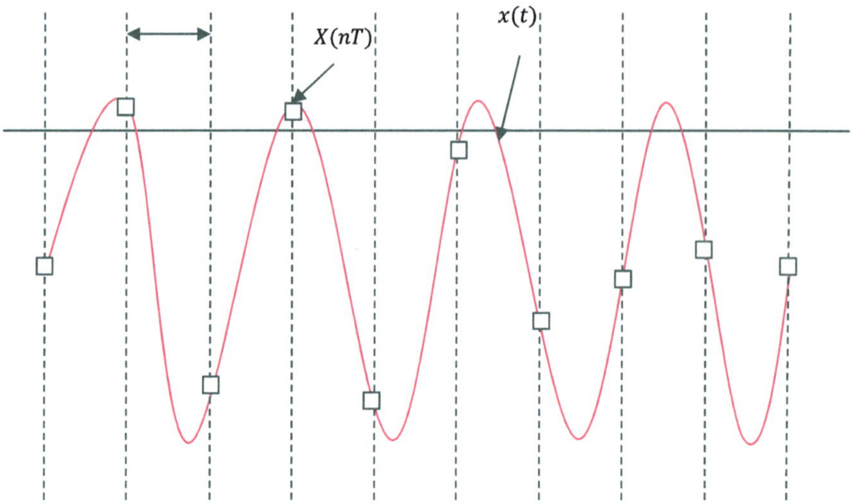

Fig. 7.5 Sampling an analog signals. A sinusoidal waveform is shown with a series of vertical dashed lines intersecting it at regular intervals. Each intersection point is marked with a square, indicating sampled data points. The horizontal distance between the dashed lines is labeled with a double-headed arrow

$$\frac{1}{3T} = 0.33 f_s$$

It is useful to consider whether the process of sampling can be carried out for any arbitrary relationship between the frequency of $x(t)$ and the sampling frequency. It will be seen that there is a rule which governs these two frequencies. Figure 7.6 shows a similar situation except that the analog signal now has a relative frequency of $1.11f_s$. If you care to draw a line joining together the square boxes in Fig. 7.5, you will end up with a rather jagged triangular waveform. The fact that your drawing looks little like a sine wave does not matter; the important issue is that it has the same frequency as $x(t)$. If you now join the boxes for Fig. 7.6, you will find your drawing displays part of a chapter of a low-frequency sine wave, i.e., it bears no relationship to the original analog waveform. If you care to check, you will find the frequency of your drawing of $0.11\,f_s$.

The effect we are seeing in Fig. 7.6 is called aliasing, since the discrete-time representation of the analog waveform no longer behaves as the original. To avoid aliasing, we have to obey a simple rule, which is called the sampling theorem. This states that an analog signal of frequency f must be sampled at a frequency greater than $2f$ to ensure a correct discrete-time representation. We can now appreciate why the discrete-time representation of Fig. 7.5 is correct, but Fig. 7.6 leads to aliasing. The general relationship between the frequency of an analog signal (*sample at fs*) and the frequency of its discrete-time representation is shown in Fig. 7.7, from which it will be apparent that there are an infinite number of frequencies for the analog signal that will result in aliasing [27, 28]. It should also be

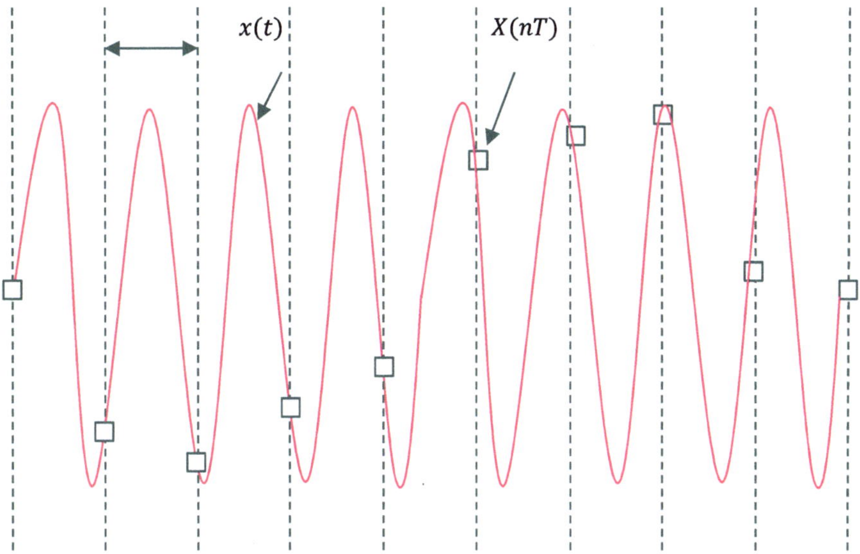

Fig. 7.6 Example of aliasing. A sinusoidal waveform is shown with vertical dashed lines indicating sampling intervals. Arrows point to the waveform and the sampled points, illustrating the concept of signal sampling

apparent why the discrete-time representation of Fig. 7.6 appears as a frequency of 0.11 f_s.

Exercise 7.7
If a digital relay samples at a frequency of 4 kHz, what is the smallest input analog frequency to the relay that could alias to 50 Hz?

The answer to this question can be calculated from Fig. 7.7 if we correctly scale the axes. Clearly, $f_s/2$ *will be* 2 *kHz*, thus showing that any analog frequency component up to 2 *kHz* can be correctly, and uniquely, represented as a discrete-time signal. If we now draw the line "*discrete time signal frequency* = 50 *Hz*," which will lie parallel to, and just above, the *x*-axis, then interchapter of this line with the curve of the graph will show the (infinite) range of analog frequencies which will alias to 50 Hz. Subsequent aliasing will occur at $(f_s + 50)Hz$, $(2f_s - 50)Hz$, $(2f_s + 50)Hz$,and so on.

7.4.3 Preventing Aliasing in a Protection Relay

Generally, in a digital protection relay, we are only interested in the power system frequency component of the relaying signals. However, under fault conditions, it is possible for the relaying signals to be corrupted transients—particularly high

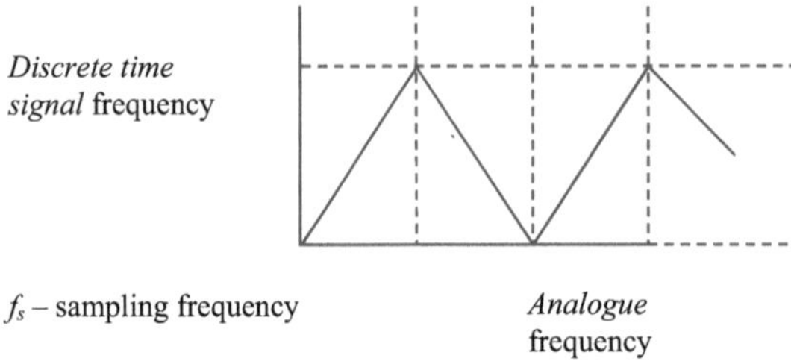

Fig. 7.7 Relationship between analog and discrete time signal frequencies. Graph illustrating the relationship between discrete-time signal frequency and analog frequency. The graph features a triangular waveform pattern with dashed vertical lines indicating sampling frequency. The left side of the image includes the text "Discrete time signal frequency" and the right side reads "Analog frequency"

frequencies. To prevent aliasing, it is necessary to ensure that the sampling theorem is obeyed. This is achieved by using a filter, prior to sampling, which removes any frequency components that are greater than, typically, one-half of the sampling frequency. This filter is called an anti-aliasing filter—see earlier.

7.4.4 Digital Filtering

If a relay samples at a frequency of 4 kHz, then the anti-aliasing filter will remove any components greater than 2 kHz. Since the relaying algorithm, almost exclusively, only processes the power system frequency ($e.\ g., 50Hz$) component of the relaying signals, there needs to be extra filtering to remove non-50 Hz components. This extra filtering is achieved by the use of a digital filter, which acts on the sampled values of the relaying signal. Since digital filters have less of a group delay than the corresponding analog types, this will lead to faster relay operating times compared to the case where a slower sampling frequency is used with a narrower passband anti-aliasing filter. Another important reason for using digital filters is that the characteristic of a digital filter used for, say, the "a" phase voltage differs from the characteristic of the digital filter used for the "b" phase voltage. If analog filters were used instead, they would have to be individual calibrated and periodically checked [29].

Conceptually, a digital filter works as shown in Fig. 7.8, where a series of sampled values are filtered to leave only the power system frequency component. The theory of digital filtering is too complicated to be explained in much detail here, so our examination will be somewhat cursory. It is useful at this point to consider normal analog filters. When describing an analog filter, we, more often than not, talk about its frequency response. Thus, in the previous paragraph, we discussed an anti-aliasing filter having a cutoff of 2 kHz. However, we rarely stop to consider the mathematical process by which the input signal is converted to the output signal. In fact, the mathematics relating an arbitrary filter input to its output are, at best, rather complicated, but it is important to appreciate that it is possible to describe the

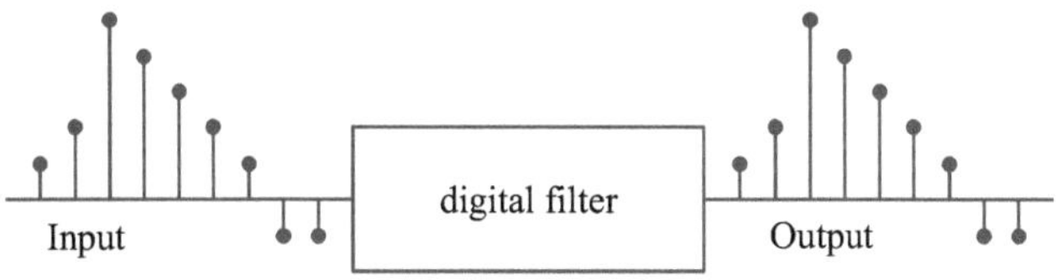

Fig. 7.8 Conceptual view of digital filter. Diagram illustrating a digital filter process. On the left, labeled "Input," a series of vertical lines of varying heights represent incoming data signals. These signals pass through a central box labeled "digital filter." On the right, labeled "Output," the filtered signals are shown as another series of vertical lines, indicating the transformation applied by the digital filter

mathematical operation which, if applied to the input, yields the output. This operation is called *convolution,* which is a branch of integral calculus—there are few people who work with concepts like this on a day-to-day basis. Fortunately, in the digital world, it is somewhat easier to understand mathematical operations since, ultimately, all operations are composed of arithmetic operations such as addition and multiplication, with which we are all familiar. Hence, it should come as no surprise that a digital filter is simply a series of mathematical operations applied to the sampled values. A simple equation fFor digital filtering is:

$$v_{filt}\left[nT\right]=v\left[nT\right]\times H_1+v\left[\left(n-1\right)T\right]\times H_2+v\left[\left(n-2\right)T\right]\times H_3+v\left[\left(n-3\right)T\right]\times H_4 \tag{7.2}$$

This shows how sampled values of $v[nT]$, which may be samples of a busbar voltage, may be combined to form the filter output $v_{filt}[nT]$. Of course, the motivation for performing this filtering is that $v[nT]$ may contain some undesirable transients which we wish the filter to remove, leaving $v_{filt}[nT]$ each value of the $v[nT]$ is multiplied by a constant H. In this equation, there are four values of H: H_1, H_2, $H_{3,}$ and H_4 In general, these values are referred to as the filter coefficients. It is the filter coefficients that make a filter perform in a predetermined manner; changing *H* will change the characteristic of the filter. In the above, if all four values of *H were set to* 1.0, then this would be equivalent to adding together four sequential values of $v[nT]$ dividing by four—a simple filter indeed. If the value of *H* were all different, then the above filtering equation would be implemented (in the microprocessor) by performing four multiplication operations and four additions. It is useful to think of the time index "*n*" in the above to represent the time "now." Every time a new sample of the relaying voltage is made, then the filtering equation needs to be applied to the four most recent samples:

$$v\left[nT\right],\ v\left[\left(n-1\right)T\right],\ v\left[\left(n-2\right)T\right],\ \text{and}\ v\left[\left(n-3\right)T\right]$$

To design a digital filter, the relay designer has to find values of H that will realize a filter that suits the given purpose. In general, digital filters used in relays have rather more than the four values of H shown above; however, it should be easy to see how this filtering equation may be extended to cater for this. This type of filter is called a finite impulse response (FIR) filter since the value of H, the filter coefficients, is also indicative of another feature of the filter called its impulse response; clearly, there is a finite number of values of H. There is also another class of digital filter called an infinite impulse response (IIR) filter. A typical equation for an IIR filter is:

$$v_{filt}[nT] = v_{filt}[(n-1)T] \times A_1 + v[(n)T] \times B_1 + v[(n-1)T] \times B_2 \tag{7.3}$$

The main difference between this equation and Eq. 7.2 is that an IIR filter equation uses previous values of the filter output as well as previous values of the filter input. In Eq. 7.3, the filter coefficients are A_1, B_1, and B_2; these coefficients are not, however, the filter impulse response—as the filter description implies. The impulse response is effectively composed of an infinite series of numbers. Generally, FIR filters are the type used in protection relays since they can be designed with a very accurate control over the filter group delay.

Exercise 7.8

The "a" phase voltage input to a digital relay is converted into discrete-time form, $v_a[nT]$.

The values of $v_a[nT]$ for $n = 0$ *to* 9 are shown.

If the relay filters, $v_a[nT]$

Using an FIR filter (Eq. 7.2) where $H_1 = H_4 = 1$ *and* $H_2 = H_3 = 4$, calculate the output of the filter for $n = 0$ *to* 9.

n	*0*	*1*	*2*	*3*	*4*	*5*	*6*	*7*	*8*	*9*
$v_a[nT]$	*392*	*422*	*449*	*473*	*495*	*513*	*528*	*540*	*548*	*553*

The FIR filters have four coefficients (H_1, H_2, H_3, and H_4); hence, we cannot expect any sensible filter output unit n = 3. Denoting the filter output as $v'_a[nT]$

For $n = 3$

$$v'_a[3T] = v[3T] \times H_1 + v[2T] \times H_2 + v[1T] \times H_3 + v[0T] \times H_4 \tag{7.4}$$

$$v'_a[3T] = 473 \times 1 + 449 \times 4 + 422 \times 4 + 392 \times 1 = 4349 \tag{7.5}$$

For $n = 4$

$$v'_a[4T] = v[4T] \times H_1 + v[3T] \times H_2 + v[2T] \times H_3 + v[1T] \times H_4 \tag{7.6}$$

$$v'_a[4T] = 495 \times 1 + 473 \times 4 + 449 \times 4 + 422 \times 1 = 4605 \tag{7.7}$$

n	*3*	*4*	*5*	*6*	*7*	*8*	*9*
$v'_a[nT]$	*4349*	*4605*	*4834*	*5033*	*5199*	*5333*	*5433*

7.4.5 Fourier Transforms

In Fig. 7.8, we see that the noisy discrete-time input signals have been filtered to yield an output that is composed only of the power system frequency component. Both the input and output of the filter are described as being in the time domain. The time domain simply means that information is referred to with respect to time; e.g., an oscillation trace of a 50 Hz waveform is a time-domain representation. A phasor is another good example of a frequency-domain value since when specifying the magnitude and angle of a phasor, there is an implicit assumption that the phasor describes only the frequency of interest, e.g., 50 Hz. There are some relay algorithms that use phasors rather than time-domain signals. We will now examine how phasors may be calculated from sampled values [29].

The process of converting information from the time domain to the frequency domain may be achieved through a mathematical procedure known as a *Fourier transform* (FT). As with convolution, which describes the filtering process, the Fourier transform uses integral calculus. However, the Fourier transform may be implemented in a microprocessor, again by using common arithmetic operations. When used in this fashion, the Fourier transform is described as being a discrete Fourier transform DFT. A conceptual view of the DFT is shown in Fig. 7.9. The DFT is basically a single equation:

$$V\left[mF\right]=\frac{1}{N}\sum_{n=0}^{N-1}v\left[nT\right]e^{\left(\frac{-j2\pi mn}{N}\right)} \tag{7.8}$$

where

$V[mF]$= component of frequency mF (in vector form)
$v[nT]$= sampled signal values (time domain)
N = number of samples of $v[nT]$ used in DFT
F = fundamental frequency of DFT = sampling frequency/N
m = the harmonic index, $0 \geq m \geq (N/2) - 1$

Equation 7.8 allows a series of N values of the discrete-time signal $v[nT]$, to be decomposed into m different frequency components. Equation 7.8 is not unlike the

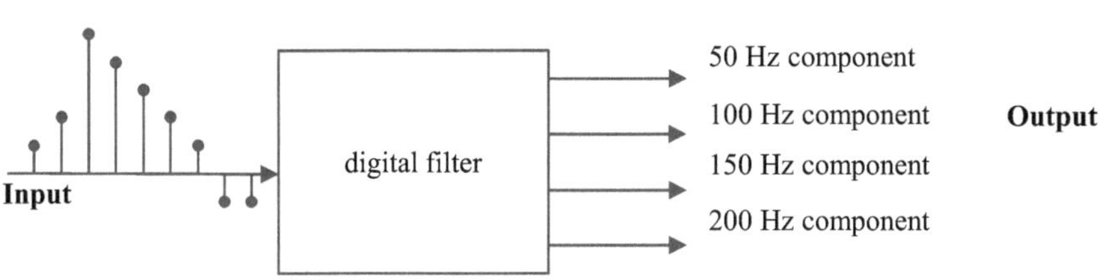

Fig. 7.9 Conceptual view of the discrete Fourier transform. Diagram illustrating a digital filter process. On the left, labeled "Input," a series of vertical lines of varying heights represents incoming data signals. These signals pass through a central box labeled "digital filter." On the right, labeled "Output," the filtered signals are shown as another series of vertical lines, indicating the transformation applied by the digital filter

filtering Eqs. 7.2 and 7.3, in that the discrete-time signal $v[nT]$ is multiplied by a series of constants, except that here the constants are defined by the complex exponential function can be expressed as a real and an imaginary component, i.e., in vector form, which can be combined to form a magnitude and angle, i.e., phasor form. In terms of Eq. 7.4 $N = 20$, the fundamental frequency $F = 100/20 = 50Hz$, and the harmonic index m, can vary between 0 *and* 9. Hence, if we apply the DFT to our 20 samples, the result will be vectors of the following frequencies ($m \times F$).

Exercise 7.9
A power system waveform has been sampled at a frequency of 1 kHz. If 20 consecutive samples are to be analyzed by using the DFT Eq. 7.8, which frequency component will be evaluated?

In protection, the DFT is usually used to form only the 50 Hz, or power system frequency, component from sampled values. The other frequency components, which are not required, are not calculated. The DFT is a widely used mathematical procedure in the protection field. It is the basis of digital current differential relays, where currents at either end of a line are compared in vector, and for fault locators, where the phasor forms of sampled busbar voltage and current signals are required. Note the time domain using a similar mathematical technique called the inverse Fourier transform (IDFT) equation, slightly different from Eq. 7.4, can be derived, which is suitable for microprocessor implementation. There has been no significant use of the IDFT in numeric protection applications to date [30].

7.4.6 The Fast Fourier Transforms

If it is desired to evaluate every frequency component of the DFT, then an alternative method of evaluating Eq. 7.4 exists that is more efficient, i.e., uses less computing time, than simply programming the equation as stated. This method is known as the fast Fourier transform (FFT), and it saves considerable computing time, particularly where the number of samples considered, N, is large. The FFT takes advantage of a numerical redundancy found in the evaluation of the complex exponential constant on the right-hand side of Eq. 7.4. Since, in a protection application, it is rare to examine any frequency component other than 50 Hz, the FFT does not offer any advantage, in this area, over the conventional form of the DFT (Eq. 7.4) evaluated at only 50 Hz. The FFT does have a wide range of applications, including calculating the frequency response of FIR filters and harmonic analysis. One drawback to using the FFT is that N, the number of samples used, must be a *power of* 2, i.e., of the form $N = 2^k$, where k is an integer.

7.5 Distance Protection

7.5.1 Introduction

In conventional distance relay, a comparison is made between the impedance presented to the relay and the boundary conditions of the characteristic. The actual impedance is not evaluated; it is only necessary to establish whether this impedance is less than or greater than the boundary value. In a numeric distance relay, a numeric evaluation of the presented impedance is calculated, which is subsequently compared against the relay characteristic. The presented impedance is the impedance between the relay busbar and the fault point.

The outline of a zone, an element of a digital distance relay, is shown in Fig. 7.10. Zone 2 and 3 elements will be discussed later. The inputs to the element are sampled voltage and current values; how these are derived is explained in Sect. 7.5.5. The first processing stage in the element is the digital filter, which removes undesirable non-50 Hz components from the signals. Next, the voltage and current signals are combined to form resistance and reactance values. The final stage of the element evaluates whether or not a fault exists by comparing the resistance and reactance values against a characteristic, thus enabling the element's trip output to be correctly set.

Each of these stages will be considered in detail in the following chapters. A sampling frequency of 4 kHz (i.e., 80 samples per 50 Hz cycle) is assumed in the following, although some relays work at a frequency as low as 300 Hz. In general, a high sampling frequency will lead to a short relay operating time [30].

7.5.2 Filtering

Signals collected from power system transducers contain many unwanted components such as exponential offsets, harmonics traveling wave, and switching noise. Since the impedance calculation technique to be explained in the next chapter is

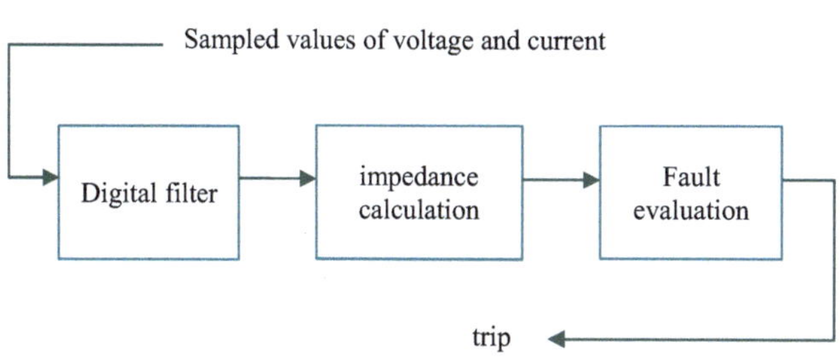

Fig. 7.10 Outlines of processing stages in a zone 1 distance element. Flowchart illustrating the process of analyzing sampled values of voltage and current. It consists of three sequential steps: "Digital filter," "Impedance calculation," and "Fault evaluation." An arrow labeled "trip" loops back from "Fault evaluation" to the beginning, indicating a feedback mechanism

based on the power system frequency component of the input signals, it is necessary to remove any non-50 Hz components by the use of a digital filter. In general, FIR filters will be used for this task and will be designed to eliminate d.c. and exponential offset as well as the higher frequency components. Atypical length of an FIR filter for this application is 10 *ms*, which, for a 4 kHz sampling frequency, corresponds to 40 values of H in Eq. 7.2 of Sect. 7.4.4.

7.5.3 Impedance Calculation

Early attempts at producing numeric distance relays used the Fourier transform approach where a DFT algorithm was used to evaluate the power system frequency component of the sampled voltage and current signals. Thus, division of the voltage vector by the current vector yielded the resistance and reactance components. Currently, evaluating resistance and reactance by solving the first-order transmission line equation is: the preferred method used in distance protection since it is immune to variation in power system frequency. This method will be described further.

Assuming that the transmission line to which the relay is connected is composed of series resistance and inductance, then the fundamental line equation is:

$$v = Ri + L\frac{di}{dt} \tag{7.9}$$

where R and L are the resistance and inductance up to the fault point. Sampled voltage and current signals taken at some point in time nT are seen to obey this equation; however, it can be seen that to solve the equation, and thus a solution to Eq. 7.9 at some other time, say K samples earlier: $(n - k)T$, is required. Hence, it may be written as:

$$v[nT] = Ri[nT] + L\frac{di[nT]}{dt} \tag{7.10}$$

$$v[(n-k)T] = Ri[(n-k)T] + L\frac{di[(n-k)T]}{dt} \tag{7.11}$$

Letting $i'[nT] = \frac{di[nT]}{dt}$, Eqs. 7.10 and 7.11 may be combined in a matrix form:

$$\begin{bmatrix} v[nT] \\ v[(n-k)T] \end{bmatrix} = \begin{bmatrix} i[nT] & i'[nT] \\ i[(n-k)T] & i'[(n-k)T] \end{bmatrix} \begin{bmatrix} R \\ L \end{bmatrix} \tag{7.12}$$

$$\begin{bmatrix} R \\ L \end{bmatrix} = \frac{1}{D} \begin{bmatrix} i'[(n-k)T] & -i'[nT] \\ -i[(n-k)T] & i[nT] \end{bmatrix} \begin{bmatrix} v[nT] \\ v[(n-k)T] \end{bmatrix} \tag{7.13}$$

where:

$$D = i[nT]i'^{[(n-k)T]} - i'^{[nT]}i[(n-k)T]$$

Equation 7.13 shows how the sampled voltage and current signals may be combined to form the resistance and inductance apparent to the relay. The derivative of the current may be calculated by using the central difference formula:

$$i'[nT] = \frac{i[(n+1)T] - i[(n-1)T]}{2T} \tag{7.14}$$

Note that since Eq. 7.13 calculates the line inductance, it is immune to any changes in the power system frequency, which would have to be corrected for if the algorithm calculated the line reactance directly. The line reactance may easily be calculated from Eq. 7.13 using a value for the nominal power system frequency, f_0:

$$X = 2\pi f_0 L \tag{7.15}$$

At the next sample interval, a new estimate of R and L will be made using new values for $v[nT]$ and $i[nT]$. A typical value of k, the number of samples difference between the two solutions of Eq. 7.9, is 6, which at 4.4 kHz sampling frequency corresponds to 1.5 *ms* time delay.

7.5.4 Fault Evaluation

The values of calculated resistance and reactance are next compared against a relay characteristic to ascertain whether or not a fault exists. In a numeric relay, virtually any shape of characteristic can be employed since the comparison between $R + jX$ and the characteristic is relatively trivial to process. We will take a quadrilateral characteristic as an example here.

Although digital filtering is employed, unwanted transient components may still be present in the voltage and current signals, which can lead to inaccuracies in the R and X calculation. Thus, the occurrence of a single impedance measurement falling within the characteristic is not a reliable indication of a fault. To ensure high accuracy, the impedance values must remain in the characteristic for a certain period of time before a fault in reliably evaluated. Figure 7.11 shows the arrangement used.

Any calculated impedance value falling within the first 80% of the characteristic, in the positive reactance direction, advances a counter by +9 (this value is referred to as the counting increment, or simply *inc*). If the impedance falls within the 80–90% area, then *inc* = 4, *whereas inc* = 1 if the impedance lies within 90–100% of the reactive reach point. Similarly, impedance values falling just outside the reach point lead to negative values of *inc*, which are biased in the opposite sense to positive increments. The counter never goes negative and thus is initially at zero if no fault is in progress. If an in-zone fault occurs, then the counter will increase at every

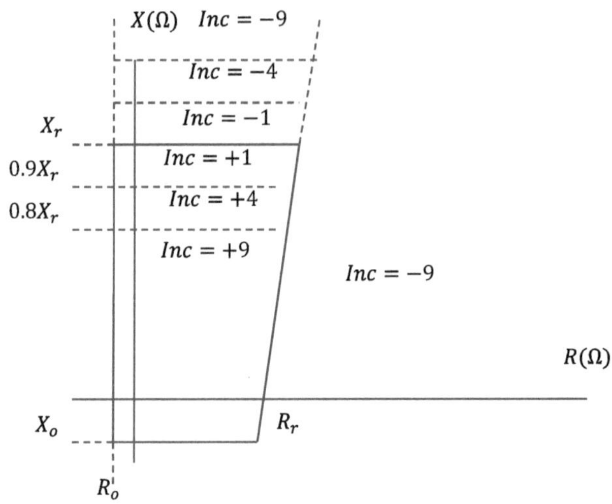

Fig. 7.11 Quadrilateral characteristic and counting strategy. Graph illustrating impedance changes with various inclination values. The horizontal axis represents resistance (R in ohms), and the vertical axis represents reactance (X in ohms). Inclination values range from −9 to +9, with corresponding horizontal lines at different reactance levels

new impedance calculation until, reaching a value of 45, it will cause the element to trip.

Exercise 7.10
Why do you think that the counting increments decrease toward the reactive reach point? What is the minimum number of consecutive in-zone impedance values that will cause the element to trip?

The choice of 9, 4, and 1 counter increments is to ensure that the relay gives exceptionally high accuracy at the reach point by slowing down the fault evaluation process. This process can be considered to be a form of filtering to supplement the filter acting on the voltage and current values. For a fault occurring well within the characteristic, there is less requirement for the filtering to act perfectly. For a fault at, say, 40% of the characteristic, the counter can reach 45 after five consecutive impedance values in the same area—this is equivalent to 1.25 ms at 4 kHz. (Note that this is not the relay operating time since this also comprises the filter and impedance calculation group delays as well as any scheme logic delays.)

If the impedance measured is not in the characteristic, then the value of *inc* is negative, and thus the counter will count down. The minimum value that the counter can assume is zero. Similarly, if a fault occurs, the maximum value that the counter can reach is 90, which ensures that the relay will easily reset after the fault has been removed.

7.5.5 Relay Elements

The independent zone (zone 1) of the relay will have six elements similar to the element described in the foregoing chapters. These elements will be responsible for detecting both earth and phase faults on the three phases. The required signal mixing is shown in Fig. 7.12, which is essentially the same approach as taken in conventional distance protection described in Chap. 4. For the phase elements, the difference between the two relevant phase signals is used; e.g., the "$a - b$" element is supplied with samples of $V_a - V_b$ voltage and $i_a - i_b$ current. For the earth elements, the relevant phase voltage is supplied, but the corresponding line current must be residually compensated. The residual current, $i_r[nT]$, is easily calculated:

$$i_r[nT] = i_a[nT] + i_b[nT] + i_c[nT] \tag{7.16}$$

This may be multiplied by the residual compensation factor K (see Sect. 4.3.1) added to the relevant line current for residual compensation, e.g., for the "a" phase element:

$$i_{ar}[nT] = i_a[nT] + Ki_r[nT] \tag{7.17}$$

where $i_{ar}[nT]$ is the residually compensated "a" phase current.

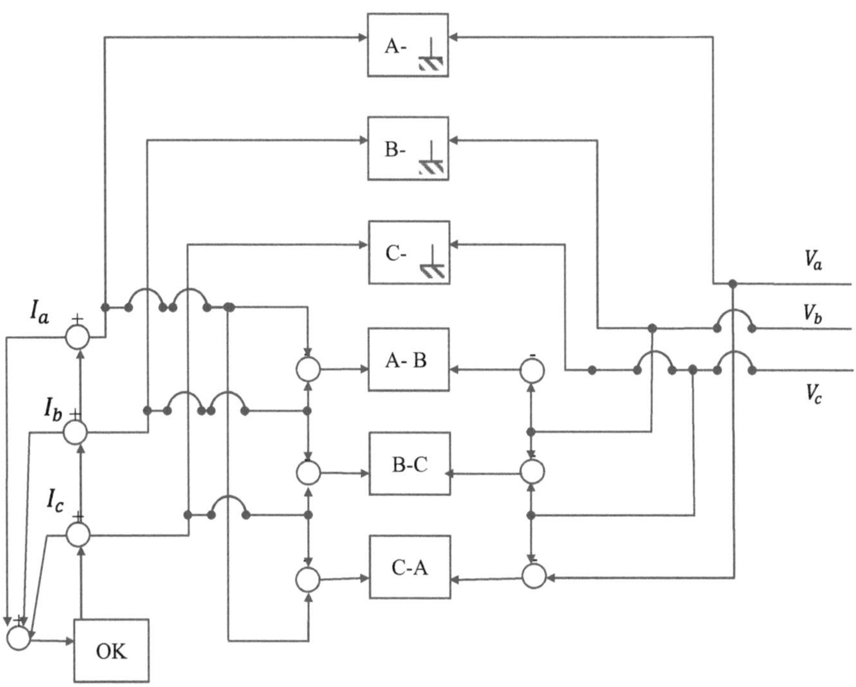

Fig. 7.12 What processing stages are required for the zone 2 and zone 3 elements

In the case of the zone 2 and 3 elements, it is not necessary to recalculate the impedance, since the corresponding zone 1 element impedance is sufficient. However, a separate fault evaluation process, including an appropriately set characteristic, will be required.

Exercise 7.11
What processing stages are required for the zone 2 and zone 3 elements?

Since the relevant impedance has already been calculated in the zone 1 elements, it is not necessary to repeat this calculation. Thus, the zone 2 and zone 3 elements consist only of "fault evaluation" stages where the relevant zone boundaries are compared against the measured impedances [31].

7.6 Differential Protection

7.6.1 Introduction

Numeric differential relays have already been mentioned in Chap. 5. This section gives a greater understanding of the operation of such relays. In a conventional differential relay, the currents at each end of the protected zone are sensed by CTs, which are directly connected using either the circulating current or balanced voltage systems. In a digital differential scheme, the currents are sensed by relays situated at each end of the protected zone, which calculate the numerical values of the locally measured current vector. These values are then transmitted along a digital communication link such that each relay possesses the current vector from its local and remote ends.

The current vectors are then compared to enable a fault decision to be made. It may now be appreciated how differential protection of multiterminal lines as described in Sect. 5.6 can easily be achieved with numeric technology since the comparison required for, say, a *teed feeder* is essentially as *Kirchoff current balance* which may easily be implemented in software.

7.6.2 Hardware

The hardware of a numeric relay differential relay is similar to that described in Sect. 7.3 and shown in Fig. 7.3, except that each relay only has three analog current inputs. The digital communications aspects of such relays are highly developed since, being of the unit type, these relays cannot operate without information from the remote end, or ends. Typical sampling rates for commercial relays are in the region of 400 Hz.

7.6.3 Current Vector Calculation

The sampled current values are converted into vector form (i.e., consisting of a real and imaginary component) by the use of the discrete Fourier transform described in Sect. 7.4.5. The DFT equation, Eq. 7.8, is used to evaluate only the power system frequency component. Equation 7.8, applied to current samples, may be adapted into the following form:

$$IIn = \frac{2}{N}\sum_{n=0}^{N-1} \sin n\omega T \times in \tag{7.18}$$

$$IIn = \frac{2}{N}\left(\frac{i[0]}{2} + \frac{i[N]}{2} + \sum_{n=1}^{N-1} \cos n\omega T \times in\right) \tag{7.19}$$

where

N = number of samples per cycle (sampling frequency divided by the power system frequency).
ω= angular power system frequency
T = sample interval = 1/(sample frequency).
$i[n]$= value of current sample at time nT
$I[n]$= value of complex current vector at time nT.

Equations 7.18 and 7.19 evaluate the vector form of the sample current. The processing is applied to each phase of the protected system, and thus three pairs of vectors are produced in each sampling interval. Comparison of each individual phase of the system can protect against earth and phase faults [31].

7.6.4 Source of Error in a Digital Differential Relay

When comparing the current vector from either end of a protected item of plant, say a transmission line, there are errors associated with the measurements as described in Chap. 5.

These are:

(a) Shunt capacitance. Some of the current entering the line will be shunted to ground, and so, under unfaulted conditions, the current entering the line will not equate the current leaving the line.
(b) CT inaccuracy. Error in the transformation ratios and saturation characteristic of the CTs will introduce errors.

The effect of the above two sources of errors is minimized by the use of biased characteristics. This is the case for all types of differential relays, numeric or otherwise.

In a numeric differential relay, there are two additional sources of error that will affect the accuracy of the measurement. These are as follows:

(c) Asynchronous sampling. The timers that provide timing for the relay sampling are not synchronous in a numeric differential scheme, i.e., on a two-ended scheme, the instants of sampling of relays at either end may not occur at the same point in time. Thus, the exact point on the wave of the samples will not be the same at each end. In the case of a relay that takes eight samples per power system cycle, there are 45 electrical degrees between each sample, and therefore, potentially a maximum error of $22\frac{1^{\circ}}{2}$ between samples made at each end. Since digital differential relays compare *vector* value of current, this error may be significant. It is seen that this source of error diminishes as the sampling frequency increases.

(d) Propagation delay. As the current vectors are transmitted across the digital communication path, there is a small delay, referred to as the propagation delay, between the message being sent and received. This can be seen to constitute an error similar to that described in c) above since the receiving relay will not know the point on the wave to which the received data is referenced. This is more of a problem in schemes using digital communication systems using digital communication system which automatically reroute data paths, thus leading to a variable propagation delay.

Compensation of sources of error c) and d) may be made automatically in a numeric relay.

7.6.5 Propagation Delay Compensation

The scheme for compensating against propagation delays is shown in Fig. 7.13. It is seen from this figure that two relays are situated at either end of a transmission line. The relay at the end P makes its samples at instant t_{p1}, t_{p2},etc., while the relay at the end Q samples at instant t_{q1}, t_{q2}, etc. At time t_{p1}, relay P sends a message to relay Q along the digital communications path. This message contains, among other information, the most recently calculated current vector at P and the time that the message was dispatched, i.e., t_{p1}. The message then proceeds along the communications path, taking t_d seconds, until it is received by relay Q at time t_q^*.

Exercise 7.12
Can relay Q now calculate the propagation delay?

The answer is no. Although $t_{p1} - t_q^*$ is exactly the communications channel propagation delay, and both these quantities are available at Q relay, Q has no time reference with P and thus is not able to make any direct use of t_{p1}.

The reason for sending t_{p1} to end Q is so that Q can retransmit this time back to P. Thus, after a delay of t_w seconds, Q sends a message back to P at time t_{q4}. Note that Q and P send messages at each sampling instant, and so Q also sends a message to P at t_{q3}; however, the t_{q3} message is not relevant to our current explanation of the

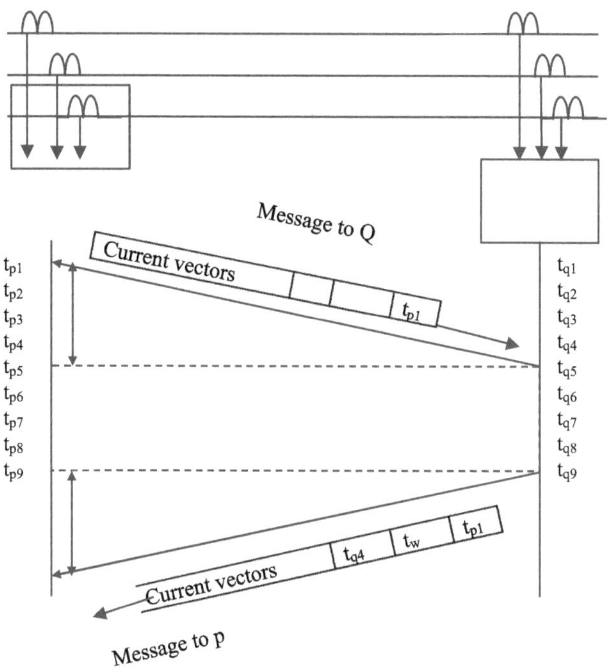

Fig. 7.13 Propagation delay compensation. Diagram illustrating a communication process between two points, labeled "p" and "Q." At the top, three parallel lines with arrows indicate a flow toward "Relay Q." Arrows labeled "Current vectors" and "Message to Q" point from left to right, while "Message to p" points from right to left, indicating bidirectional communication

propagation delay measurement. The t_{q4} message to P contains: t_{p1}, t_w, and the most recent current vector. When received by P, at time t_p^*, P can now calculate the propagation delay since it is in possession of the time at which the message was transmitted, t_{p1}, the delay at Q, t_w, and the time of receipt, t_p^*.

Exercise 7.13
How does P calculate the propagation delay? What assumption is being made?

Assuming that the propagation delay is equal in both directions, *P to Q and Q to P*, the delay t_d, may be calculated as:

$$t_d = \frac{\left(t_p^* - t_w - t_{p1}\right)}{2}.$$

Exercise 7.14 How can the relay at P now relate the most recent current vector from Q to its own most recent current vector?

Within the time reference of relay P, Q's most recent current vector (which originated from Q at time t_{q4}) occurred at time $\left(t_p^* - t_d\right)$, whereas P's most recent current vector occurs at time t_{p6}. Thus, the time existing between these two vectors is $\left(t_{p6} - t_p^* - t_d\right)$. Hence the angular separation of the vectors, θ, is:

$$\theta = \left(t_{p6} - t_p^* - t_d\right) \times 360° \times f$$

where f is the power system frequency. Thus, the channel propagation delay and lack of relay synchronous have artificially lagged Q's vector by θ behind P's vector, in addition to any other phase relationship that exists between the two vectors. Thus, if Q's current vector is brought forward in phase by θ, then both current vectors may be compared without any of the errors described in c) and d) of Sect. 7.6.4. Hence, Q's vector needs to be compensated for the phase shift:

$$I'[n] = I[n] \times (\cos\theta + \mathrm{j}\sin\theta)$$

where $I'[n]$ is the compensated vector from Q. The sine and cosine constants are stored in a look-up table in the memory for varying values of θ; thus, these values do not need to be calculated and do not incur much processing overhead. After compensation, Q's current vector may now be compared with P's most recent current vector [32].

7.6.6 Differential Characteristics

We will now refer to the two current vectors from ends P and Q as I_p and I_q, respectively. I_q is assumed to have been compensated for propagation delay effects as described in the previous section. In order to implement a biased differential characteristic, it is necessary to calculate bias and differential currents. The differential current, I_{diff}, is formed from the vector summation of I_p and I_q since I_p and I_q are composed of real and imaginary components, this may readily be achieved [33, 34]:

$$RI_{diff} = RI_p + RI_q,$$

and

$$II_{diff} = II_p + II_q$$

The magnitude of I_{diff} may be formed as:

$$I_{diff} = \sqrt{RI_{diff}^{\,2} + II_{diff}^{\,2}}.$$

In practice, taking a square root is a time-consuming task for the microprocessor, so this process is usually performed by a linear approximation technique. The bias current I_{bias} may be similarly formed:

$$\left|I_{bias}\right| = \frac{1}{2}\left(\left|I_p\right| + \left|I_q\right|\right).$$

Using a percentage biased characteristic, the condition for relay tripping may be expressed as:

$$|I_{diff}| > K|I_{bias}|$$

$$|I_{diff}| > I_s$$

where K is the percentage biased setting and I_s is the minimum differential current setting—see Chap. 5.

Self-Assessment Questions

? Self-assessment question 7.1a

A digital current differential relay converts analog current signals into two's complement binary form, where each binary step is equivalent to 0.0375 A. Find the original analog input current for the following samples:

(i) 0000 0000 1000 1101
(ii) 0000 1100 0110 0011
(iii) 1111 1100 0110 0011.

? Self-assessment question 7.1b

Deduce how the following data are stored in a digital protection relay:

1. Programming instruction for the relaying algorithm.
2. Sampled values of voltage and current.
3. Relay setting.

? Self-assessment question 7.2a

A digital protection relay samples six analog inputs:

Three-phase voltages and three-line currents. Describe the hardware used to convert the analog input signals into digital values. Your solution must be cost-effective and capable of converting all six inputs at the same instant in time.

? Self-assessment question 7.2b

Software written for the same protection relay is composed of the following activities shown with their execution times:

Activity	Execution time (instruction cycles)
Analog conversion	153 (per input)
Digital filtering	135 (per input)
Signal formation	961
Fault evaluation	720
Housekeeping tasks	10% of the total execution time for the above activities

All the above activities are executed on a digital signal processor, which executes one instruction cycle per clock cycle. If the processor is clocked at 15 MHz, calculate the maximum relay sampling frequency.

? Self-assessment question 7.3

A faulted current waveform is sampled, the results are shown in the table below:

n	*0*	*1*	*2*	*3*	*4*	*5*	*6*	*7*	*8*	*9*
i[*nT*]	*0.0*	*0.029*	*0.301*	*0.717*	*1.141*	*1.440*	*1.513*	*1.326*	*0.911*	*0.368*

n	*10*	*11*	*12*	*13*	*14*	*15*	*16*	*17*	*18*
i[*nT*]	*−0.171*	*−0.572*	*−0.736*	*−0.630*	*−0.289*	*−1.189*	*1.669*	*1.017*	*0.135*

Calculate the effect of filtering this waveform for the period n = 5–18, using finite impulse response filters having the following impulse responses:

$$H_1 = H_2 = H_3 = H_4 = H_5 = H_6 = 1.0$$

$$-H_1 = -H_3 = H_4 = H_5 = H_6 = 0.5; H_2 = H_5 = 1.0$$

Draw a graph of the waveforms and comments on the result.

? Self-assessment question 7.4a

Describe a numerical technique for converting sampled current values into phasor form.

? Self-assessment question 7.4b

The "a" phase currents measured from two ends of a transmission line are shown in the table below, where the sampling interval is 2.5 ms, and the power system frequency is 50 Hz. Using your result to part a), calculate the phasor values of the current.

n	*0*	*1*	*2*	*3*	*4*	*5*	*6*	*7*	*8*
End S	*1732*	*1932*	*1000*	*−518*	*−1732*	*−1932*	*−1000*	*518*	*1732*
End R	*−1060*	*−1949*	*−1696*	*−450*	*1060*	*1949*	*1696*	*450*	*−1060*

? Self-assessment question 7.5a

Describe how resistance and reactance values may be calculated from sampled values of voltage and current, assumed to be at power system frequency.

? Self-assessment question 7.5b

SAQ 7.5 Figure 7.14 shows the counting strategy used in a digital distance relay; the relay will trip if the counter reaches a value of 45. The table below shows the resistance and reactance values calculated by the "a" to earth element of the relay. Draw a graph of the "a" to earth element of the relay. Draw a graph of the "a" to earth element trip counter versus time, and thus predict the operation of the relay:

Time (ms)	R	X
0.0	430	200.0
0.5	110.5	86.5
1.0	5.0	25.5
1.5	2.3	13.0
2.0	−4.3	6.0
2.5	−1.5	6.5
3.0	1.5	9.0
3.5	2.0	10.0
4.0	2.5	12.5
4.5	2.5	14.5
5.0	2.5	14.5
5.5	2.5	14.5
6.0	2.5	14.5

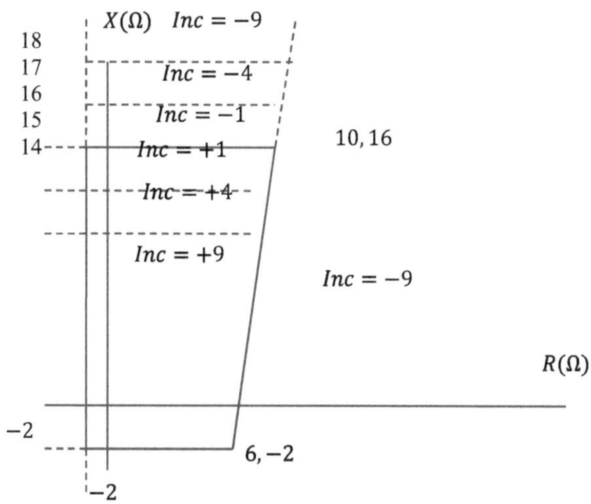

Fig. 7.14 A graph illustrating a series of horizontal lines with varying increments labeled "Inc" ranging from −9 to +9. The vertical axis is labeled "X(Ω)," and the horizontal axis is labeled "R(Ω)." The lines represent different increments, showing changes in values across the graph

Bibliography

1. Cleberton Reiz; Jonatas Boas Leit "Impact Analysis of Distributed Generation on Protection Devices Coordination in Power Distribution Systems" 2021 IEEE PES Innovative Smart Grid Technologies Conference - Latin America (ISGT Latin America) in 2021
2. Hana A. Halim; Melaty Amirruddin; Nur S. Noorpi; Nurhakimah M. Mukhtar "An Improved Protection Scheme for Smart Distribution Grid published in 2013 1st International Conference on Artificial Intelligence, Modelling and Simulation.
3. Jamil Abdo & Tariq Masood "New Technique to develop Intelligent, highly reliable & Efficient Unmanned Power Generation Modular" 22nd International Conference on Renewable Energies and Power Quality (ICREPQ'24) Bilbao (Spain), 26th to 28th June 2024. https://www.icrepq.com/icrepq24/392-24-abdo.pdf
4. Jamil Abdo & Tariq Masood "Operational Reliability- Production Best Practice Pragmatic Case Studies & Improvement" published in Proceedings of the 9th North American Conference on Industrial Engineering and Operations Management Washington D.C., USA, June 4-6, 2020
5. Jamil Abdo & Tariq Masood "Techno Economics Analysis to Reshape Global Clean and Green Energy" 22nd International Conference on Renewable Energies and Power Quality (ICREPQ'24) Bilbao (Spain), 26th to 28th June 2024. https://www.icrepq.com/icrepq24/415-24-abdo.pdf
6. Jane Courtney; Aaron McDonnell "Impact on Distribution System Protection with the Integration of EG on the Distribution Network" published in 54th International Universities Power Engineering Conference (UPEC) in 2019
7. L. Ndahepele; S. Chowdhurym "Impact of Distributed Generation on Traditional Protection in Distribution and Transmission Systems: A Review" Published in IEEE PES/IAS PowerAfrica in 2020
8. Lü Zhining; Ning Baifeng; Luo Weifeng; Liu Wei "Distribution networks ground fault protection method based on fuzzy measure fusion criterion"published in IEEE 8th International Conference on Advanced Power System Automation and Protection (APAP) in 2019
9. Luyao Pei; Cheng Xu; Jinsong Zhang; Bo Wang; Yining Ma "Artificial Intelligence Algorithm in Intelligent Power Supply and Distribution Network Design System published in IEEE 3rd International Conference on Electronic Technology, Communication and Information (ICETCI) in 2023
10. M.Saeed Ullah, Tariq Masood, Umair Zahid, H. Abuzar Ahsan "Electrochemical Oxidation of Dissolved organic Dye in Aqueous Solution' Published in IMASCON-2019
11. Nasser Alemadi, Tariq Masood, R Aggarwal, Suhail A Qureshi, D Kothari "Novel selection approach of STATCOM device tailored and optimized engineering configuration to implement in between Qatar and Bahrain" published in Int. J. Autom. Power Eng. https://doi.org/10.14355/ijape.2014.0302.01
12. Qin Lijun; Wang Ying; Zheng Shenrui; Guo Qing "Situation of study on distribution network protection containing distributed generation and application scheme of intelligent protection" 2011 4th International Conference on Electric Utility Deregulation and Restructuring and Power Technologies (DRPT)
13. Rani Kumari; Bhukya K. Naick "Protection Coordination Impact on Reliability of Distribution Systems With Distributed Energy Resources" 2024 Third International Conference on Power, Control and Computing Technologies (ICPC2T).
14. Seongil Kim; Byungchul Sung; Jeong-Hyeok Heo; Duck-Su Lee; Protection against Transient Overvoltages Caused by Intermittent Earth Faults in Unearthed Marine Power Distribution Network" published in IEEE International Conference on Power Systems Technology (POWERCON) in 2022
15. Tariq Masood & Jamil Abdo & "Maintenance Strategy Production Best Practices—Pragmatic Case Studies and Operations Improvement" published in Proceedings of the 9th North American Conference on Industrial Engineering and Operations Management Washington D.C., USA, June 4-6, 2020

16. Jamil Abdo & Tariq Masood "Operational Reliability- Production Best Practice Pragmatic Case Studies & Improvement" Proceedings of the 9th North American Conference on Industrial Engineering and Operations Management Washington D.C., USA, June 4-6, 2024. https://doi.org/10.46254/NA09.20240065
17. Tariq Masood & Jamil Abdo "Maintenance System—Production Best Practice pragmatic case studies and Operations Improvement" Proceedings of the 9th North American Conference on Industrial Engineering and Operations Management Washington D.C., USA, June 4-6, 2024 https://doi.org/10.46254/NA09.20240064
18. Tariq Masood & Jamil Abdo "New Technique to Enhance Hybrid Renewable Energy Resources' Operational Flexibility and Reliability" 22nd International Conference on Renewable Energies and Power Quality (ICREPQ'24) Bilbao (Spain), 26th to 28th June 2024. https://www.icrepq.com/icrepq24/396-24-masood.pdf
19. Tariq Masood and Jamil Abdo "Review: Uninterrupted Liquid Hydrogen (E-Methanol) Production framework to deliver Clean & Green Energy" Published 22nd International Conference on Renewable Energies and Power Quality (ICREPQ'24) Bilbao (Spain), 26th to 28th June 2024. https://www.icrepq.com/icrepq24/414-24-masood.pdf
20. Tariq Masood, Muhammad Tajammal, Samer Kareem, Suhail Aftab Qureshi, Abdul Jabbar Khan, DP Kothari, Nair Jameel" Unified power Flow Controller Modeling and Analysis Technique on the GCC Power Grid Operations and Control" published IEEE Energy Con, 2018 June 7–9 Limassol Cyprus. https://ieeexplore.ieee.org/document/8398834/
21. Tariq Masood, Muhammad Tajammal, Samer Kareem, Suhail Aftab Qureshi, Murtaza Hashmi, Nasir Jameel "Unified power Flow Controller (UPFC) Modeling and Analysis Technique by using Matlab and EMTP" published (ICREQP, 18 Salamanca Spain 21–23 March, 2018). http://www.icrepq.com/icrepq18/250-18-masood.pdf
22. Tariq Masood, Muhammad Tajammal, Samer Kareem, Suhail Aftab Qureshi, Murtaza Hashmi, Naser Jameel "STATCOM Controller Performance Enhancement and Optimization Technique" published (ICREQP,18 Salamanca Spain 21-23 March, 2018). http://www.icrepq.com/icrepq18/249-18-masood.pdf
23. Tariq Masood, Muhammad Tajammal, Samer Kareem, Suhail Aftab Qureshi, Murtaza Hashmi, DP Kothari, "Smart Grid Operational functions and Control Challenges by Implementing SSSC Tailored to Optimize performance in between Qatar and KSA on the GCC Electrical-power grid" published in the ASTESJ Journal Advances in Science, Technology and Engineering Systems Journal.
24. Vassilis C. Nikolaidis; Evangelos Papanikolaou; Anastasia S. Safigianni "A Communication-Assisted Overcurrent Protection Scheme for Radial Distribution Systems With Distributed Generation" IEEE Transactions on Smart Grid
25. Xiaohui Song; Yu Zhang; Song Zhang; Shizhan Song; Jing Ma; Weibo Zhang "Active distribution network protection mode based on coordination of distributed and centralized protection" publish in 2017 China International Electrical and Energy Conference (CIEEC).
26. Yu Deng; Dingkang Liang; Jiahao Wang; Xinyi Song; Zehua Li; Zhetong Hong "Reliability Evaluation of Distribution System Based on Dynamic Bayesian Network" Published in 3rd Power System and Green Energy Conference (PSGEC) in 2023
27. Yuwei Xiong; Wei Chen "Positive sequence current differential protection based on least square method for di stribution network with distribution generation" 2021 IEEE 4th International Electrical and Energy Conference (CIEEC).
28. Zong Guo; Jiarui Li; Hongbo Lian; Shihua Zhao; Yang Chen; Renjie Dai; Xiuneng Xu; Yufeng Gu; Yangyang Liu "Optimization Method for Protection of Distribution Network with MW-PV Based on Multi-level Cooperative Protection Strategy" 2019 IEEE Sustainable Power and Energy Conference (iSPEC).
29. Ping Zeng; Zhaoxin Du; Desheng Zhou; Xiaofeng Zou; Qian Feng; "A Study on Adaptive Protection System of Distribution Network Based on Passive Protection Terminal of Ring Network Cabinet" Published in IEEE 6th Conference on Energy Internet and Energy System Integration (EI2) in 2022

30. Tariq Masood, Jamil Abdo and Kaimiao Liu, "Facilities Integrity Management System – Production Best Practice pragmatic case studies for improving Production Operations in the Energy Sector" IEOM 2025 Conference in Orlando. June 17 to June 19, 2025. Online paper link will be available end of September.
31. Tariq Masood, Jamil Abdo and Kaimiao Liu, "Risk Assessment and Management – Production Best Practice Case study for improving production Operations in the Energy Sector" published in IEOM 2025 Conference in Orlando. June 17 to June 19, 2025. Online paper link will be available end of September.
32. Tariq MASOOD(1), Jamil Abdo(1), Nasser j. AL-Kuwari(2) Tahir Imran Qureshi(3), Kaimiao Liu(1) "Nuclear ammonia as a renewable energy source: an innovative approach to advancing sustainable fuel solutions for the global energy crisis" published in 23rd International Conference on Renewable Energies and Power Quality (ICREPQ'24) Tenerife (Spain), 25-27th June 2025. https://repqj.com/index.php/repqj/article/view/4513.
33. Tariq Masood, Jamil Abdo, Atif Iqbal "Protection of Electrical Power Distribution Systems: Smart-grid, Microgrid, AI and Cybersecurity" ISBN: 978-1-394-32764-5, published by Wiley-IEEE Press 2025 https://mitpressbookstore.mit.edu/book/9781394327645.
34. Nasser J.Al-Kuwari(1) Tariq MASOOD(2), Jamil Abdo(2), Tahir Imran Qureshi(3), Kaimiao Liu(2) "Advancing Sustainable Hydrogen Production: Cost, Environmental, and Technological Innovations in Biomass, Pyrolysis, and SMR" published in 23rd International Conference on Renewable Energies and Power Quality (ICREPQ'24) Tenerife (Spain), 25-27th June 2025 www.enerqj.com/E&Q/eqj3/213-25.pdf.

Smart Grid Operation and Control

8

Contents

8.1 Introduction.......... 174
8.2 The Smart Grid Goal.......... 175
  8.2.1 Smart Grid (Evolutionary Development).......... 176
8.3 Smart Grid Driving Factors and Their Influences.......... 177
  8.3.1 Fundamental Characteristics of Smart-Grid Operations.......... 177
  8.3.2 STATCOM Operations.......... 178
  8.3.3 STATCOM: DC Capacitor Control.......... 179
  8.3.4 STATCOM: DC Capacitor Control.......... 180
8.4 Understanding Electric Grid Performance.......... 180
  8.4.1 Development of Transmission Grid Infrastructure.......... 180
  8.4.2 Power System Complexity.......... 181
  8.4.3 Power Flow Limits on AC Transmission System.......... 182
  8.4.4 Consideration of Reactive Power to Maintain Seamless AC Power Transmission Operations.......... 183
  8.4.5 Reactive Power Balance at Different Load Levels with Respect to "SIL".......... 184
  8.4.6 Complexity of Managing Reactive Power.......... 186
8.5 Technology Option for a Smart Transmission Grid.......... 189
  8.5.1 Transmission Technologies.......... 189
  8.5.2 FACTS Controller for Power Transmission Technology.......... 191
  8.5.3 HVDC Power Transmission Technology.......... 198
8.6 State-of-the-Art Smart Technology.......... 201
  8.6.1 Case Study.......... 201
  8.6.2 Cost Analysis of HVDC and FACTS Technology.......... 202
8.7 Types of Transmission Protection.......... 203
  8.7.1 Smart Protection Settings for a Ring Distribution System.......... 208
  8.7.2 Smart Distance Protection.......... 209
  8.7.3 Unit Protection.......... 213
8.8 Self-Assessment Questions.......... 216
Bibliography.......... 218

Supplementary Information The online version contains supplementary material available at https://doi.org/10.1007/978-3-031-94312-6_8.

T. Masood et al., *Protection of Electrical Power Transmission Systems*, Power Systems, https://doi.org/10.1007/978-3-031-94312-6_8

Abstract

When you have completed this chapter, you will be able to understand electric grid performance—conventional transmission technologies, smart transmission technologies, state-of-the-art (FACTS and HVDC) device operations and control, and the future outlook of smart device operations. You will be able to make smart and rigorous approach to calculate the operating conditions for an overcurrent protection system used for the protection of radial feeders and calculation for the settings of directional overcurrent relays in a ring distribution system. You will be able to use smart calculate conditions in distance protection system using stepped characteristics and calculate conditions in a current differential protection network used for unit protection.

Objectives

- Understanding electric grid performance.
- Conventional transmission technologies.
- Smart transmission technologies.
- State of the art.
- Future outlook.
- Calculate the operating conditions for an overcurrent protection system used, for example, for the protection of radial feeders.
- Make a simple calculation for the settings of directional overcurrent relays in a ring distribution system.
- Calculate conditions in the distance protection system using stepped characteristics.
- Calculate conditions in a current differential Protection network used for unit protection.

8.1 Introduction

The smart grid represents a significant advancement beyond mere smart metering; it is a transformative technology that is reshaping the landscape of power distribution systems. Encompassing vital components such as wires, substations, transformers, and switches, the smart grid is designed to modernize the way electricity is distributed and managed. By adopting this innovative approach, we are not only enhancing the reliability, quality, and efficiency of power distribution systems but also preparing them to meet the demands of the future.

At the heart of the smart grid are bi-directional electric grids and sophisticated communication networks, which work together to enhance the reliability, security, and efficiency of electric systems. These improvements cater to both small-scale and large-scale operations in generation, transmission, distribution, storage, and consumption. Integrating advanced software and hardware applications, the smart grid facilitates dynamic, integrated, and interoperable optimization of system operations, maintenance, and planning. It also enables effective interconnection and integration of distributed energy resources, providing consumers with real-time feedback and controls that allow them to monitor and manage their energy usage with precision and confidence.

In the realm of power system operations, occasional equipment failures are an unfortunate reality. This is where smart protection measuring equipment comes into play, acting as a crucial safeguard. This cutting-edge protection technique continuously monitors the condition of the equipment and is programmed to swiftly remove any faulty components from the system at the first sign of trouble. The importance of smart protection cannot be overstated, as it offers essential benefits:

(a) It efficiently eliminates faulted equipment, thus preserving the stability of the remaining power system operations.
(b) It curtails damage to equipment caused by excessive fault currents, safeguarding our infrastructure. Smart protection systems come with several indispensable features:
   - They are designed to trigger tripping only when a genuine fault is detected, focusing on the removal of only the affected equipment—this capability is known as discrimination.
   - They are engineered to actuate the circuit breaker rapidly, which plays a critical role in maintaining system stability.
   - Their reliability is paramount, ensuring that they function as intended without fail.

This chapter provides an in-depth exploration of the fundamental principles underpinning various types of smart protection used in power transmission and distribution systems. It includes comprehensive calculations for protection relays utilized in specific sections of the power system network. In practical applications, achieving precise relay settings is vital for obtaining the required protection discrimination with a high level of accuracy, ensuring the safe and efficient operation of our power infrastructure [1].

8.2 The Smart Grid Goal

A smart grid represents a revolutionary advancement in our energy delivery network, utilizing automated systems that facilitate a bidirectional flow of electricity and information. This innovative framework is capable of monitoring and adapting to fluctuations in power generation, consumer preferences, and even the operation of individual appliances. By transforming the entire energy conversion chain into a smart infrastructure, we can unlock significant benefits across several essential areas of control and operations. Decentralized energy management systems are revolutionizing communities by significantly enhancing energy efficiency. With advanced communication solutions, we ensure seamless connectivity and effortless data exchange, making energy management more intuitive. Intelligent substation

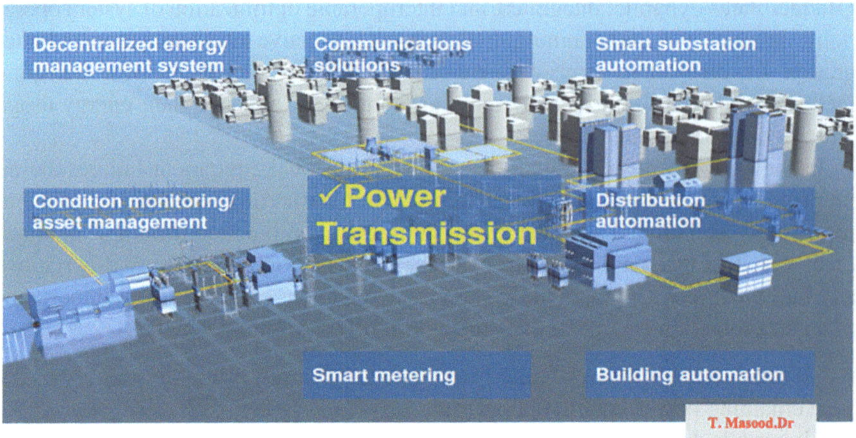

Fig. 8.1 Smart infrastructure. A digital illustration of a cityscape highlighting various aspects of power transmission and management. Key areas are labeled: "Decentralized energy management system," "Communications solutions," "Smart substation automation," "Condition monitoring/asset management," "Distribution automation," "Smart metering," and "Building automation." The central focus is on "Power Transmission" in bold yellow text. The image conveys interconnected systems within an urban environment, emphasizing modern energy solutions

automation not only boosts reliability but also maximizes operational effectiveness, leading to a more resilient energy infrastructure. By employing condition monitoring and asset management strategies, we can extend the life of our infrastructure while dramatically reducing maintenance costs. Robust power transmission systems guarantee quick and reliable delivery of electricity, ensuring that no household is left in the dark.

Distribution automation systems effectively optimize power distribution, minimizing outages and keeping the lights on when it matters most. Smart metering offers consumers real-time insights into their energy usage, empowering them to make better-informed decisions about their consumption. Finally, building automation systems seamlessly integrate technology to foster energy conservation and enhance comfort, creating a sustainable future for us all. These components work in harmony to create a more resilient and responsive energy landscape, as depicted in Fig. 8.1. Embracing this smart infrastructure is not just a choice; it is a necessary step toward a sustainable energy future [2].

8.2.1 Smart Grid (Evolutionary Development)

This digitally enabled approach transforms the electrical grid into a time-based system, allowing it to gather, distribute, and act on information about the behavior of all participants (Conventional, Distributed Renewable Generation, Transmission, and Distribution). This transformation significantly enhances the efficiency, reliability, economics, and sustainability of electricity services. A smart grid, in this context, is a powerful tool that not only increases the efficiency of the entire electrical distribution system but also plays a crucial role in reducing greenhouse gas emissions, addressing one of the most pressing environmental concerns of our time.

8.3 Smart Grid Driving Factors and Their Influences

Grid Conditions and Requirements

- Variable power supply.
- Integration of distributed resources.
- Increasing distances between resources.
- Loads.

Operational Factors

- Greater transmission network complexity.
- Aging infrastructure and lack of experts.
- High-power delivery quality.

External Influences

- Legal and regulatory framework.
- Integration of renewable energy resources.
- CO_2 reduction.
- Energy efficiency.

This is a complex grid structure that requires "smart solution."

8.3.1 Fundamental Characteristics of Smart-Grid Operations

The self-healing grid system is an advanced infrastructure that rapidly identifies, analyzes, and addresses disturbances, effectively restoring its functionality with minimal disruption. This innovative approach not only empowers consumers by actively involving them in the grid's design and operation but also tailors the system to accommodate various types of consumer equipment and behaviors. Significantly, this grid boasts a remarkable capacity for resilience, exhibiting strong tolerance against both physical and cyber threats. Its design incorporates robust mechanisms for mitigating attacks, ensuring that the entire system remains secure and reliable, safeguarding the interests of all users.

Exercise 8.1
How can power quality be achieved and monitored?

This groundbreaking system is uniquely designed to empower users in the twenty-first century by delivering the superior power quality essential for both consumer and industrial success. It skillfully adapts to a wide array of supply and

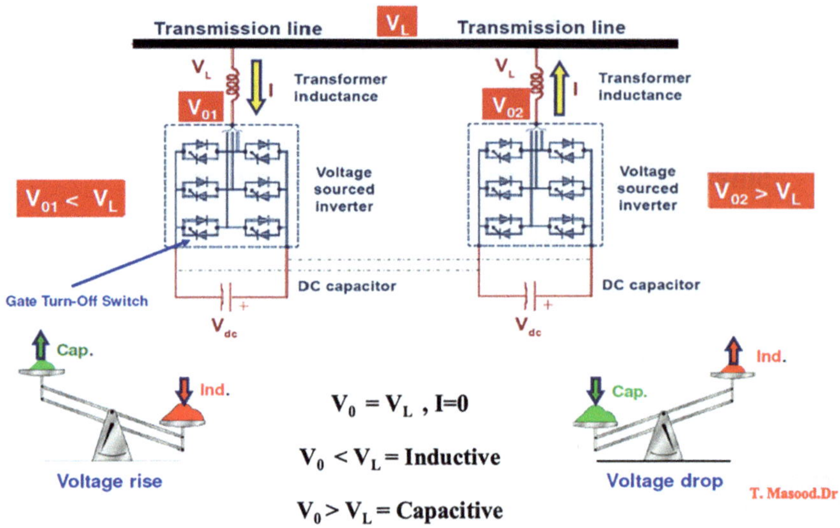

Fig. 8.2 Self-healing power delivery. Diagram illustrating a voltage sourced inverter system with two transmission lines. Each line connects to a transformer with inductance, leading to a voltage-sourced inverter and a DC capacitor. The left side shows a condition where voltage is less than line voltage, labeled as inductive, with a gate turn-off switch. Scales depict voltage rise and drop, with capacitive and inductive labels

demand scenarios by leveraging various resources, including demand response mechanisms as well as the combined capacitance and inductance response of the transmission line.

When operating in inductive mode, if V01 is less than VL (V01 < VL), reactive current flows back to the STATCOM (static synchronous compensator), ensuring optimal performance. In contrast, in capacitive mode, when V01 exceeds VL (V01 > VL), the reactive current flows from the STATCOM to the transmission line, maximizing efficiency (Fig. 8.2). Furthermore, this system thrives within a dynamic framework of competitive markets, fostering reliability, driving continuous innovation, and significantly enhancing energy sustainability for a better future [3].

8.3.2 STATCOM Operations

A STATCOM (static synchronous compensator) is an advanced power electronics device integral to modern power systems, specifically designed for reactive power regulation and voltage support. It comprises a static voltage source converter (SVS) that is energized by a direct current (DC) storage capacitor labeled Vdc. The SVS interfaces directly with the alternating current (AC) system by connecting in parallel (shunt) to the AC bus through a coupling transformer, which has a characteristic leakage reactance denoted as XT, as illustrated in Fig. 8.3.

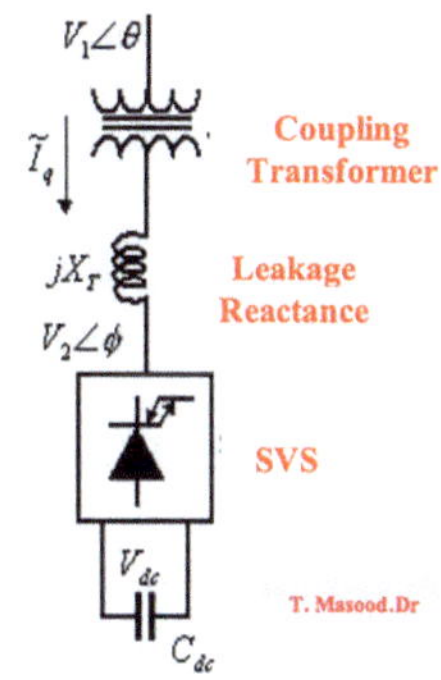

Fig. 8.3 STATCOM operations. Diagram illustrating a power transmission line with labeled components: a coupling transformer, leakage reactance, and SVS (Static Var System). The diagram includes electrical symbols and notations

The operation of the STATCOM heavily depends on the relationship between the phase angles ∠θ and ∠φ of the voltages connected to it. When these phase angles are equal, it indicates a balanced condition where the purely reactive current, represented as *Iq*, is influenced primarily by the voltage magnitudes V1 (the voltage at the AC system) and V2 (the voltage at the STATCOM).

In scenarios where V1 is greater than V2 (V1 > V2), the reactive current flows from the AC system into the STATCOM. This condition prompts the converter to absorb reactive power in an inductive manner, effectively providing voltage support to the grid and helping to stabilize the system. Conversely, if the situation reverses and V2 exceeds V1 (V2 > V1), the reactive current flows from the STATCOM back into the AC system. This action allows the converter to generate reactive power in a capacitive manner, thereby enhancing the reactive power resources available to the AC system.

The need for precise monitoring of these voltage conditions is critical, as it ensures optimal performance and reliability of the STATCOM. Adjustments in reactive power compensation are essential to maintain voltage stability in the system, particularly during conditions of varying load demands or disturbances. This operational flexibility of the STATCOM plays a significant role in improving power quality and supporting the overall performance of electrical networks [4].

8.3.3 STATCOM: DC Capacitor Control

The approach of purely reactive injection or absorption proves to be both impractical and undesirable for effective system operation. In this context, the converter is supported by a DC capacitor, which plays a crucial role in maintaining voltage stability. If the STATCOM fails to operate as a lossless device, the voltage across the DC capacitor will inevitably decline, jeopardizing system performance.

To counteract these losses—stemming from switching and other operational factors—it becomes essential to replenish the energy by drawing power from the AC system through the coupling transformer circuit. Consequently, it is imperative that the phase angle ∠φ lags slightly behind the phase angle ∠θ. This careful relationship is vital; if it is not upheld, the voltage of the DC capacitor may fluctuate unpredictably, leading to potential instability within the entire system.

Exercise 8.2
Describe the STATCOM DC capacitor control and operation.

8.3.4 STATCOM: DC Capacitor Control

In a multi-step converter, the output voltage's fundamental component plays a vital role and is fundamentally determined by the voltage magnitude across the DC capacitor. This relationship enables us to manipulate the voltage magnitudes, specifically V2 in relation to V1, by carefully charging or discharging the DC capacitor through precise control of the phase angle (φ). However, this method often results in a sluggish control loop, which can hinder the overall performance and responsiveness of the system. To overcome this limitation, pulsewidth modulation (PWM) can be effectively employed in a multilevel converter. This technique significantly enhances control response and contributes to better overall efficiency and adaptability. In sinusoidal PWM, the fundamental component of the output voltage magnitude can be adeptly adjusted by modifying either the voltage of the capacitor or the modulation index. While the diode-clamped topology has proven to be effective for various applications, it faces a serious challenge in the form of capacitor voltage imbalance, which can affect system reliability and performance. On the other hand, the flying capacitor topology presents an innovative solution to this pressing issue, showcasing its ability to balance capacitor voltages more effectively. This advancement offers a promising outlook for the future of multi-step converters, emphasizing their potential for enhanced efficiency and performance in modern applications.

8.4 Understanding Electric Grid Performance

8.4.1 Development of Transmission Grid Infrastructure

First, the expansion of interconnected power transmission systems will greatly enhance the efficiency and reliability of power exchange among various regions. This initiative will facilitate the transmission of substantial power blocks over extensive distances, particularly harnessing the potential of hydroelectric resources that thrive in mountainous and river-rich areas. Additionally, it will enable the effective utilization of renewable energy sources located in optimal locations, supporting a sustainable and greener energy future.

Exercise 8.3
Describe the power system complexity elements.

8.4.2 Power System Complexity

The first and foremost challenge we face in the energy sector is the suboptimal control of power flows throughout the grid. This situation can lead to inefficiencies and even potential disruptions. To address this, it is essential to manage reactive power and reactive reserves effectively, as they play a critical role in maintaining voltage levels and overall system stability. Without proper management, we risk compromising the reliability of our electrical networks. Power complexity attributes (Fig. 8.4):

- Reactive power management.
- Stabilities.
- Power flow.

Moreover, it is important to highlight that many planners and operators often receive inadequate training, which limits their ability to make informed decisions in real-time situations. It is imperative that investment in training programs be prioritized to equip these professionals with the necessary skills and knowledge.

In addition, effective data integration and seamless information exchange between various stakeholders remain areas that require significant improvement. Enhanced communication and access to real-time data will empower operators and planners to make more accurate assessments and timely decisions. Finally, the inherent complexity of the power system contributes significantly to the challenges we face. Therefore, a focused effort on advancing our approaches to reactive power management, ensuring stability, and optimizing power flow is crucial to the success

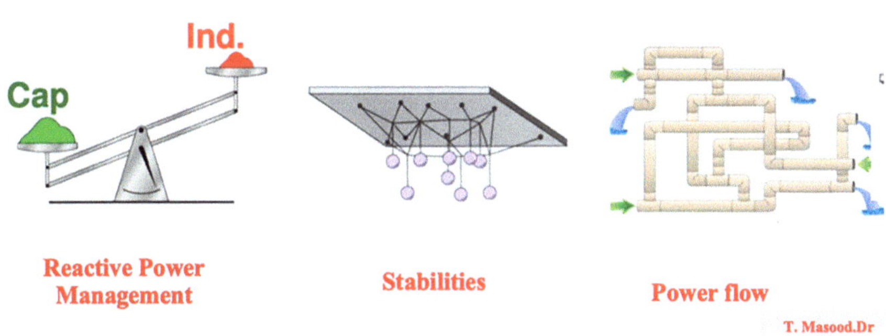

Fig. 8.4 Power system complexity. The image consists of three diagrams. The first diagram, labeled "Reactive Power Management," shows a seesaw with "Cap" in green on one side and "Ind." in red on the other, illustrating the balance between capacitive and inductive power. The second diagram, labeled "Stabilities," depicts a suspended platform with weights hanging below, suggesting equilibrium and stability. The third diagram, labeled "Power flow," features a complex network of interconnected lines with arrows indicating the direction of flow

of our energy infrastructure. By addressing these areas, we can create a more resilient and efficient power system for the future.

Exercise 8.4
Describe the uncontrolled power flow consequences in the power transmission network.

8.4.3 Power Flow Limits on AC Transmission System

Uncontrolled Power Flows

Diminished Power Transfer Capability: Uncontrolled power flows can significantly reduce the ability of transmission lines to carry electricity efficiently. This limitation can lead to an inadequate supply to meet demand, causing potential shortages and reliability issues (Fig. 8.5).

Presence of Bottlenecks: When power flows are not effectively managed, specific areas of the grid may become bottlenecks. These bottlenecks restrict the movement of electricity, resulting in congestion that can hinder overall system performance and prolong delivery times for consumers.

Creation of Loop Flows: Loop flows occur when power takes unintended paths through the grid, often due to differences in generation and load conditions. These flows can create instability in the network, leading to complications in maintaining a balanced and reliable power supply.

Overall, addressing these issues is crucial for the efficient operation of our electrical grid and to ensure a consistent energy supply for all users (Figs. 8.6 and 8.7).

Exercise 8.5
Describe the electrostatic field and electromagnetic field.

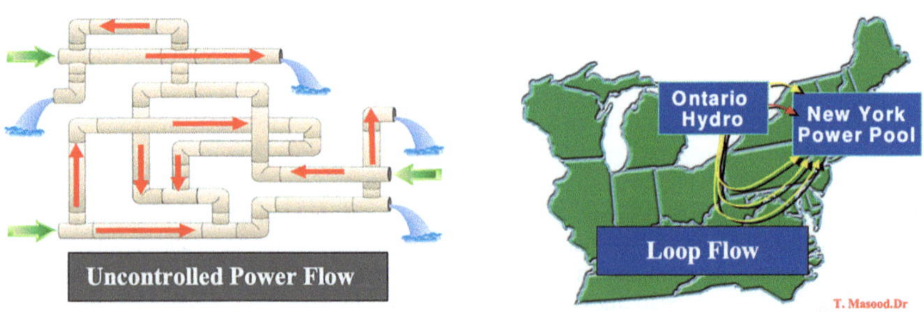

Fig. 8.5 Uncontrolled flows and loop flows. The image consists of two diagrams. The left diagram illustrates "Uncontrolled Power Flow" using a network of pipes with red arrows indicating the direction of flow and water spilling out at various points. The right diagram shows a map of the northeastern United States and parts of Canada, highlighting "Loop Flow" with labeled regions: "Ontario Hydro" and "New York Power Pool." Yellow arrows depict the flow of power between these regions

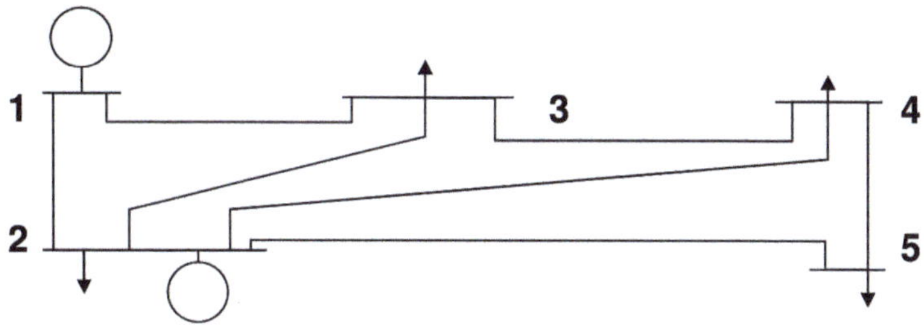

Fig. 8.6 Diagram of a mechanical system with five labeled points. Point 1 is at the top left, connected to a horizontal line leading to a downward arrow at point 2. A diagonal line extends from point 2 to a vertical line with an upward arrow at point 3. This connects to another horizontal line leading to point 4, which has an upward arrow. A final horizontal line extends to point 5, which has a downward arrow. Two circles are positioned near points 1 and 2

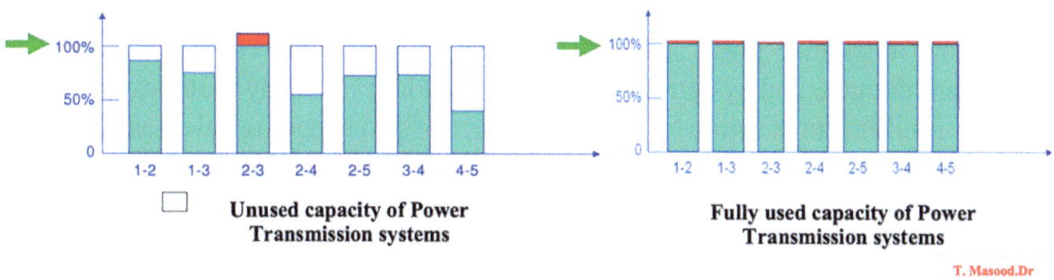

Fig. 8.7 The image shows two bar graphs comparing power transmission system capacity. The left graph illustrates unused capacity, with bars for connections 1–2 to 4–5 showing varying levels of unused space, indicated in white. The right graph depicts fully used capacity, with all bars reaching 100% for the same connections, marked in teal. A legend explains the color coding: white for unused capacity and teal for used capacity

8.4.4 Consideration of Reactive Power to Maintain Seamless AC Power Transmission Operations

In the realm of AC power transfer systems, the effective measurement of power is vital for maintaining a reliable electrical grid. Here, real power is expressed in megawatts (MW), reflecting the actual energy consumed by electrical devices, while reactive power is quantified in megavars (Mvar), which accounts for energy that oscillates between the source and the load without being consumed. Two pivotal forms of reactive power compensation are employed to enhance system performance.

First, capacitive reactive power compensation, which is closely associated with electrostatic fields (represented as $f(v)$), serves to counteract the lagging current often observed in inductive loads, thus improving voltage stability and overall efficiency. Second, inductive reactive power compensation, linked to electromagnetic fields (denoted as $f(I)$), is essential for managing the energy that flows back into the system, ensuring that equipment operates smoothly without overloading the infrastructure. By focusing on these compensatory strategies, we can significantly optimize energy systems, ensuring that they deliver reliable and high-quality power to end users [5].

Surge impedance loading (SIL) is $f(V, L, C)$

$$SIL = V^2 \cdot \sqrt{C/L} \tag{8.1}$$

8.4.5 Reactive Power Balance at Different Load Levels with Respect to "SIL"

When the power transmission network operates under what is described as "Light" loading—characterized by conditions below the short-circuit current level (SIL)—the system experiences an excess of capacitive elements compared to inductive ones. This imbalance poses a significant risk of overvoltage, which can lead to equipment damage and system failures. If the loading precisely matches the SIL, a delicate equilibrium is achieved between inductive and capacitive elements, but this results in an underutilized power transmission network, failing to harness its full potential (Fig. 8.8).

On the other hand, during periods of "Heavy" loading, where conditions exceed the SIL, the inductive components dominate, creating an environment ripe for

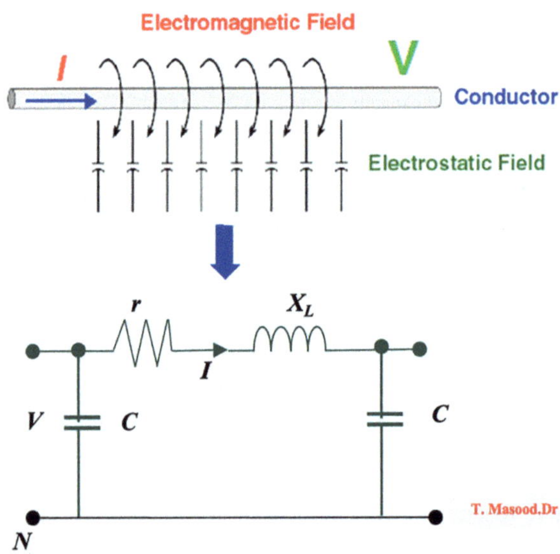

Fig. 8.8 Transmission line operations. Diagram illustrating the relationship between electromagnetic and electrostatic fields in a conductor. The top section shows a conductor with current "I" generating an electromagnetic field, depicted with curved arrows, and an electrostatic field, shown with straight arrows. The bottom section presents an equivalent circuit diagram with components: a resistor "r," an inductor "X_L," and capacitors "C," connected between voltage "V" and neutral "N"

voltage instability. Such instability can severely disrupt power delivery and threaten the reliability of the entire network. These potential risks make it abundantly clear that meticulous management of the balance between inductive and capacitive elements is essential. By strategically adjusting reactive power within the power transmission network, we can safeguard against these risks and ensure a robust, efficient system that operates effectively under varying load conditions [6].

Reactive Power Balance (System Voltage)

1. **Significance of Reactive Power Management**

   Maintaining a smooth and uniform voltage profile is crucial for the overall performance and reliability of our electric transmission system. When reactive power is insufficient, it can lead to noticeable voltage drops, creating potential disruptions in the flow of electricity. Conversely, a surplus of reactive power, often generated by lightly loaded transmission lines or cables, can result in an unintended rise in voltage levels. This imbalance can create serious challenges, contributing to phenomena like voltage collapse, which has been a significant factor behind several major power outages worldwide in recent years (Fig. 8.9).

   On the other hand, achieving an adequate supply of reactive power is not merely a technical necessity; it opens the door to substantial improvements in transmission capacity and enhances the efficiency of real power delivery. This balanced approach to managing reactive power not only stabilizes our energy systems but also positions us for a brighter and more reliable future in power distribution, ensuring that the electricity we depend on is both dependable and sustainable.

2. **How to Manage Reactive Power**

   Generators: Generators serve a crucial dual function within energy systems; they can either generate or absorb reactive power, which is vital for maintaining the stability and reliability of the electrical grid. When in an overexcited state, generators actively supply reactive power, bolstering voltage levels that are essential for the proper operation of electrical equipment. Conversely, when they are

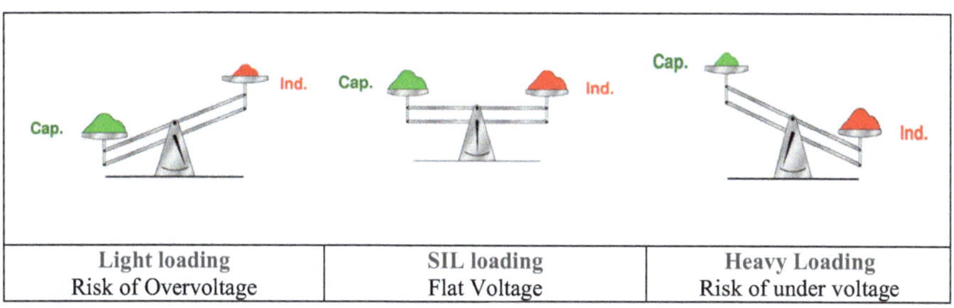

Fig. 8.9 Reactive power management. Three balance scales illustrate different weight distributions. The left scale tilts right, with a green "Cap." weight on the left and a red "Ind." weight on the right. Risk of overvoltage. The center scale is balanced with equal green "Cap." and red "Ind." Weights, Flat voltage. The right scale tilts left, with a green "Cap." weight on the left and a red "Ind." weight on the right, Risk of under voltage

underexcited, they absorb reactive power, helping to mitigate excess voltage and prevent potential overloads. The reactive power capability curve is an essential tool that outlines the specific limits on both the production and absorption of reactive power, guiding operators to optimize the generators' performance and ensure operational efficiency [7].

Overhead Lines: These robust lines play a significant role in power transmission, adeptly managing reactive power based on varying loading conditions. When the loading exceeds the Surge Impedance Loading (SIL), overhead lines absorb reactive power, acting as a buffer that alleviates strain on the network. Conversely, when the loading falls below the SIL, they transform into a source of reactive power, actively producing vars that enhance voltage stability within the system. This dynamic capability of overhead lines is essential for maintaining the overall health of the power grid.

Transformers: Operating continuously, transformers are the workhorses of electrical systems. They are designed to absorb reactive power at all times, regardless of the loading conditions. This consistent absorption helps to maintain optimal voltage levels across the grid, ensuring that all devices and systems receive the power they require to function effectively. Their role is indispensable in safeguarding against fluctuations and ensuring the reliability of the power supply.

Underground Cables: These specialized cables bring a unique advantage to the table by generating reactive power when operating below the SIL. This characteristic makes them an efficient choice for various applications, enhancing the overall performance of electrical systems by providing reactive power where it is needed most. Their ability to generate vars under specific conditions contributes to improved system efficiency and helps in managing voltage levels effectively.

8.4.6 Complexity of Managing Reactive Power

Many generating plants that supply reactive power play a critical role in supporting substantial loads throughout the electrical system, and it is vital that they are accurately modeled to reflect their true capabilities. As the operating conditions fluctuate, whether transitioning from light load scenarios to peak demand or experiencing variations in their power factor, the demand for reactive power alters significantly.

The challenge is compounded by the high X/R ratio present in transmission lines, which hinders the efficient long-distance transfer of reactive power. Much of this power becomes trapped or dissipated because I^2X losses, reducing its availability for effective voltage regulation. Therefore, it is essential to avoid relying on long-distance transfers of reactive power. Instead, implementing localized means of voltage control emerges as a highly effective strategy. By focusing on localized solutions, we can effectively manage reactive power and significantly enhance the overall stability of the electrical system.

Understanding Reactive Power
<u>Mvar support is exchangeable with MW.</u>
The exchange rate could be up to 0.5 MW/Mvar.

1. Power Angle and Voltage Stability ISSUES

Examine the critical issues related to power angle and voltage stability within the power system network. Power angle refers to the difference in phase angle between the voltage at the generator and the voltage at the load. This difference can significantly impact system stability. A suitable power angle is vital for preventing system overloads and maintaining the balance between generation and consumption.

In contrast, voltage stability is essential for ensuring that voltage levels across the network remain within acceptable limits during normal operations and under various disturbances. Poor voltage stability can lead to voltage collapse, resulting in widespread outages and compromising the reliability of the entire power system. These two issues are interconnected and require proactive management and monitoring to ensure a robust and resilient power network. Addressing these challenges is crucial for enhancing the overall performance and security of our energy infrastructure (Figs. 8.10 and 8.11).

2. AC Transmission Challenges

Thermal limits, stability limits, and uncontrolled power flow are essential concepts in managing power network operations, each playing a crucial role in ensuring reliable and efficient electricity delivery as described in Fig. 8.12.

Thermal limits refer to the maximum temperature that electrical equipment—such as transformers, transmission lines, and generators—can safely handle without sustaining damage. These limits are based on the physical properties of the materials used in the equipment. Exceeding these temperatures can lead to overheating,

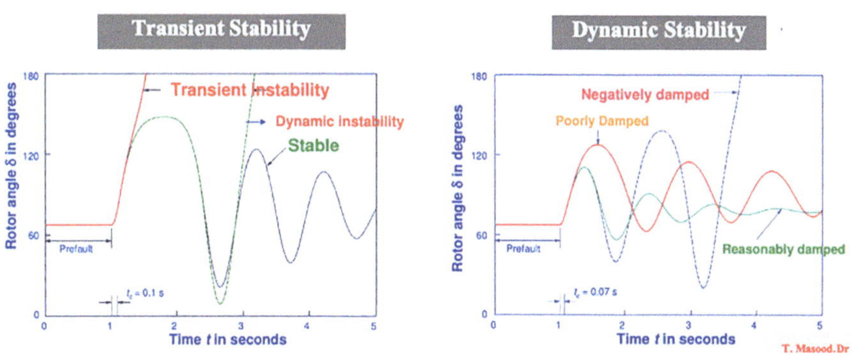

Fig. 8.10 Transient and dynamic angle stability issues. The image consists of two graphs comparing the transient and dynamic stability of rotor angles over time. The left graph, titled "Transient Stability," shows three curves: transient instability in red, dynamic instability in blue, and stability in green. The right graph, titled "Dynamic Stability," displays three curves: negatively damped in red, poorly damped in orange, and reasonably damped in green. Both graphs have rotor angle in degrees on the vertical axis and time in seconds on the horizontal axis, with a prefault period indicated

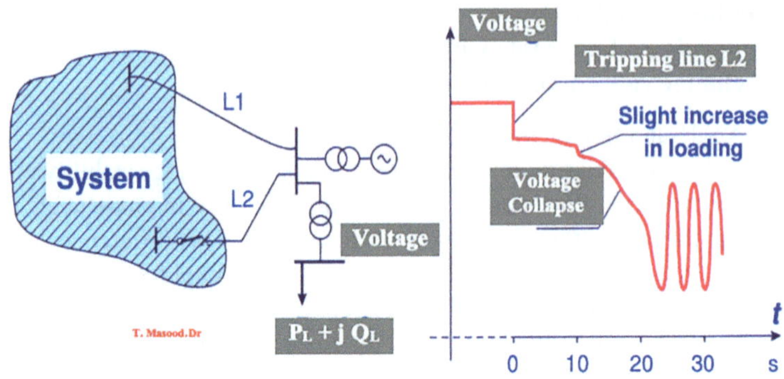

Fig. 8.11 Voltage stability issues. Diagram illustrating a power system with two lines, L1 and L2, connected to a load. The left side shows a schematic of the system with labeled components: "System," "Voltage," and "P_L + j Q_L." The right side features a graph with time on the x-axis and voltage on the y-axis. The graph shows a voltage drop labeled "Tripping line L2," followed by a "Slight increase in loading," leading to "Voltage Collapse" with oscillations

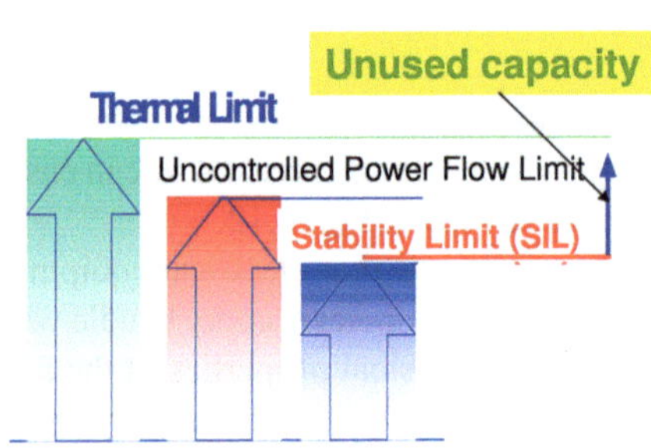

Voltage (kV)	SIL (MW)	Typical Thermal Rating (MW)
230	150	400
345	400	1200
500	900	2600
765	2200	5400
1100	5200	24000

Fig. 8.12 AC transmission challenges. The image illustrates power transmission limits with a focus on stability and thermal constraints. On the left, arrows represent different limits: the thermal limit is the highest, followed by the uncontrolled power flow limit, and the stability limit (SIL) is the lowest. A label indicates "Unused capacity" above the stability limit. On the right, a table lists voltage levels in kilovolts (kV) alongside their corresponding Surge Impedance Loading (SIL) in megawatts (MW) and typical thermal ratings in MW

decreased equipment lifespan, and even catastrophic failures. Therefore, careful monitoring and management of power loads are essential to prevent thermal overloads.

Stability limits define the operational boundaries of the power system, particularly concerning the balance between load and generation. These limits are vital during both normal operations and transient conditions, such as sudden changes in load or generation due to generator outages or spikes in demand. If these limits are exceeded, the power system may drift into unstable conditions, resulting in oscillations or blackouts. To maintain stability, operators must implement robust control strategies and continuously assess system performance against these stability thresholds.

Uncontrolled power flow occurs when power transfers within the network exceed established control measures, potentially leading to overloads and instability. This situation can arise from improper coordination between generation and load or from unexpected events such as line faults or sudden changes in generation capacity. The consequences can be severe, potentially resulting in cascading failures across the grid. To address uncontrolled power flow, operators need to utilize advanced monitoring tools, dynamically adjust generation resources, and establish effective communication protocols among grid operators.

In summary, thoroughly understanding and actively managing thermal limits, stability limits, and uncontrolled power flow are imperative for the safe and efficient operation of power networks. Prioritizing these elements not only protects the infrastructure but also ensures a consistent and reliable supply of electricity to consumers [8].

8.5 Technology Option for a Smart Transmission Grid

Technology serves as the fundamental building block for enhancing transmission capacity and effectively managing both real and reactive power flows. In the realm of electrical power system operations, it is indispensable for achieving efficient and reliable performance, significantly boosting our capability to ensure uninterrupted service. To develop a smart transmission grid, we have an array of innovative technology options at our disposal for managing power delivery and its dynamics. These solutions include dynamic circuit rating, synchrophasors, and broad-area monitoring, protection, and control systems. It is imperative to embrace these cutting-edge technologies to tackle the root causes of issues within power systems, rather than merely responding to their effects.

Leveraging power electronics-based controllers, such as high-voltage direct current (HVDC) systems and flexible AC transmission systems (FACTS), with or without energy storage, empowers us to optimize our power delivery infrastructure for the future. Adopting these advancements is not just beneficial; it is essential for ensuring a resilient and efficient power system.

8.5.1 Transmission Technologies

Our initial technology-driven collaborative efforts are dedicated to effectively managing thermal limits. By adopting dynamic thermal circuit rating technology and tackling the issues of high-temperature, low-sag conductors, we are making significant strides in this critical area.

Line sagging can lead to reduced clearance, diminished conductor strength, and increased annealing problems, as depicted in Fig. 8.13 with the traditional ACSR (aluminum core steel reinforced) conductors. However, our innovative high-temperature, low-sag composite conductors not only reduce weight but also enhance ampacity and strength, as illustrated in Fig. 8.14. This advancement represents a transformative step toward more efficient and reliable power transmission.

Management of Power Delivery and Dynamics,	HVDC & FACTS Technologies	Monitor, dynamic circuit rating

Fig. 8.13 Technology option for smart transmission. The image consists of three sections. The first section shows a control room with multiple computer monitors and operators working at desks, indicating a monitoring or operations center. The second section features a person looking through a telescope with the text "Use of new Technologies," suggesting innovation or exploration. The third section is a diagram of power lines with labels for "Line Tension Monitor," "Weather Station," and "Temperature Monitor," illustrating remote monitoring of electrical infrastructure

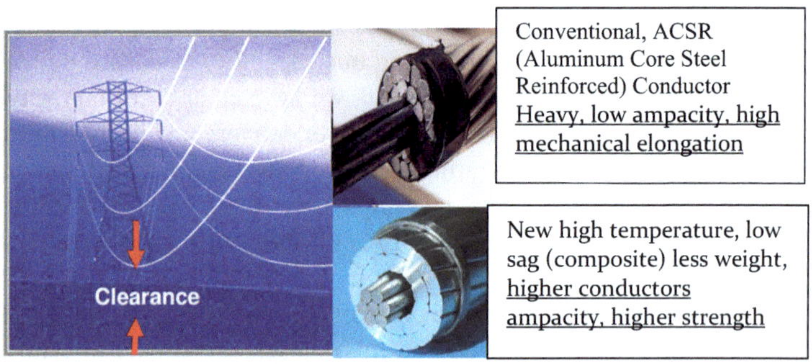

Fig. 8.14 ACSR (Aluminum Core Steel Reinforced) The image compares two types of electrical conductors. On the left, a diagram shows power lines with arrows indicating "Clearance" beneath them. On the right, two photos depict conductor cables. The top photo shows a conventional ACSR (Aluminum Core Steel Reinforced) conductor, described as "Heavy, low ampacity, high mechanical elongation"

As power system operators, you hold a vital responsibility in the effective deployment of innovative smart technologies that are reshaping the energy landscape. These cutting-edge technologies involve sophisticated methods of power flow control and system dynamics, primarily utilizing power electronics-based transmission systems designed for optimal performance. This includes the deployment of advanced controllers such as FACTS (Flexible AC Transmission Systems) and HVDC (High Voltage Direct Current) technologies. Together with segmentation strategies and the implementation of grid shock absorbers, these components play a crucial role in bolstering the resilience and efficiency of our power grids. In addition, the integration of wide-area monitoring, protection, and control systems with energy storage solutions and synchrophasor technology enhances our ability to

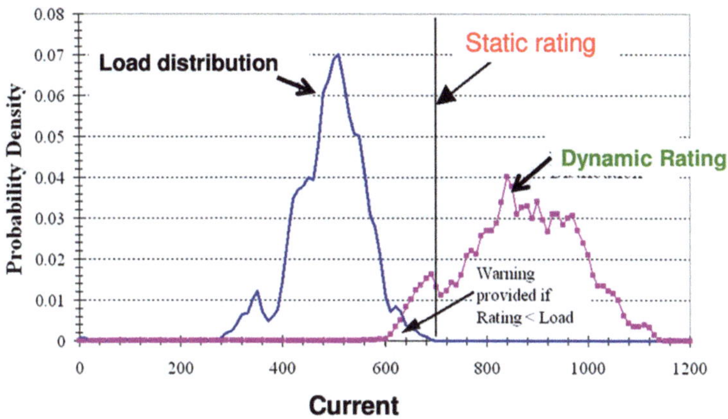

Fig. 8.15 Load distribution static and dynamic rating. A graph showing probability density versus current. The blue line represents load distribution, peaking around 600 units. The pink line indicates dynamic rating, peaking near 800 units. A vertical line marks the static rating at approximately 700 units. Annotations highlight "Load distribution," "Static rating," "Dynamic Rating," and a warning note stating "Warning provided if Rating < Load"

oversee and manage power delivery on a larger scale. The significance of HVDC and FACTS becomes increasingly evident as they provide essential support for the optimization of transmission operations and the dynamic management of power flow. A visual representation of a typical probability distribution of load and dynamic rating can be seen in Fig. 8.15, highlighting the transformative potential of these advanced technologies in advancing our energy systems and ensuring a more sustainable future [9].

8.5.2 FACTS Controller for Power Transmission Technology

8.5.2.1 First-Generation Power Transmission Line Control Devices

Figure 8.16 provides a comprehensive illustration of the control mechanisms for critical power flow parameters, which include voltage magnitude, transmission line reactance, and power angle. This control process initiates with the adoption of two distinct operational modes: the switched capacitor mode and the switched reactor mode, both of which are elaborated in Fig. 8.17.

In the switched capacitor mode, reactive current is purposefully injected into the transmission line. This is achieved through the use of a mechanical switch that can seamlessly toggle the capacitor on and off as needed. When this injection occurs, it raises the voltage levels, referred to as V1 and V2, at both the sending and receiving terminals of the transmission network. This action is particularly crucial when there are fluctuations or deviations from the expected operating conditions, ensuring that the system remains within the prescribed operational envelope.

Conversely, in the switched reactor mode, when the voltage levels surpass the predetermined operating limits, the reactor switch is activated mechanically. This

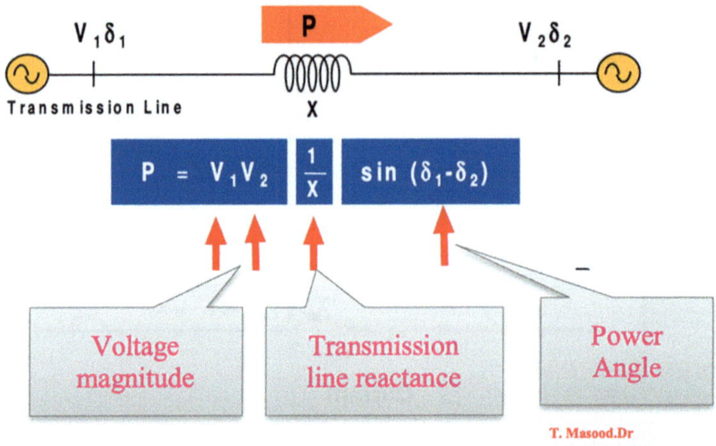

Fig. 8.16 Controlling power transfer parameters. Diagram illustrating power flow in a transmission line. It shows a transmission line with voltage sources at each end, connected through a reactance is represented by an arrow and is calculated using the formula. Labels indicate "Voltage magnitude," "Transmission line reactance," and "Power Angle"

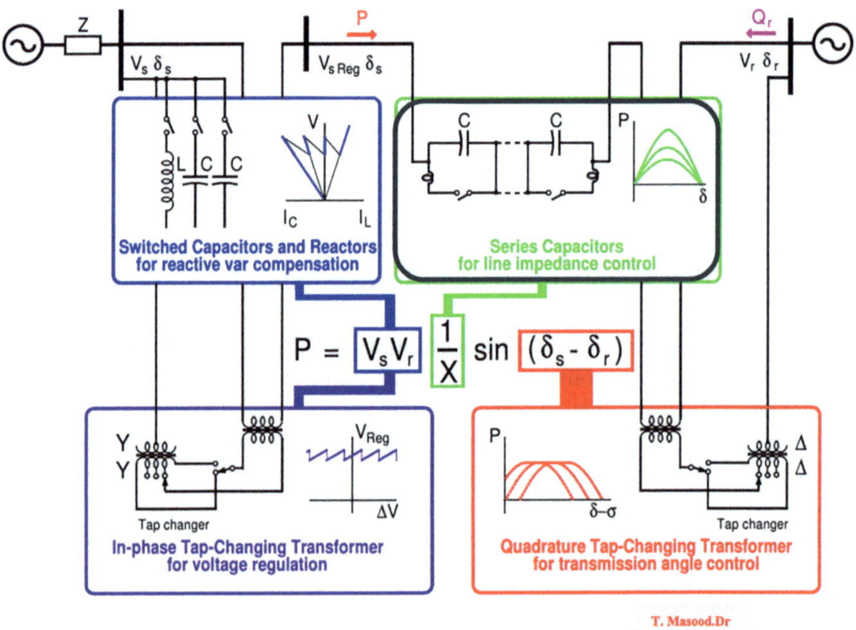

Fig. 8.17 Mechanical switched-based passive control devices. Diagram illustrating an electrical power system with components for voltage and impedance control. The system includes switched capacitors and reactors for reactive var. compensation, series capacitors for line impedance control, and two types of tap-changing transformers. The in-phase tap-changing transformer is used for voltage regulation, while the quadrature tap-changing transformer manages transmission angle control. Mathematical expressions and graphs depict relationships between variables such as voltage, impedance, and phase angles

switch absorbs the excess reactive current from the transmission line, effectively managing and stabilizing the voltage profile at both ends of the pipeline. This two-mode approach allows for dynamic management of the network's voltage, ensuring consistency and reliability throughout the system.

Subsequently, we transition to a series capacitive mode of operation, also illustrated in Fig. 8.17. This mode plays a key role in injecting reactive current into the transmission line by adjusting the line reactance. By utilizing mechanical switching, we can significantly enhance the power flow between the sending and receiving ends of the network. Should the power flow exceed the defined operating limits, the reactor switches engage in a mechanical operation to absorb reactive current from the line in series, thereby preventing potential overload and ensuring adherence to the operating envelope.

The quadrature tap-changing transformer stands out as a pivotal component within the power transmission control framework. By adjusting the power transmission angle, this transformer optimizes the power flow within the interconnected area network, as depicted in Fig. 8.16. Achieving this optimization is essential for maximizing the network's capacity, as demonstrated in Fig. 8.17. The operational benefits include increasing power flow in areas experiencing lower power availability, alleviating bottlenecks caused by uneven load distribution, and effectively managing loop flows in the system, all based on the network's inherent load-carrying capability [10].

In addition to these operational modes, mechanical switched capacitors and reactors perform their functions with precision, taking between 120 and 180 ms to complete their transitions, as shown in Fig. 8.17. These components, known for their reliability, utilize mechanically driven circuit breakers that operate on capacitive and inductive principles, thereby characterizing them as forms of passive transmission technology. This thoughtful integration of advanced technologies is crucial for maintaining a resilient and efficient power transmission system. Figure 8.18

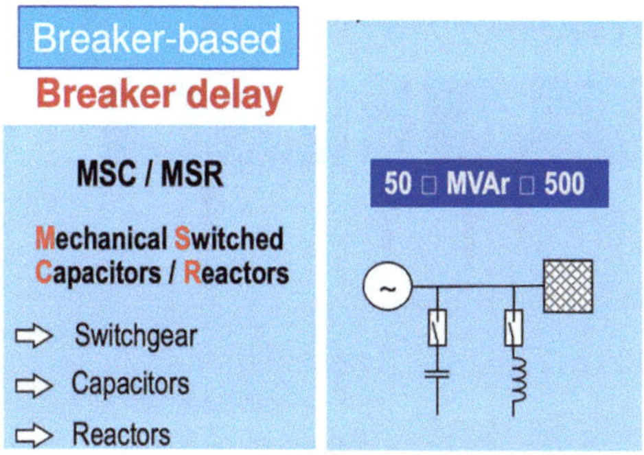

Fig. 8.18 First-generation time-based passive control devices operations and transmission line capacitive and inductive compensation. Diagram illustrating a breaker-based system with a focus on breaker delay. The left section includes the text "MSC/MSR" with highlighted letters for "Mechanical Switched Capacitors/Reactors." The icons labeled "Switchgear," "Capacitors," and "Reactors." The right section shows a schematic with electrical symbols, labeled "50 MVAr 500," representing the system's capacity

describes the components and operational time period (120–180 ms) of the first-generation devices.

8.5.2.2 Second-Generation Power Transmission Line Control Devices

Figure 8.19 provides a comprehensive illustration of the essential functions performed by the static var. compensator (SVC) and the thyristor-controlled series capacitor (TCSC) in regulating key power flow parameters, including voltage magnitude, transmission line reactance, and power angle. The management of these parameters is critical for the stability and efficiency of power transmission networks.

The process initiates with the implementation of the SVC mode, as depicted in Fig. 8.19. In its capacitive mode, the SVC plays a vital role in improving voltage levels within the transmission line. It achieves this by injecting reactive current into the line using a mechanical switch that can be adjusted based on system requirements. When the current flow deviates from the expected operational envelope, this capacitive action works to enhance the voltage at both the sending end (V1) and the receiving end (V2) of the transmission network. Conversely, during the inductive mode of operation, a reactor switch is activated through the thyristor when voltage levels surpass the predetermined operating limits. This action absorbs reactive current from the transmission line, ensuring that the voltage profile remains stable and operates within the defined envelope at both ends of the pipeline.

Following the SVC process, the TCSC mode is employed, as shown in Fig. 8.20. The operation of the TCSC focuses on the controlled injection of reactive current into the transmission line. By executing thyristor switching at precise intervals, operators can effectively adjust the transmission line reactance between the sending and receiving ends. This adjustment is instrumental in increasing power flow in the

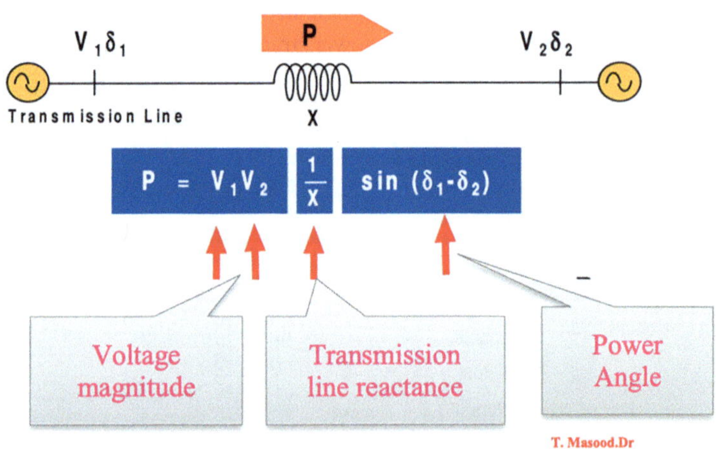

Fig. 8.19 Controlling power transfer parameters. Diagram illustrating power flow in a transmission line. The line connecting two voltage sources is represented by an arrow and calculated using t. Three labeled components highlight “Voltage magnitude,” “Transmission line reactance,” and “Power Angle”

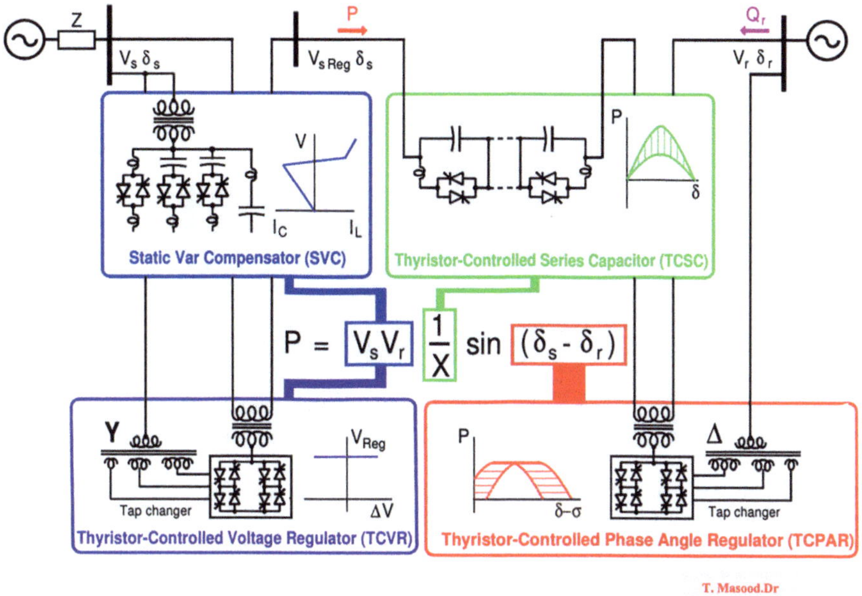

Fig. 8.20 Thyristor-based active control devices. Diagram illustrating a power system with components for voltage and phase regulation. The static var compensator (SVC) is shown in blue, featuring a schematic of capacitors and inductors. The thyristor-controlled series capacitor (TCSC is highlighted in green, with a diagram of thyristors and a sinusoidal graph. The thyristor-controlled voltage regulator (TCVR) is in purple, showing a tap changer and voltage graph. The thyristor-controlled phase angle regulator (TCPAR) is marked in red, with a phase angle graph and tap changer

system. In instances where the power flow exceeds the specified operational limit, the reactor is promptly switched on or off via the thyristor to absorb excess reactive current from the transmission line, thereby preventing overloading and ensuring safe operation.

A critical component in this power transmission control framework is the quadrature tap-changing transformer, as illustrated in Fig. 8.20. This transformer is responsible for adjusting the power transmission angle and optimizing power flow throughout the inter-tie area network. Such adjustments are crucial for fully utilizing the maximum available capacity of the network, as depicted in Fig. 8.20. The overall process involves several key actions [11]:

- **Increasing Power Flow**: Enhancing power delivery, especially in low-power areas of the network, facilitates overall system efficiency.
- **Reducing Bottlenecks**: This involves redistributing electrical load across neighboring networks to prevent congestion, ensuring a balanced distribution of power resources.
- **Mitigating Loop Flows**: Loop flows can create inefficiencies and potential overloads within the network; managing these flows is essential for maintaining system stability.

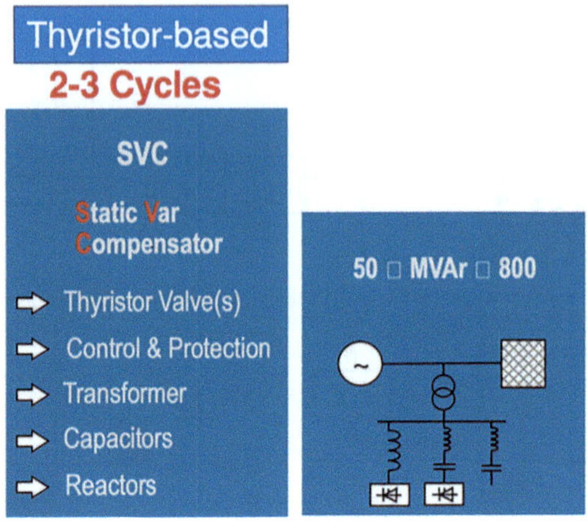

Fig. 8.21 Second-generation time-based active control devices operations and transmission line capacitive and inductive compensation. Diagram illustrating a thyristor-based static var compensator (SVC) system, which operates in two to three cycles. The left section lists components: thyristor valves, control and protection, transformer, capacitors, and reactors. The right section shows a schematic with symbols representing electrical components, labeled with "50 MVar 800"

This entire operational framework is deeply rooted in the concept of loadability, which defines the maximum quantity of power a network can accommodate without risk of overload scenarios. Efficient power distribution not only optimizes resource utilization but also underscores the significant responsibility held by operators in managing network capacity effectively to maintain system reliability.

As evidenced in Fig. 8.20, the use of thyristor-based capacitors and reactors is integral to modern power transmission systems. These devices typically require two to three cycles to execute their operational functions, highlighting their responsiveness in dynamic situations. Additionally, mechanically driven circuit breakers that operate based on capacitive and inductive modes are categorized as passive transmission technologies. These components are indispensable in the overall power transmission process, accentuating the critical nature of effective network capacity management to ensure uninterrupted power delivery to consumers and industries alike. Figure 8.21 describes the components and operational time period (two to three cycles) of the second-generation devices.

8.5.2.3 Third-Generation Power Transmission Line Smart Control Devices

Figure 8.22 decisively illustrates the control of critical power flow parameters, including voltage magnitude, transmission line reactance, and power angle. This process is initiated with the STATCOM (static synchronous compensator) mode of operation, as depicted in Fig. 8.23. This mode is essential because the switched STATCOM capacitor injects reactive current into the line using a converter-based switch that can be activated or deactivated quickly. This capability effectively enhances the voltages V1 and V2 at both the sending and receiving ends of the

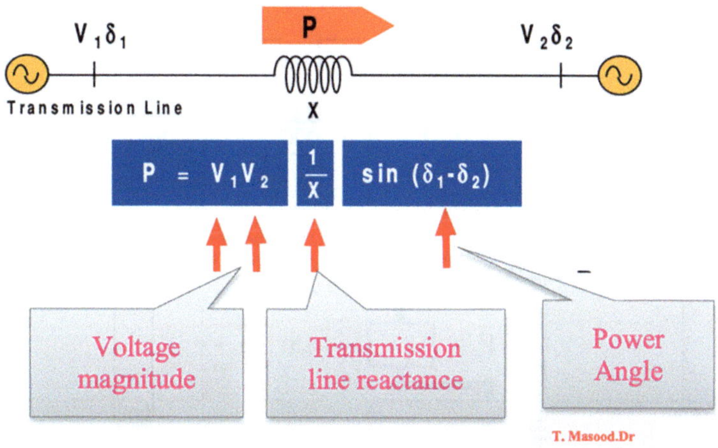

Fig. 8.22 Controlling power transfer parameters. Diagram illustrating power flow in a transmission line. It shows two voltage sources connected by a transmission line with reactance \(X\). The power flow \(P\) is represented by an arrow and calculated. Annotations highlight key components: "Voltage magnitude," "Transmission line reactance," and "Power Angle"

transmission network, even when there is a deviation from the predetermined operating envelope—an established range of acceptable operating conditions.

In situations where voltage exceeds the operating limit during the STATCOM reactor mode, the reactor switch is reliably turned on or off through converter-based technology, absorbing reactive current from the transmission line. This proactive measure maintains the voltage profile within the operating envelope on both ends of the pipeline.

Furthermore, we implement the SSSC (static synchronous series compensator) mode of operation to inject reactive current, as shown in Fig. 8.23. By employing converter-based switching, we can decisively enhance power flow by adjusting the line reactance between the sending and receiving ends of the transmission network. If power flow surpasses the operating limit, the reactor switches effectively engage in series to absorb reactive current from the transmission line. This control action is fundamental in ensuring that power flow remains well within the operating envelope.

The converter-based phase angle regulator serves as a vital component in power transmission control. It adjusts the power transmission angle and optimizes the power flow within the interconnected transmission network, as demonstrated in Fig. 8.22. This function is crucial for maximizing the network's available capacity, as highlighted in Figs. 8.6 and 8.7. The process effectively increases power flow in low-power areas of the network, removes bottlenecks by sharing the load with neighboring systems, and manages loop flows efficiently. By relying on the network's load capacity, we ensure optimal power distribution, emphasizing the critical responsibility of managing network capacity to maintain a stable and reliable power supply.

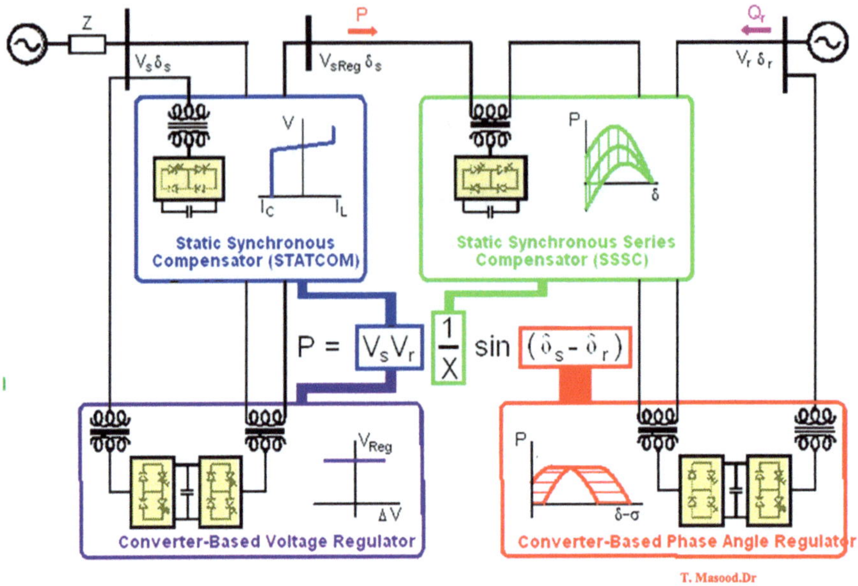

Fig. 8.23 Converter-based smart control devices. Diagram illustrating a power system with three main components: Static synchronous compensator (STATCOM) in blue, static synchronous series compensator (SSSC) in green, and converter-based phase angle regulator in red. Each component is represented with symbols and graphs indicating voltage and power flow. The STATCOM shows a voltage graph, the SSSC displays a power curve, and the phase angle regulator includes a phase angle graph. Equations and symbols are used to denote power (P), voltage (V), and phase angles (δ). The system also includes a converter-based voltage regulator in purple, showing voltage regulation

Thyristor-based devices such as STATCOMs, SSSCs, voltage regulators, and phase angle regulators are designed to operate swiftly within one to two cycles to complete their functions. This real-time control and adjustment of power flow, known as active transmission technology, guarantees the efficient and reliable operation of the network, reinforcing confidence in the effectiveness of our power management strategies. Figure 8.24 describes the components and operational time period of the third-generation devices.

8.5.3 HVDC Power Transmission Technology

The HVDC Link is an innovative and efficient power transmission system specifically designed for connecting offshore wind farms located far from existing AC transmission infrastructure. Its primary advantage is the ability to transmit large amounts of electricity over long distances with exceptional efficiency, showcasing advanced engineering capabilities. The system uses robust cables to transport power over moderate to long distances, as illustrated in Fig. 8.25, highlighting the engineering excellence that supports modern energy needs (Table 8.1).

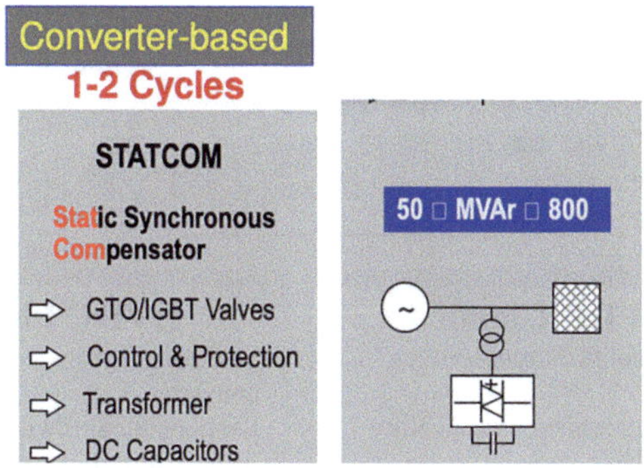

Fig. 8.24 Third-generation time-based active control devices operations and transmission line capacitive and inductive compensation. The image illustrates a converter-based system labeled "1–2 Cycles" with a focus on STATCOM, which stands for static synchronous compensator. Key components listed include GTO/IGBT valves, control and protection, transformer, and DC capacitors. A diagram on the right shows a schematic with a label "50 MVar 800," indicating the system's reactive power capacity

HVDC controllers are crucial for the system's operation, enabling seamless power transfer between asynchronous systems that operate at different frequencies or have varying phase angles. This functionality allows for efficient management of electricity flow, strategically directing power to specific areas and effectively reducing loop currents while addressing congestion in transmission lines that serve densely populated regions.

Additionally, the HVDC power delivery method significantly enhances overall system stability and reliability, serving as an effective safeguard against the risks of cascading outages. By embracing HVDC technology, we are paving the way for a more resilient and dependable energy infrastructure, ensuring that future energy demands are met with a focus on innovation and sustainability [12].

8.5.3.1 Application of HVDC Based on LCC and VSC

1. Line commutated converter (LCC)-based technologies.
2. Voltage sourced converter (VSC)-based technologies.

Explore the advanced technologies related to line commutated converters (LCCs) and voltage sourced converters VSCs, both of which play vital roles in modern power electronics.

LCC-based technologies use thyristors and are mainly utilized in high-voltage direct current (HVDC) transmission systems. This approach allows for efficient power transfer over long distances with minimal losses, making it ideal for connecting large power grids and integrating renewable energy sources. LCCs are particularly effective in scenarios requiring stable voltage and current control; however, they depend on a robust AC grid for commutation.

In contrast, VSC-based technologies utilize insulated-gate bipolar transistors (IGBTs) or similar devices to convert DC to AC, offering greater flexibility and

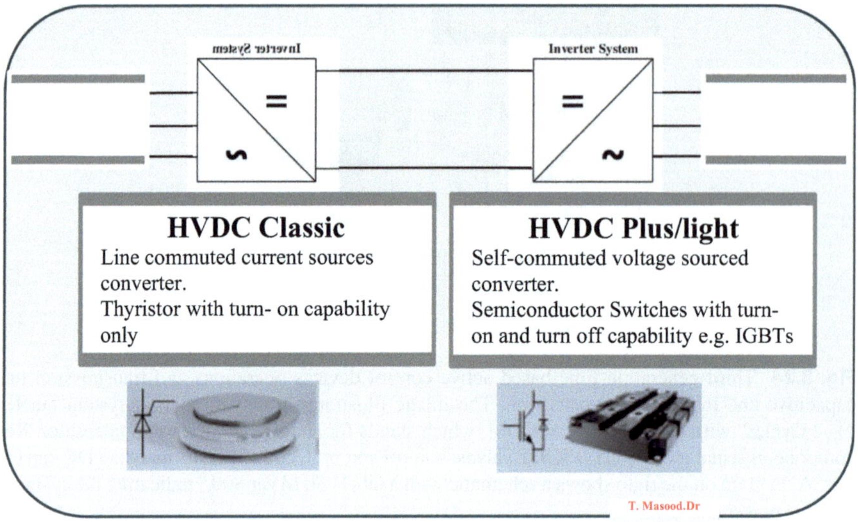

Fig. 8.25 HVDC controllers performance analysis. Diagram illustrating two types of HVDC systems connecting AC Grid System 1 and AC Grid System 2. On the left, "HVDC Classic" uses a line-commuted current source converter with thyristors, which have turn-on capability only. An image of a thyristor is shown below. On the right, "HVDC Plus/light" features a self-commutated voltage source converter with semiconductor switches like IGBTs, capable of both turn-on and turn-off

Table 8.1 Comparative analysis of thyristor and converter-based technology. Comparison image showing two types of technology. On the left, "Thyristor based Technology capacitors/reactors" with bullet points: Limited performance, Limited functionality, and Large footprint. The image shows outdoor electrical equipment. On the right, "Voltage source Converter based Technology" with bullet points: Superior performance, Versatile functionality, and Smaller footprint. Image shows indoor electrical equipment with two people inspecting it

Thyristor based technology capacitors/reactors	Voltage source converter based technology
Thyristor switched	Converter-based controllers
• Limited performance	• Superior performance
• Limited functionality	• Versatile functionality
• Large footprint	• Smaller footprint

Table 8.2 Comparative analysis of LCC and VSC (need vs. technology)

Moving large amounts of power long distances	Interconnecting weak AC networks
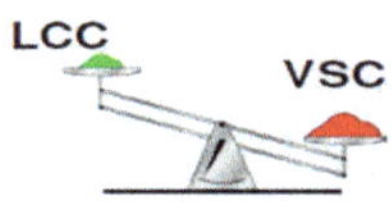	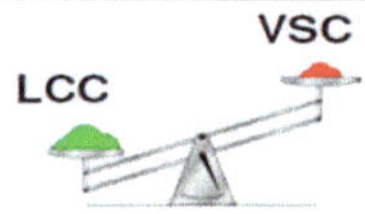
Improving system stability and reliability	HVDC technology of the future
	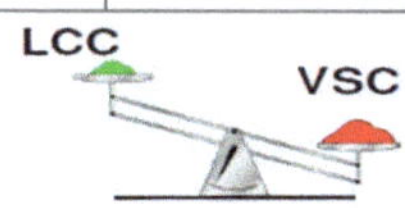

control in power systems. VSCs enable bidirectional power flow management, which not only facilitates the integration of renewable energy sources but also enhances grid stability. They are especially beneficial in applications that require rapid response times and precise control, such as offshore wind farms and microgrids.

Together, these technologies create a strong framework for improving the efficiency, reliability, and sustainability of electricity transmission and distribution systems worldwide. Understanding their capabilities and applications is essential for navigating the future of energy management as described in Table 8.2.

8.6 State-of-the-Art Smart Technology

8.6.1 Case Study

NYPA's Marcy Convertible Static Compensator: Transforming Transmission Capacity

Figure 8.26 illustrates the significant impact of innovative technologies on addressing critical transmission bottlenecks within the grid. The implementation of robust dynamic voltage support at the Marcy facility has successfully enhanced the overall transmission capacity by approximately 200 megawatts (MW). This increase in capacity translates to sufficient energy to supply around 200,000 homes, demonstrating its substantial effect on energy distribution. What truly sets this advancement apart is the incorporation of static synchronous compensator (STATCOM) operation. This integration not only allows for improved voltage stability but also introduces remarkable levels of controllability and flexibility to the transmission networks. The ability to adjust and manage reactive power dynamically enables operators to respond swiftly to fluctuating demand and maintain optimal voltage levels across the system.

Moreover, Fig. 8.26 underscores the sophisticated converter-based technology that underpins the Marcy convertible static compensator. This state-of-the-art technology is engineered for rapid and stable control of both voltage (V) and reactive power (Q). By leveraging the features of the STATCOM FACTS (flexible AC transmission systems) device, this system exemplifies a vital component of

Fig. 8.26 STATCOM Implemented Marcy New York. Two workers wearing hard hats stand beside a large industrial machine in a factory setting. The machine features multiple rows of complex components and wiring. Overhead lights illuminate the spacious room. The workers appear to be inspecting or discussing the machinery

contemporary, efficient operational technology. These advancements play a crucial role not only in bolstering grid reliability but also in empowering utilities to meet the surging energy demands of the future. By enhancing transmission capabilities and system responsiveness, the Marcy facility sets a leading example in modern energy management and distribution (Fig. 8.27) [13].

8.6.2 Cost Analysis of HVDC and FACTS Technology

Table 8.3 presents a comprehensive analysis of four scenarios for transmitting power at 200 MW, 500 MW, 1000 MW, and 2000 MW. This analysis is conducted using advanced HVDC (High-Voltage Direct Current) two-terminal systems alongside FACTS (Flexible AC Transmission Systems) devices, ensuring a thorough understanding of the topic.

Let's delve into the first scenario, where the investment for transmission ranges from $40 million to $50 million. The accompanying FACTS devices, a crucial part of the process, require an additional $5 million to $10 million to transmit 200 MW.

The second scenario further demonstrates the potential for significant returns, with total costs estimated to be between $75 million and $100 million, and FACTS devices costing approximately $10 million to $20 million to transmit 500 MW of power.

For scenarios three and four, Table 8.3 provides detailed insights that highlight the strategic benefits and cost-effectiveness of these modern transmission solutions. This information should reassure you about the value of these technologies.

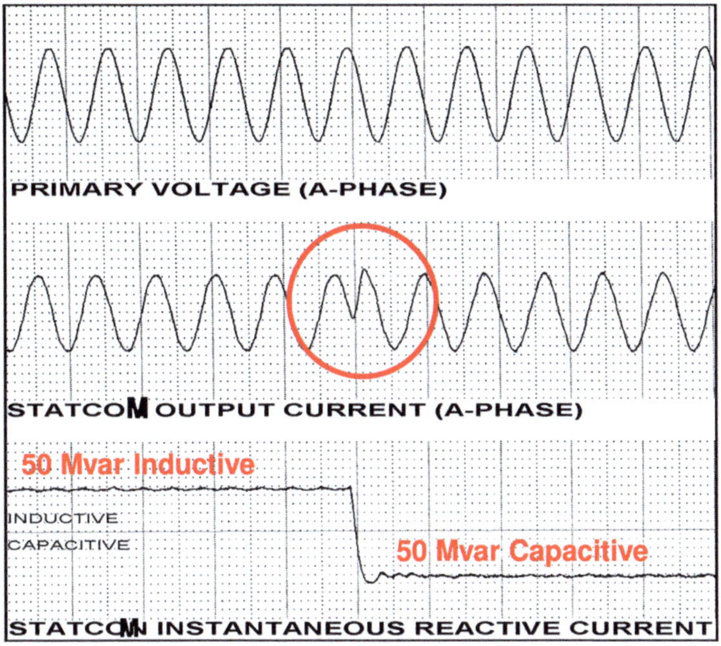

Fig. 8.27 Dynamic voltage support

Table 8.3 Comparative cost analysis

S.N°	Throughput (MW)	HVDC 2-terminal (million dollars)	FACTS (million dollars)
1	200	40–50	5–10
2	500	75–100	10–20
3	1000	120–170	20–30
4	2000	200–300	30–50

8.7 Types of Transmission Protection

Protection devices may be classified into two basic types: non-unit systems and unit systems.

Non-Unit Systems

These are time-graded systems that utilize information (currents and voltages) derived from a particular point in the system. Overcurrent and distance protection are prime examples. Time grading is applied through a sequence of equipment so that when a fault occurs, a number of protection devices respond, but ideally only those closest to fault complete the tripping function. The others make incomplete operations and then reset.

Overcurrent Protection

The most common non-unit scheme, which is widely applied in both distribution and transmission systems, is the overcurrent device.

Exercise 8.6

What is the basic difference between a transmission and a distribution system, in terms of protection requirements?

A transmission system is a high-voltage interconnected system in which generators are connected via transmission lines for the purpose of bulk power transfer. Voltage levels are very high (275 kV and 400 kV in the United Kingdom; 500 kV and above used in other parts of the world). Very fast, expensive SMART, and highly sophisticated protection is employed on such systems in order to minimize the risk of losing a lot of power. A distribution system is used for distributing power to the consumers. Normally, it comprises lines and cables that are fed from one end only with load taps at various points along the lines. Voltage levels are relatively low (33 kV and 11 kV) for large industrial consumers and 415 V (line voltage) and 230 V (phase voltage) for the domestic consumers. Relatively slow and less expensive protection is normally employed on systems since a loss of line would affect only a very small number of consumers.

Overcurrent protection is applied as the primary and often the only protection scheme at distribution voltages and is also applied in the transmission system as *back-up* protection.

Exercise 8.7

What is the fundamental difference between primary and back-up protection?

Primary (or main) protection has the fastest measuring speed and is designed to be absolutely discriminative. Back-up protection often operates at a relatively slow speed and has relatively low discriminative properties. It is provided as backup in case of primary protection failure.

The protection device relies upon a comparison of the current measured via the current transformer (CT) at a particular time. This comparison is often arranged to produce particular current versus tripping time response curves according to the requirements of the system. In the simplest case, tripping is initiated when the measured current exceeds a pre-set value.

Exercise 8.8

What is essential to employ a current transformer for current measurements? In a particular overcurrent protection system, a CT with a 500:1 ratio is employed. If the level of fault current is 2.5 kA, what will be the current passing through the overcurrent relay?

Under fault conditions, the level of fault current flowing through the distribution line becomes very high and is therefore not suitable for the protection device. A CT is thus employed to step down this current to a more suitable level.

With CT ration = 500:1 and I_{fault} = 2.5 then the current through the device is

$$I_{fault} / CT\ ratio = \frac{2500}{500} = 5A$$

These relays have inverse operating-time/ current characteristics and a definite minimum operating time. These can be arranged to provide for discriminative tripping by suitably adjusting the individual relay so that their operating times versus current curves bear a definite relationship with respect to one another.

In terms of relay settings, there are essentially two control parameters:

(i) Plug setting that defines the relay sensitivity.
(ii) Time-multiplier setting (TMS) for coordinating the time grading of relays on system to attain discrimination.

Relay Plug Setting

Plug setting, i.e., current in the relays, depends upon the relay current setting, which, in turn, is given as a percentage plug setting. Typical plug settings for minimum current operations are 40% of full load current for earth fault and 125% of full load current for other types of fault. The current in the relay is calculated as a multiplication of the plug setting and is given by;

$$Plug\ setting\ multiplier = 100 I / \left[CT\ ratio \times \%plug\ setting \times nominal\ rating\right]$$

where I is the actual fault current.

Exercise 8.9

A current transformer with a CT ratio of 500:1 feeds an overcurrent relay with the following settings:

I. 100%
II. 40%
III. 125%.

Determine the current in the relay (as a multiple of plug setting) if the fault current is 2 kV, assuming a nominal relay rating of 1 A.

$$I_1 = \frac{100 \times 2000}{500 \times 100} = 4\ plug\ settings$$

$$I_2 = \frac{100 \times 2000}{500 \times 40} = 10\ plug\ settings$$

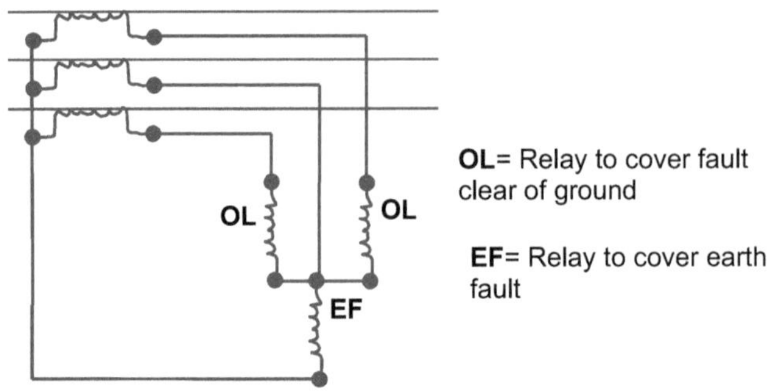

Fig. 8.28 A typical overcurrent relay arrangement. A schematic diagram of an electrical relay system. The diagram shows three relays labeled "OL" and one labeled "EF." The "OL" relays are designed to cover faults clear of ground, while the "EF" relay is for covering earth faults. Lines connect the relays, indicating their configuration in the system

$$I_3 = \frac{100 \times 2000}{500 \times 125} = 3.2\ plug\ settings$$

It is apparent from the above that a lower percentage setting makes the relay more sensitive. In practice, this facility is very useful (and desirable) for making the relay operating times comparable for earth and other types of faults. Level of fault currents for earth faults is generally lower than for other types, such as a three-phase system involving the three elements of the relay, as shown in Fig. 8.28, from which it can be seen that the aforementioned differing sensitivities for different types of faults can be achieved by making only one relay element responsive to earth faults.

Exercise 8.10
In practice, it is desirable to have nearly the same relay operating times for all types of faults.

In a feeder arrangement of the type shown in Fig. 8.29a in which a number of overcurrent relays are involved, an abnormally long operating time of the faulted chapter relay for a particular type of fault could very easily upset the relay coordination, thus affecting relay discrimination.

Relay Time-Multiplier Setting

Time multipliers enable the coordination of the time grading so that sufficient discrimination may be provided to isolate a faulty feeder correctly. In the simple radial feeder system of Fig. 8.29a, we would have to grade the time/current characteristics with differing time multiplier settings typical as shown in Fig. 8.29b. It should be noted that all overcurrent relays of the type discussed above are, in a sense, universal relays with a universal relationship between operating time and relay current in multiples of plug setting typically as shown in Fig. 8.30.

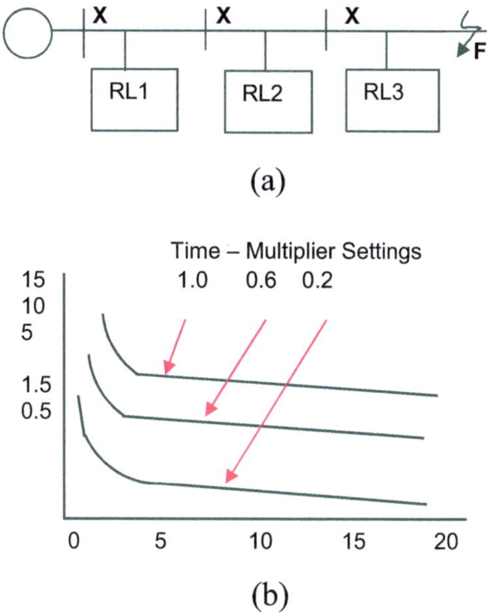

Fig. 8.29 Protection of a radial feeder system: (**a**) system arrangement and (**b**) relative time-multiplier settings. (**a**) (Operating relay) Diagram of a beam supported at three points marked with "X" and labeled RL1, RL2, and RL3. An arrow labeled "F" indicates a force applied at the right end of the beam. (**b**) (Operating current in multiples of plug setting) Graph showing three downward-sloping curves representing time multiplier settings of 1.0, 0.6, and 0.2. The x-axis ranges from 0 to 20, and the y-axis ranges from 0 to 15. Pink arrows point to each curve, indicating the direction of decrease

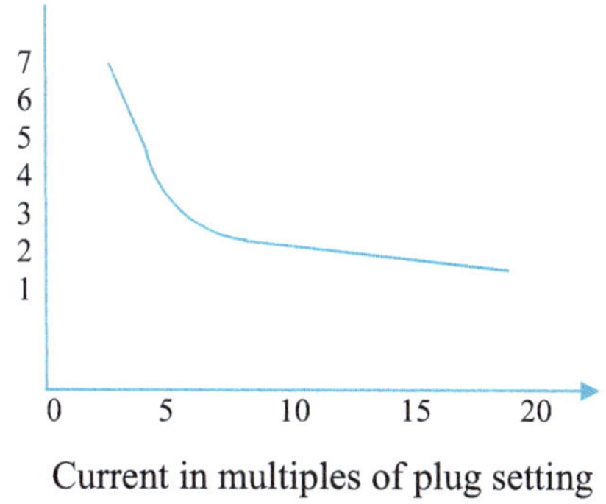

Fig. 8.30 Universal relay characteristic. This characteristic is better described by its formula. A line graph showing a curve that starts at a high point on the left and decreases sharply before leveling off toward the right. The y-axis ranges from 1 to 7, and the x-axis ranges from 0 to 20. The curve indicates a rapid decline initially, followed by a gradual decrease

$$t = \frac{0.14}{psm^{0.02} - 1}$$

where **psm** is the plug setting multiplier.

Furthermore, in any application, the problem of deciding suitable time-multiplier settings is governed by breaker operating time.

Exercise 8.11

An overcurrent relay with an inverse characteristic shown in Fig. 9.3 sees a fault current on the feeder of four plug settings. If the breaker operating time is 0.2 s and an overall fault clearance time of 1 s is required, determine the relay time multiplier setting.

For a plug setting of 4, the operating time from Fig. 9.3 using the characteristics equation is 4.98 s. The relay must operate in $(1 - 0.2) = 0.8\ s$. Therefore, time multiplier setting =0.8/4.98 = 0.16.

8.7.1 Smart Protection Settings for a Ring Distribution System

You will see from the foregoing that radially fed power systems have very low integrity, as a high fault in the first chapter of Fig. 8.29a will result in the loss of supply to Chaps. 2 and 3. A better alternative power supply is to use a further interconnection between the remote subchapter and the supply substation, as shown in Fig. 8.31.

Unfortunately, this means that a simple time-graded overcurrent system of the type described before cannot be used, as power and fault current can flow in both directions. To provide discrimination, directional relays must be used. These are arranged to operate when fault current flows out of the feeder. The contacts of the relays are arranged in conjunction with the over-current relays to allow measurement when closed and prevent measurement when open.

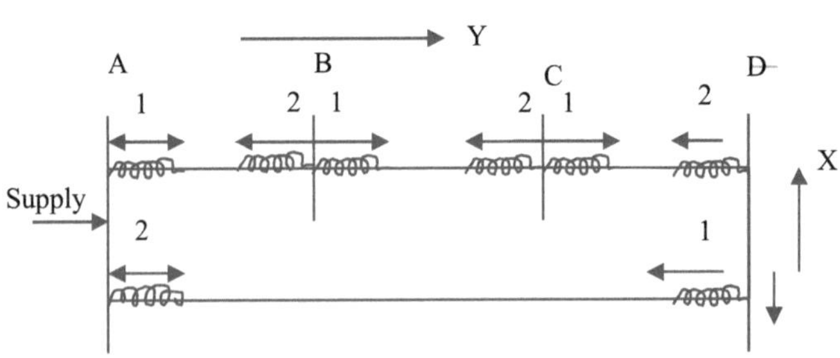

Fig. 8.31 Ring distribution system. Diagram illustrating a network of interconnected nodes labeled A, B, C, and D, with directional arrows indicating flow. The supply enters at node A with a value of 2. Arrows between nodes show flow values: A to B (1), B to C (1), and C to D (2). Additional arrows indicate flow between nodes and external points, with values labeled 1 or 2. Axes X and Y are marked, showing directionality

At the supply substation, there is no necessity to use directional relays as the only possible way that fault current can flow is into the feeder.

The arrows on the system diagram in Fig. 8.31 indicate the direction of current flow, which will operate the relay at that location. Examination of the system shows that, for protection purposes, the system can be regarded as two radial systems. One concerned with relays A1, B1, C1, and D1 and the other involving relays A2, D2, C2, and B2. The fault level at the end of each feeder would be calculated by assuming the circuit breaker at A2 is open when calculating the fault level for relays A1, B1, C1, and D1. And the circuit breaker at A1 is open when calculating the fault level for relays A2, B2, C2, and D2 [14].

In terms of relay coordination for the system shown in Fig. 9.4, this can best be achieved by treating the system as two radial systems, one comprising relays A1, B1, C1, and D1 and the other comprising relays A2, B2, C2, and D2.

In the respect of radial systems, relays may conveniently be coordinated by starting at the farthest load point and working toward the source. Each relay is successfully set such that downstream relays operate before upstream relays. With reference to Fig. 8.31, for a particular fault level at the incoming busbar, a typical relay coordinating in terms of time multiplier settings (TMSs) will be:

Relay TMS			Relay TMS		
D1	0.2	**Direction X**	D2	0.2	**Direction Y**
C1	0.35		C2	0.35	
B1	0.51		B2	0.51	
A1	0.67		A2	0.67	

8.7.2 Smart Distance Protection

The distance or impedance measuring principle represents the most widely applied non-unit protection as primary (or main) protection to high-voltage transmission lines. The basic measuring principle is simple in that, from a knowledge of the various voltages and currents at one end of the line, it is possible, from the ratio of the measured voltage to current, to determine the impedance between the relay and a particular fault point on the line. Figure 8.32 illustrates the basic principles.

With reference to Fig. 9.5, if the distance from the relaying point to the fault is X, then the impedance measured for a zero-resistance fault is given as

$$\frac{V}{I} = xZ_L$$

where Z_L is the positive phase impedance/meter.

The measured impedance follows the typical locus shown in Fig. 8.33.

It is important that the distance relay, say at end A in Fig. 8.32, sees fault only on the feeder AB, i.e., forward faults, and does not operate for reverse faults behind A. Furthermore, in practice, there is very often a small resistance present in the fault. The distance protection is thus arranged to compare the circuit impedance

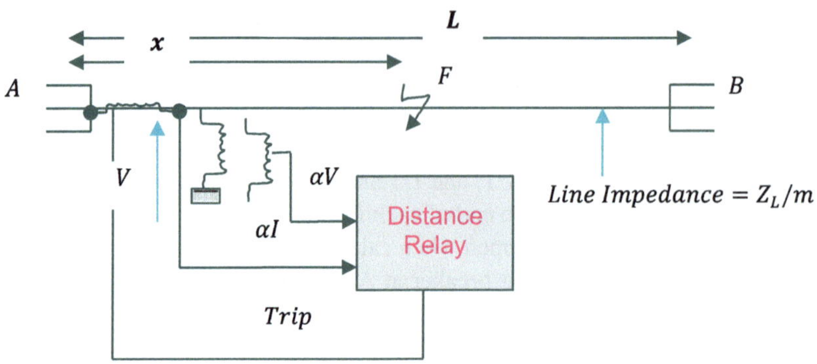

Fig. 8.32 Basic distance relay principle. Diagram of an electrical circuit showing a distance relay system. The circuit runs from point A to point B with a fault at point F. The distance from A to F is labeled as "x" and from A to B as "L." Voltage (V) and current (I) are measured and fed into the distance relay, which calculates impedance to determine if a trip is necessary. The line impedance is noted as \(Z_L/m\)

Fig. 8.33 Typical impedance locus. Graph illustrating the relationship between measured reactance and resistance, labeled as I(xZL) and R(xZL), respectively. The plot features a solid fault impedance locus, marked by a curved line. Points A and B are indicated, with point xZL highlighted. The line LZL is shown intersecting the locus

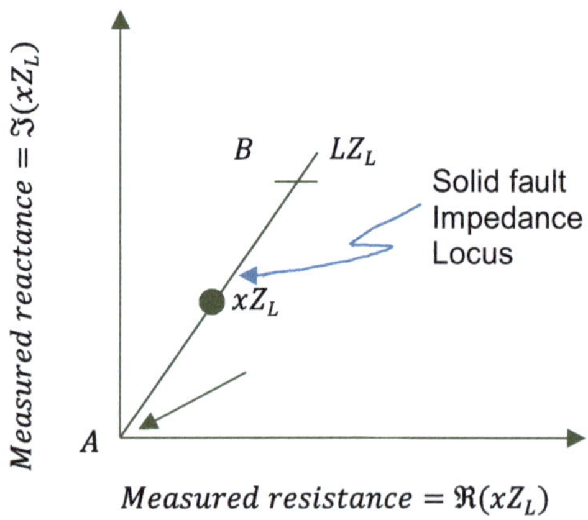

with a defined area of the measured impedance plane so that any fault within the defined area causes tripping of the circuit breaker. A very common form of arranging the distance protection characteristic is shown in Fig. 8.34a (also known as the MHO characteristics), which very adequately meets the two aforementioned requirements. The primary circle shown in Fig. 8.34a is constructed on the diameter AB of Fig. 8.33. AB is the impedance of the protected length of line. If the value of line impedance to the point of fault plus any fault resistance lies inside this circle, then the protection will operate [15].

It is normally practice to arrange the protection characteristics in zones as shown in Fig. 8.34b.

Exercise 8.12
Why is it necessary to cater for a small fault resistance? What will happen if the fault resistance is very large?

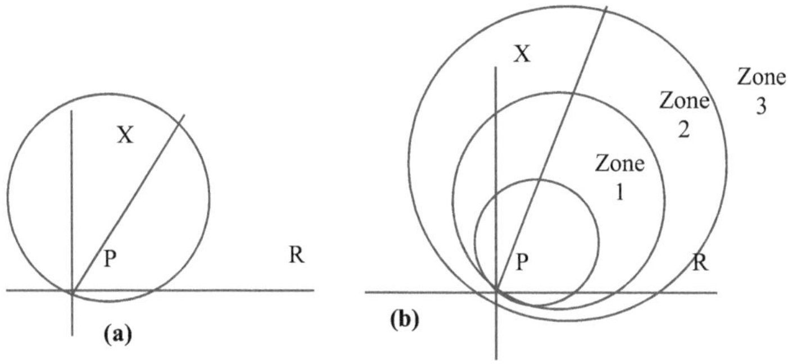

Fig. 8.34 Relay MHO characteristics. The image consists of two diagrams labeled (**a**) and (**b**). Diagram (**a**) shows a circle with a center point P, a vertical line labeled X, and a horizontal line labeled R intersecting the circle. Diagram (**b**) features three concentric circles centered at point P, with the same vertical line X and horizontal line R. The circles are divided into three zones labeled Zone 1, Zone 2, and Zone 3, moving outward from the center

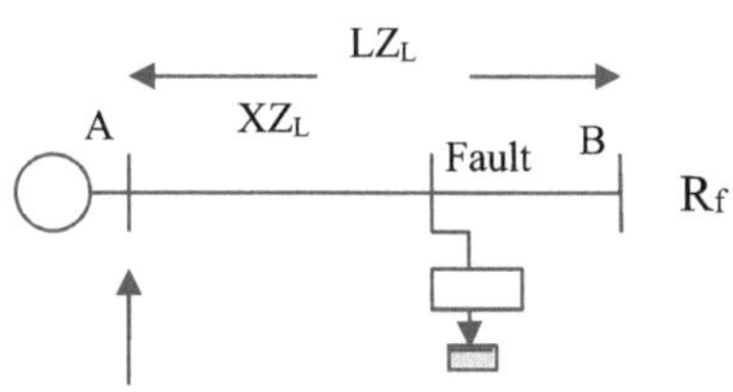

Fig. 8.35 Diagram illustrating an electrical transmission line with points A and B. A fault is indicated between these points. The line is labeled with impedance values "XZL" and "LZL," with arrows showing direction. A box below the fault suggests a connection to a monitoring or protection device

Consider the following simple feeder arrangement, on which a fault involving a small resistance R_f has occurred, as shown in Fig. 8.35.

If $R_f = 0$, the measured impedance

$$Z_M = \frac{V}{I} = xZ_L$$

Otherwise

$$Z_M = \frac{V}{I} = xZ_L + R_f$$

This means that if the relay characteristics are confined only to the solid fault locus as shown in Fig. 8.34a and not extended to that shown in Fig. 9.7a, then the relay will fail to see this as a fault. For very large fault resistances, the measured impedance can fall well outside the extended characteristics shown in Fig. 8.34a.

The zone 1 relay is set to protect about 80% of the line AB and trips for all faults that present an impedance inside this range. The zone 1 operating time corresponds

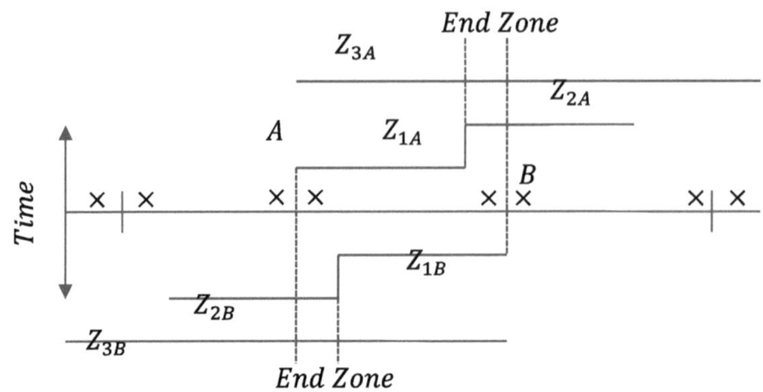

Fig. 8.36 Time distance characteristic of practical distance protection scheme. Diagram illustrating a time-based sequence with two main paths labeled A and B. Path A includes zones Z1A, Z2A, and Z3A, while path B includes zones Z1B, Z2B, and Z3B. Both paths have an "End Zone" marked. The sequence is represented with steps and crosses along a horizontal time axis, with vertical lines indicating transitions between zones

to the fastest measuring speed that the particular distance relay will achieve and is typically between 1 and 30 ms in modern equipment(s).

The zone 2 characteristic operates after a delay and is seen to have a reach in excess of the remote line terminal (this reach is typically 120%). The delay is typically 0.3–0.4 s and is arranged to give satisfactory tripping at A in an acceptable time for faults beyond B, which, due to local failure, are not cleared by the switchgear beyond end B. This is the previously referred to remote back-up function. Zone 3 is seen to encompass the relay location.

This zone operates in a time greater than zone 2 and thus provides some further remote back-up extending further into the system. It is also seen to provide some local reverse back-up that is useful in the event of the circuit breaker behind relay location A failing to clear the fault. Figure 8.36 shows how, ideally, the operating time of the protection would vary with fault position relative to end A.

Exercise 8.13
Why zone 1 relay set to only 80% of the line?

In practice, there is always an element of uncertainty in the protection relay measuring process near the boundary to which it is set. The relay accuracy is affected by prevailing system and fault conditions, and this means that if it is set to cover the full 100% of the feeder, then it could "see" faults in the next feeder chapter—clearly a situation to be avoided [16].

8.7.3 Unit Protection

These are schemes that respond only to faults inside a clearly defined zone. They utilize information derived from two (or occasionally more) points in a system. A comparison of quantities on the boundaries of a protected zone is the usual technique. Examples of unit protection are differential current relays, where the current entering a zone is compared with that which leaves it.

Exercise 8.14
What are the fundamental differences between unit and non-unit protection?

A non-unit scheme makes measurements from information derived only at one end of the protected zone, whereas a unit scheme utilizes information from both ends of the zone. Secondly, the former provides back-up protection through time grading, whereas the latter has no such property. All unit schemes derive from the Merz and Price circulating system. In its simplest form, the CTs at each end of the line are connected as shown in Fig. 8.37 so that, in theory, no current flows in the relay for faults outside the zone.

An alternative, but similar, technique hinges upon connecting the CTs in opposition so that, under external fault conditions, no current flows in the series-connected relays. This technique, often referred to as the balanced voltage principles, is shown in Fig. 8.38.

Circulating current schemes are rather limited in feeder applications because the voltage developed across the CT secondaries increases in proportion to the length of the feeder. Prohibitively large CTs are required when applying the principles to feeders of longer than 2 km, and the balanced voltage principles find more widespread use. The series-connected relays are of relatively high impedance, and the system may be satisfactorily applied in principle for the protection of a feeder up to about 50 km. It should be noted that, in such applications, pilot wire circuits (which are normally rented telephone cables) have to be employed for transmitting currents from one end to the other for the purpose of comparison.

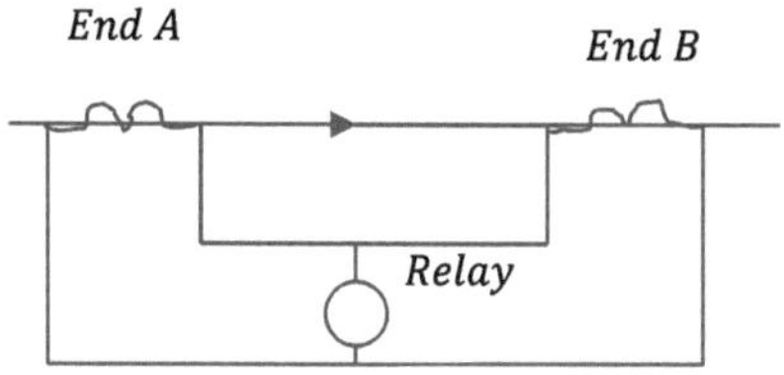

Fig. 8.37 Circulating current system. Diagram illustrating a circuit with two ends labeled "End A" and "End B." A line connects the two ends with an arrow pointing from End A to End B, indicating direction. Below the line, a relay is depicted as a circle with the label "Relay." The circuit includes wavy lines near both ends, suggesting signal or current flow

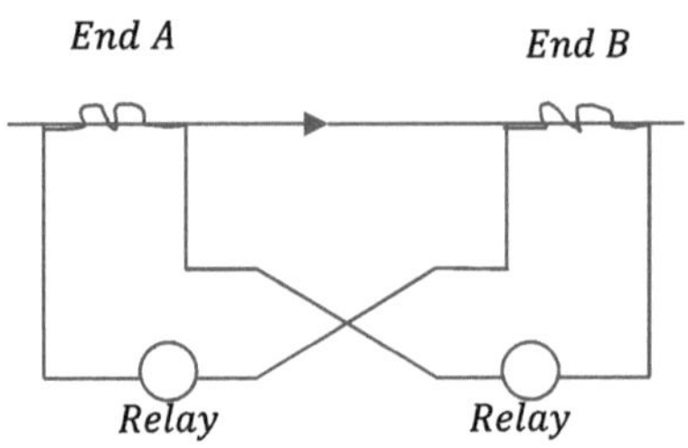

Fig. 8.38 Balanced voltage system. Diagram illustrating a telegraph circuit with two ends labeled "End A" and "End B." The circuit includes two relays, each represented by a circle, connected by wires. The wires cross in the middle, and an arrow indicates the direction of the current flow from End A to End B

Some modification to the basic scheme is required for two main reasons:

(i) Under heavy through fault conditions, the secondary CT currents are not exactly equal due to the inevitable difference (however slight) in the CT ratios.
(ii) The pilot wire circuit has shunt capacitance so that, even if the voltages at either end are identical, some current will flow in the relays. This will tend to cause tripping under healthy conditions.

There are various methods of overcoming these foregoing problems, the most common of which is the use of bias. Biasing of relays consists of restraining the operation of a relay if the magnitude of the summed local and remote currents exceeds a certain percentage (which can be set according to the application requirements) of the sum of the magnitude of the individual local and remote currents. Figure 8.39 shows a typical bias curve in which [17, 18]:

$$\left|I_{diff}\right| = \left|I_A + I_B\right|$$

and

$$\left|I_{bias}\right| = k\left\{\left|I_A\right| + \left|I_B\right|\right\}$$

Summation Devices

So far, the differential principle has been discussed as though it were applied to single-phase systems. In dealing with three-phase systems, the principles could be applied to each phase, but this is not economically justified because three pilot wire pairs would be required. It is therefore common practice to sum the currents at the feeder ends. A common arrangement using a summation transformer is shown in Fig. 8.40.

Here, the output is proportional to

$$\left(n+2\right)I_a + \left(n+1\right)I_b + nI_c$$

where n is typically 3.

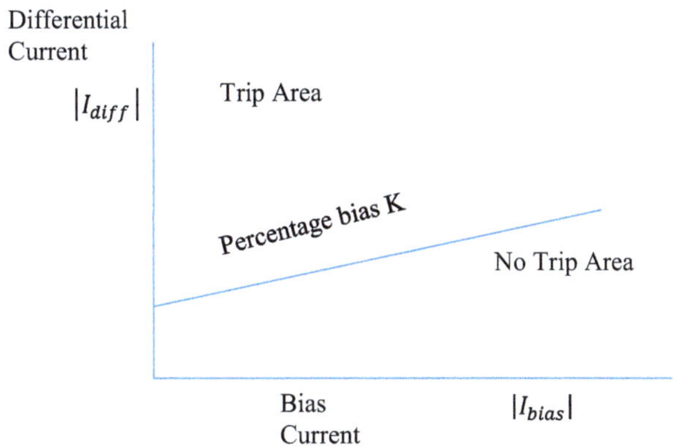

Fig. 8.39 Typical bias characteristic. Graph illustrating the relationship between differential current and bias current. The y-axis is labeled "Differential Current" with the variable, and the x-axis is labeled "Bias Current" with the variable. A diagonal line divides the graph into two regions: "Trip Area" above the line and "No Trip Area" below. The line represents "Percentage bias K"

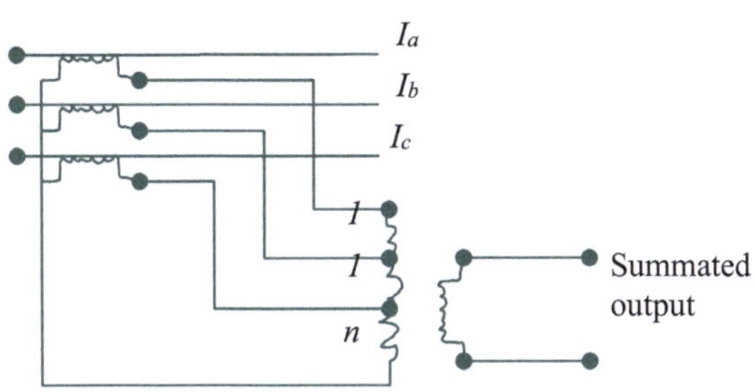

Fig. 8.40 A typical summation arrangement. Diagram of a summation transformer circuit. Three input currents labeled \(I_a\), \(I_b\), and \(I_c\) are connected to coils. These inputs are combined with two additional lines labeled "1" and "n" into a single output coil, resulting in a "Summated output"

Note that, in practice, there are a number of different types of current differential protection schemes in use, and their usage depends upon the type of application. For example, static relays employing metallic pilot wires as a means of sending power frequency analog signals between the two ends of the protected zone are suited for circuit lengths of only up to about 50 km, principally because of the limitation imposed by the resistance and capacitance of the pilot wire. An alternate scheme is the phase comparison carries the protection in which high frequency carrier signal, modulated by a power frequency signal, is sent over the protected line itself. This type of scheme is suitable for the protection of a long line up to about 500 km in length.

More recently, the development of high-capacity, long-haul digital communication systems *(such as fiber optical links)* has made possible the development of digital current differential relays [19, 20].

Here, the power frequency analog signals are first coded into binary data, and because of the speed and high capacity of the communication link, it is then possible to send the current signals of each phase (rather than a signal summated quantity) to the other end of the protected zone. More details of these various types of practical schemes can be found in Module 3, Chap. 5 on differential protection.

8.8 Self-Assessment Questions

? Self-assessment question 8.1

Figure 8.41 shows a feeder arrangement on which overcurrent protection relays R1 and R2 are employed with a CT ratio.

Each breaker has an operating time of 0.2 s, and the plug setting on both relays is 125% for normal overcurrent. If each relay has the inverse characteristics as shown in Figs. 8.37 and 8.38, determine the following:

(a) The time multiplier setting for relay R2 if the three-phase fault current is 2 kV and the corresponding fault clearance time is 1 s.
(b) The time multiplier setting for relay R1 necessary for discrimination.
(c) The fault clearance time for a 4-kV three-phase fault occurring beyond breaker B2 and also for a similar fault occurring between the two breakers.
(d) Will the same time multiplier§ settings be suitable when dealing with earth fault currents of 1 kA if the plug setting for the earth fault element is 40%.

? Self-assessment question 8.2

(a) Outline the principles of distance protection and explain why it is useful for performance remote back-up protection functions.
(b) A distance relay with MHO characteristics is situated at end A of a 400-kV single circuit transmission line AB and has a first zone setting of 80%. The positive phase sequence impedance of the line Ab is (2.0 + j36.0), and the relay is

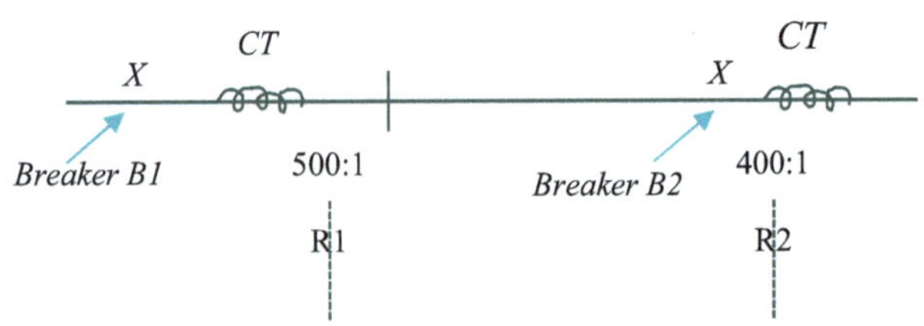

Fig. 8.41 Diagram of an electrical circuit with two sections. The left section shows Breaker B1 with a current transformer labeled "CT" and a ratio of 500:1, connected to resistor R1. The right section shows Breaker B2 with a current transformer and a ratio of 400:1, connected to resistor R2

set to protect 80% of this line. If a single-phase earth fault of resistance 20 Ω occurs three-quarters of the way along the line from end A, show that the relay at end A under-reaches to such an extent that it will not "see" this as a fault in zone 1 *(the remote end infeed from end B can be neglected).*

? Self-assessment question 8.3

(a) Outline the principle of a simple differential protection scheme and explain why it is necessary to use a bias-type characteristics.
(b) A low-voltage distribution line is protected by a bias current differential protection as shown in Fig. 8.42.

$$I_A = (40 - j30)A$$

$$I_B = (20 - j15)A$$

If the tripping criterion for the relay is such that

$$|I_{diff}| = |I_A + I_B| \geq |I_{bias}| = k\{|I_A| + |I_B|\} \tag{8.2}$$

Deduce a value of the constant k for which the relay would "just" operate.

? Self-assessment question 8.4

A 30 MVA transformer feeds several feeders through an 11-kV busbar. The transformer circuit breaker is provided with a 250:1 current transformer and each feeder is provided with 100:1 current transformer. Each set of current transformers feeds and induction-type overcurrent relay and each relay has a time-current characteristic given by:

$$t = \frac{0.14}{\left(psm^{0.02} - 1\right)} s \tag{8.3}$$

where psm is the operating current in multiples of plug setting.

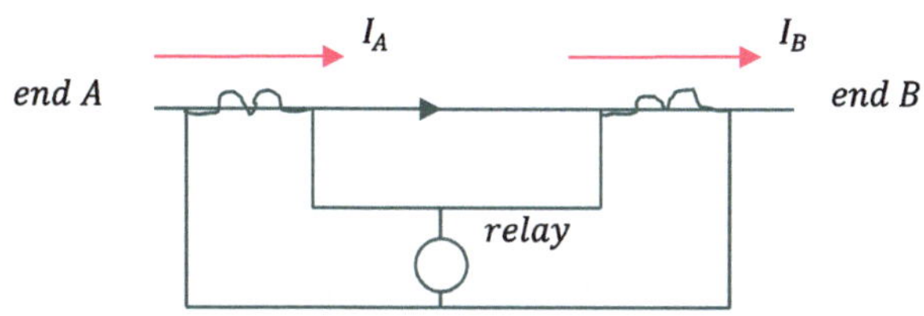

Fig. 8.42 Inter-fault occurs on the feeder. Diagram illustrating a relay circuit with two ends labeled "end A" and "end B." Current flows from end A to end B, indicated by arrows labeled "I_A" and "I_B." A relay is connected in the circuit, depicted as a circle with a line

All relays have plug settings of 125%.
A fault on one feeder results in a phase current of 1000 A.
Determine:

(i) Operating time of the feeder relay if its time multiplier setting is 0.5.
(ii) The time setting required on the transformer relay to provide 0.5 s discrimination time margin.

? Self-assessment question 8.5

A distance protection relay with an MHO characteristic is positioned at the sending end of a 400-kV single circuit transmission line and set to protect 80% of the line. The line has a positive phase sequence impedance of $(2 + j36)\Omega$.

A single-phase to earth fault of impedance $(10 + j3.75)\Omega$ occurs at a point 60% of the way along the line from the sending end.

(a) Construct the MHO characteristic for the relay and use it to show that the relay will not operate for this fault.
(b) Determine the maximum value for a resistive fault at the point concerned for the relay just to detect the fault.

Bibliography

1. Nasser Alemadi, Tariq Masood, R Aggarwal, Suhail A Qureshi, D Kothari "Novel selection approach of STATCOM device tailored and optimized engineering configuration to implement in between Qatar and Bahrain" published in Int. J. Autom. Power Eng. https://doi.org/10.14355/ijape.2014.0302.01
2. Tariq Masood & Jamil Abdo "New Technique to Enhance Hybrid Renewable Energy Resources' Operational Flexibility and Reliability" 22nd International Conference on Renewable Energies and Power Quality (ICREPQ'24) Bilbao (Spain), 26th to 28th June 2024. https://www.icrepq.com/icrepq24/396-24-masood.pdf
3. Tariq Masood and Jamil Abdo "Review: Uninterrupted Liquid Hydrogen (E-Methanol) Production framework to deliver Clean & Green Energy" Published 22nd International Conference on Renewable Energies and Power Quality (ICREPQ'24) Bilbao (Spain), 26th to 28th June 2024. https://www.icrepq.com/icrepq24/414-24-masood.pdf
4. Tariq Masood, Abdel Aty Edris, Suhail Aftab Qureshi, Muhammad Tajammal, Murtaza Hashmi, DP Kothari, Samer Kareem "SSSC tailored to optimization performance in Between United Arab Emirates and Oman on the GCC Power Grid" published in the 13th IET International conference on AC and DC power Transmission Conference on 14–16 February, 2017. http://ieeexplore.ieee.org/document/7934934/
5. Tariq Masood, Abdel Aty Edris, Suhail Aftab Qureshi, Muhammad Tajammal, Murtaza Hashmi, DP Kothari, Samer Kareem "Smart Grid Operations and Control Challenge by Implementing STATCOM Tailored to Optimization Performance at the GCC Power Grid" published in the 13th IET International conference on AC and DC power Transmission Conference on 14–16 February, 2017. http://ieeexplore.ieee.org/document/7934988/
6. Tariq Masood, Muhammad Tajammal, Samer Kareem, Suhail Aftab Qureshi, Abdul Jabbar Khan, DP Kothari, Nair Jameel "Unified power Flow Controller Modeling and Analysis Technique on the GCC Power Grid Operations and Control" published IEEE Energy Con, 2018 June 7–9 Limassol Cyprus. https://ieeexplore.ieee.org/document/8398834/
7. Tariq Masood, Muhammad Tajammal, Samer Kareem, Suhail Aftab Qureshi, Murtaza Hashmi, Nasir Jameel "Unified power Flow Controller (UPFC) Modeling and Analysis Technique by

using Matlab and EMTP" published (ICREQP, 18 Salamanca Spain 21–23 March, 2018). http://www.icrepq.com/icrepq18/250-18-masood.pdf

8. Tariq Masood, Muhammad Tajammal, Samer Kareem, Suhail Aftab Qureshi, Murtaza Hashmi, Naser Jameel "STATCOM Controller Performance Enhancement and Optimization Technique" published (ICREQP, 2018 Salamanca Spain 21–23 March, 2018). http://www.icrepq.com/icrepq18/249-18-masood.pdf
9. Tariq Masood, Muhammad Tajammal, Samer Kareem, Suhail Aftab Qureshi, Murtaza Hashmi, DP Kothari, "Smart Grid Operational functions and Control Challenges by Implementing SSSC Tailored to Optimize performance in between Qatar and KSA on the GCC Electrical-power grid" published in the ASTESJ Journal Advances in Science, Technology and Engineering Systems Journal Vol. 2, No. 1, 30.09.2017. https://astesj.com/special-issue-4/
10. Tariq Masood, Raj K Aggarwal, Suhail Aftab Qureshi, RAJ Khan "STATCOM model against SVC control model performance analyses technique by MATLAB" published in International Conference on Renewable Energies and Power Quality (ICREPQ'10) https://icrepq.com/icrepq'10/655-Masood.pdf
11. Tariq Masood, N Al-Emadi, SA Qureshi, YY Almulla, AJ Khan, SK Shah "FACTS Control Devices (Statcom, SSSC and UPFC) Re-Configuration Techniques By PSIM/MATLAB" Published in: 2007 International Conference on Electrical Engineering. https://ieeexplore.ieee.org/abstract/document/4287295
12. Tariq Masood, Raj K Aggarwal, SA Qureshi, RAJ Khan "STATCOM and SVC control operations and optimization during network fault conditions" https://ieeexplore.ieee.org/document/5636868
13. Tariq Masood, RK Aggarwal, SA Qureshi, RAJ Khan "Statcom control reconfiguration technique for steady state and dynamic performance optimization during network fault conditions" published in International Conference on Renewable Energies and Power Quality (ICREPQ'10) Granada (Spain) https://www.icrepq.com/icrepq'10/656-Masood.pdf
14. Tariq Masood, Samer Karim Shah, Muhammad Tajammal, "Unified power flow controller modeling and analysis technique on the GCC power grid" published in 2018 IEEE International Energy Conference (ENERGYCON) https://ieeexplore.ieee.org/abstract/document/8398834
15. Tariq Masood, Suhail A Qureshi, RK Aggarwal, DP Kothari "Multiple FACTS devices integration through distributed control system and its application on the GCC interconnection" published in 2013 7th IEEE GCC Conference and Exhibition (GCC) https://ieeexplore.ieee.org/abstract/document/6705750
16. Tariq Masood, Jamil Abdo and Kaimiao Liu, "Facilities Integrity Management System – Production Best Practice pragmatic case studies for improving Production Operations in the Energy Sector" IEOM 2025 Conference in Orlando. June 17 to June 19, 2025. Online paper link will be available end of September
17. Tariq Masood, Jamil Abdo and Kaimiao Liu, "Risk Assessment and Management - Production Best Practice Case study for improving production Operations in the Energy Sector" published in IEOM 2025 Conference in Orlando. June 17 to June 19, 2025. Online paper link will be available end of September
18. Tariq MASOOD(1), Jamil Abdo(1), Nasser j. AL-Kuwari(2) Tahir Imran Qureshi(3), Kaimiao Liu(1) "Nuclear ammonia as a renewable energy source: an innovative approach to advancing sustainable fuel solutions for the global energy crisis" published in 23rd International Conference on Renewable Energies and Power Quality (ICREPQ'24) Tenerife (Spain), 25-27th June 2025.Online paper link…
19. Tariq Masood, Jamil Abdo, Atif Iqbal "Protection of Electrical Power Distribution Systems: Smart-grid, Microgrid, AI and Cybersecurity" ISBN: 978-1-394-32764-5, published by Wiley-IEEE Press 2025 https://mitpressbookstore.mit.edu/book/9781394327645
20. Nasser J.Al-Kuwari(1) Tariq MASOOD(2), Jamil Abdo(2), Tahir Imran Qureshi(3), Kaimiao Liu(2) "Advancing Sustainable Hydrogen Production: Cost, Environmental, and Technological Innovations in Biomass, Pyrolysis, and SMR" published in 23rd International Conference on Renewable Energies and Power Quality (ICREPQ'24) Tenerife (Spain), 25-27th June 2025 www.enerqj.com/E&Q/eqj3/213-25.pdf

Microgrid Operation and Control 9

Contents

9.1 Introduction........ 222
9.1.1 Classification of Microgrid........ 223
9.2 AC Microgrid Operations and Control........ 225
9.3 DC Microgrid Operations and Control........ 226
9.4 Hybrid Microgrid Operations and Control........ 229
9.5 Microgrid Control........ 230
9.5.1 Off-Grid and Islanded Microgrid........ 231
9.5.2 Different Types of Islanding........ 233
9.5.3 Grid Following Inverters........ 234
9.5.4 Grid-Forming Inverters........ 236
9.6 Short-Circuit Study........ 237
9.6.1 Advantages of Short Circuit Ratio Study........ 237
9.6.2 Traditional Short Circuit Ratio (SCR) Method........ 237
9.7 Strong and Weak Microgrid........ 237
Bibliography........ 238

Abstract

At the end of this chapter, you will learn about the Microgrid, a topic directly relevant to the field of study in electrical engineering. This includes independent active and reactive power control, correction of voltage sag and system imbalances, and meeting the grid's load dynamics requirements. This chapter also covers the classification of microgrids, the merits and demerits of AC and DC microgrid functionality, and the functionality of hybrid microgrids. Finally, this

Supplementary Information The online version contains supplementary material available at https://doi.org/10.1007/978-3-031-94312-6_9.

T. Masood et al., *Protection of Electrical Power Transmission Systems*, Power Systems, https://doi.org/10.1007/978-3-031-94312-6_9

chapter will discuss the performance comparison of DC and AC microgrids, providing insights directly applicable to the work and studies in electrical engineering. Cyber and information programs are also discussed in this chapter.

9.1 Introduction

The U.S. Department of Energy (DOE) describes a microgrid as a sophisticated and dynamic network that integrates various interconnected loads and distributed energy resources (DER) within specifically defined electrical boundaries. This cutting-edge system is designed to operate as a cohesive and controllable unit in relation to the larger main power grid. One of the key features of a microgrid is its ability to seamlessly connect and disconnect from the grid, which allows for versatile operations in both grid-connected and island modes. This capability significantly enhances energy resilience, making it possible for communities to maintain power during outages or disruptions, while also improving the overall reliability of the energy supply. The microgrid model supports a sustainable energy future by utilizing a range of local energy sources, thereby promoting greater energy independence and efficiency (Fig. 9.1).

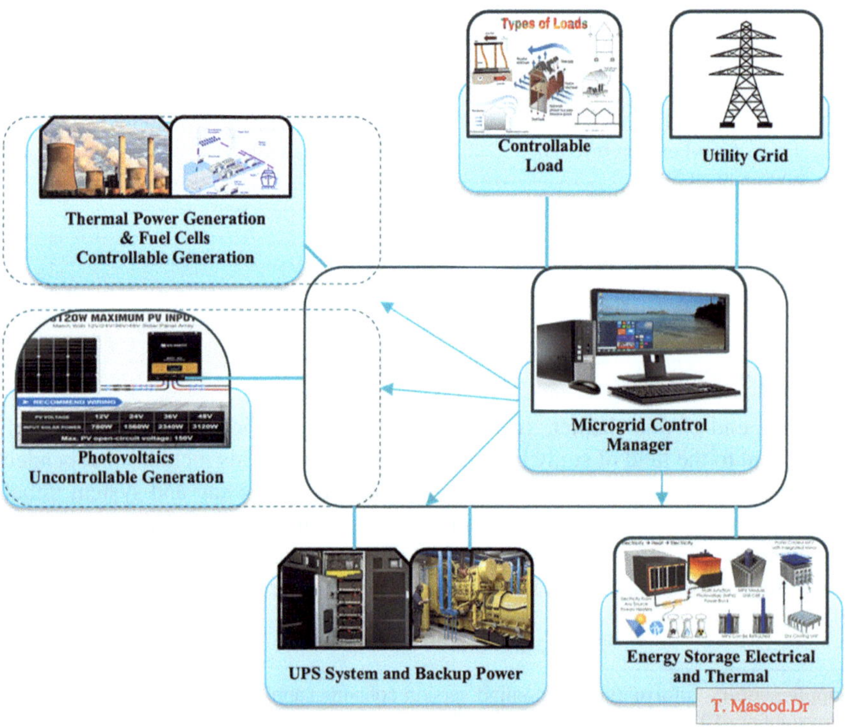

Fig. 9.1 (Microgrid schematic) Diagram illustrating a microgrid system. Central to the image is the "Microgrid Control Manager," depicted with a computer setup. To the left, "Thermal Power Generation & Fuel Cells" and "Photovoltaics" are shown as controllable and uncontrollable generation sources, respectively. Below, "UPS System and Backup Power" and "Energy Storage Electrical and Thermal" are depicted. At the top, "Controllable Load" and "Utility Grid" are connected to the system

9.1.1 Classification of Microgrid

In the first layer of our exploration, we take a deep dive into the intricate workings of microgrids, providing a nuanced analysis of their various scales. These power assets can be categorized into three distinct sizes: small microgrids, which operate at less than 10 kW; medium microgrids, with capacities up to 1 MW; and large microgrids, exceeding 1 MW. Our thorough examination reveals the diverse applications of microgrids, focusing on three essential components: performance optimization, loss reduction, and enhanced resilience in power delivery. By cultivating a solid understanding of these concepts, we empower stakeholders to make well-informed decisions. Furthermore, this chapter delves into the complexities of grid connections and navigates through both transient and autonomous operational scenarios, ensuring that every critical aspect of microgrids is meticulously addressed, as illustrated in Fig. 9.2.

In the second layer, we broaden our focus to encompass distributed energy systems as they relate to microgrid operations. Here, we present innovative solutions through the implementation of DC microgrids, AC microgrids, and hybrid microgrid systems. Each of these architectures is uniquely designed with specific principles and operational advantages that significantly enhance the overall performance [1, 2]. Notably, the microgrid configuration may include a radial grid connection,

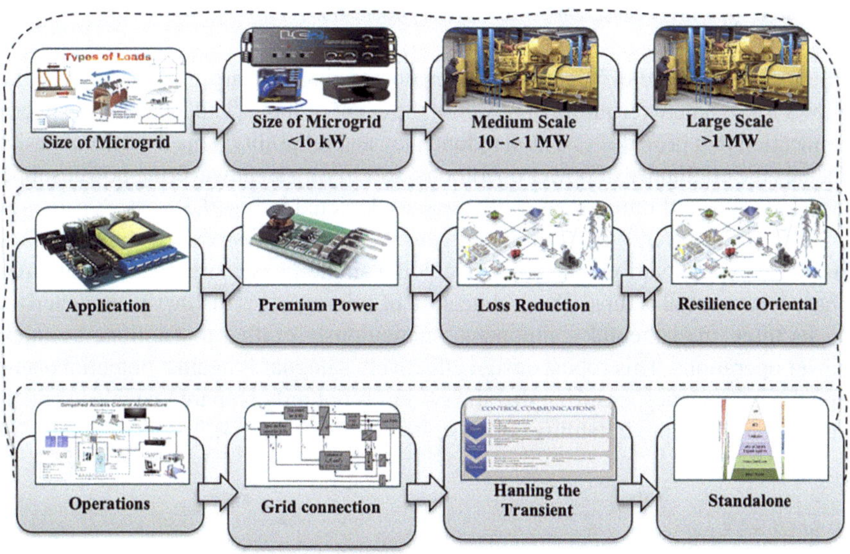

Fig. 9.2 (Microgrid size, application, and operations) Diagram illustrating a microgrid system. Central to the image is the "Microgrid Control Manager," depicted with a computer setup. To the left, "Thermal Power Generation & Fuel Cells" and "Photovoltaics" are shown as controllable and uncontrollable generation sources, respectively. Below, "UPS System and Backup Power" and "Energy Storage Electrical and Thermal" are depicted. At the top, "Controllable Load" and "Utility Grid" are connected to the system

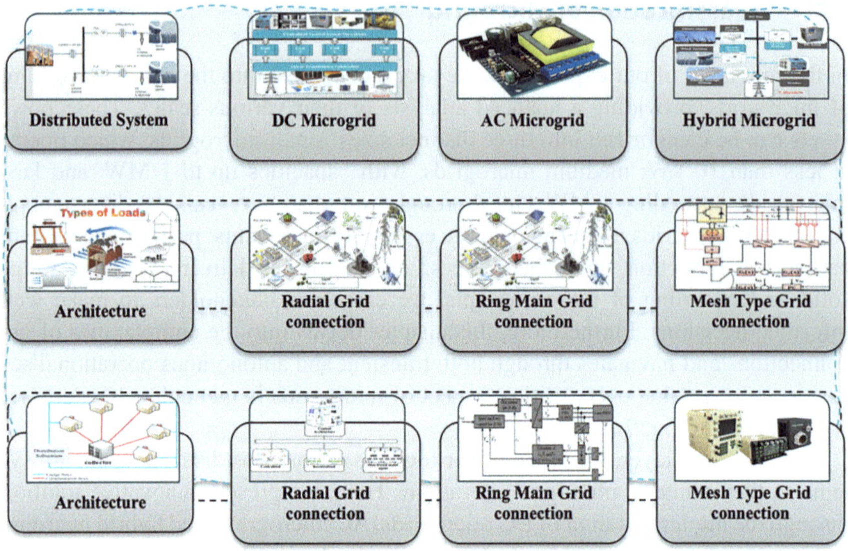

Fig. 9.3 (Microgrid distributed system and architecture) Diagram illustrating various power grid systems and architectures. The top row shows "Distributed System," "DC Microgrid," "AC Microgrid," and "Hybrid Microgrid," each with corresponding images. The middle row depicts grid connection types: "Architecture," "Radial Grid connection," "Ring Main Grid connection," and "Mesh Type Grid connection," with visual representations

which optimizes the efficiency of power flow, a ring primary grid connection that allows for dynamic rerouting of power in the event of faults, and a mesh-type hybrid connection that provides essential redundancy and reliability. This thoughtful design ensures the continuity of power operations, solidifying an unwavering power supply even in the face of unforeseen challenges, as depicted in Fig. 9.3.

In the third layer, our attention shifts to the realm of distributed configurations. Here, we explore various scenarios, including single-phase, three-phase, and three-phase with neutral setups. The architecture of microgrids is distinctly characterized by its three fundamental components, meticulously crafted to facilitate seamless power operations. This robust design effectively safeguards against potential operational contingencies, ensuring that the system is not only prepared but also resilient in confronting unpredictability, as illustrated in Fig. 9.4 [3, 4].

Exercise 9.1
Describe AC microgrid operational and control challenges. Describe the classification of the microgrid and its main components.

Exercise 9.2
Describe AC microgrid operational and control challenges.

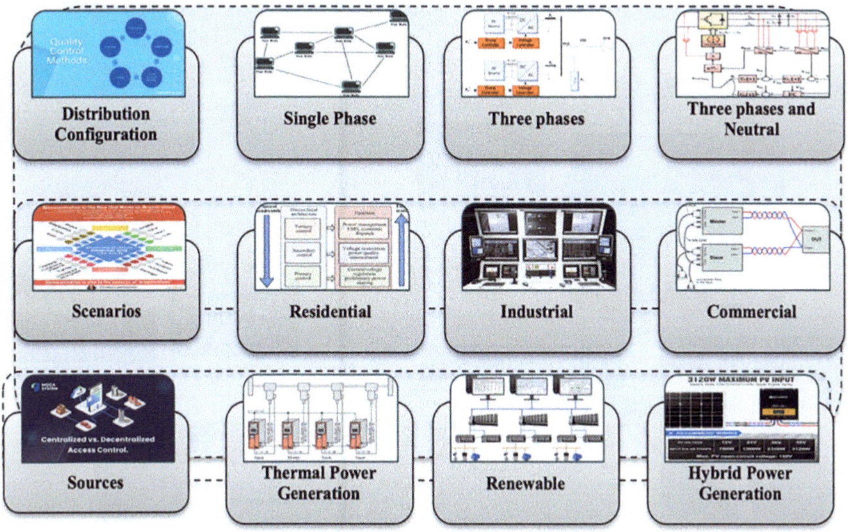

Fig. 9.4 (Microgrid distributed configuration, scenarios, and resources) A grid layout with 12 labeled sections related to power systems. Top row: "Distribution Configuration" with a diagram of quality control methods, "Single Phase" showing a network diagram, "Three Phases" with a flowchart, and "Three Phases and Neutral" displaying a circuit diagram. Middle row: "Scenarios" with a colorful flowchart, "Residential" featuring a hierarchical diagram, "Industrial" showing control room monitors, and "Commercial" with a block diagram

9.2 AC Microgrid Operations and Control

An AC microgrid serves as a sophisticated network that interconnects various energy generation sources and consumer loads through a robust AC bus system, presenting a myriad of advantages. One of its foremost benefits is the capability to seamlessly integrate with the traditional electricity grid, or alternatively, operate independently in islanded mode during emergencies. This versatility enhances the stability and reliability of the energy supply.

Cost efficiency stands out as a significant advantage, positioning AC microgrids as vital components in safeguarding power systems and maintaining their overall functionality. Furthermore, these microgrids are engineered to accommodate higher loads, ensuring that they can meet increasing energy demands without compromising performance.

However, it is essential to acknowledge some inherent challenges. The initial investment associated with converters can be substantial, and the complexity involved in control methods requires careful management and expertise. Additionally, AC microgrids may experience lower transmission efficiency compared to other systems. Despite these drawbacks, their potential to revolutionize

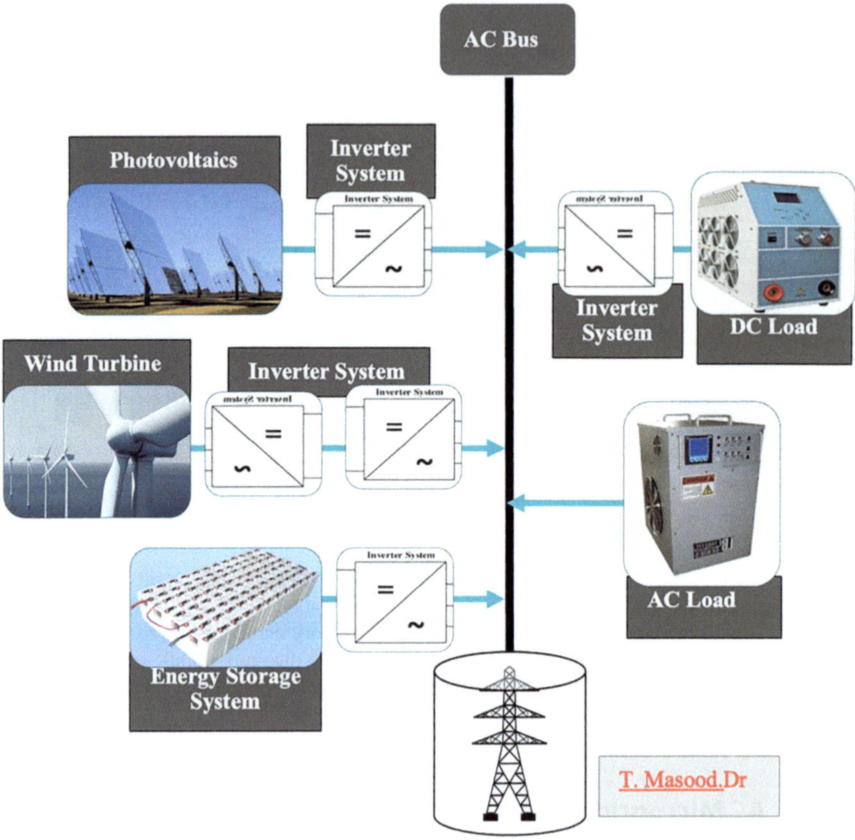

Fig. 9.5 (AC microgrid operations and control) Diagram illustrating a renewable energy system connected to an AC bus. The system includes three main sources: photovoltaics, wind turbines, and an energy storage system. Each source connects to an inverter system, converting DC to AC. The AC bus distributes power to a DC load and an AC load. The diagram emphasizes the integration of renewable energy sources into a unified power system

energy distribution makes AC microgrids an intriguing and promising option for the future of power systems (Fig. 9.5) [5].

Exercise 9.3
Describe DC microgrid operational and control challenges.

9.3 DC Microgrid Operations and Control

The hallmark of a DC microgrid is its distinct utilization of a DC bus network for interconnection, marking a significant shift away from the traditional AC bus found in conventional microgrids. This specialized DC bus network acts as a vital conduit,

Table 9.1 (Performance comparison of DC and AC microgrid)

Factor affecting	DC microgrid	AC microgrid
Series-connected power converters	Yes	No
Possibility of reconfiguration of existing facilities	No	Yes
Stability	Free from external effects	Affected by external disturbances
Microgrid control	Simple control approach	Control process is complex due to frequency
Grid-compliant devices	No	Yes
Required number of power electronics interfaces	Medium	High
Complexity of the required power electronics interfaces	Simple	Complex
Quality of the energy in the microgrid	High	Low
Easy management of the energy storage	Yes	No
Protection system	Complex, costly, and immature protection components	Simple, cheap, and mature protection schemes

efficiently linking various distributed generators and loads, thus facilitating coordinated energy sharing throughout the system.

One of the most compelling advantages of a DC microgrid lies in its remarkable transmission efficiency, which can lead to significant energy savings. Additionally, it boasts a straightforward control system that simplifies operations, ensuring that power delivery remains reliable and of high quality, even under varying load conditions. While it is true that the initial investment can be higher, this upfront cost should be viewed in the context of the long-term gains in efficiency and performance [6, 7].

However, it is important to acknowledge some challenges. The lack of reactive power sources in a DC power network can complicate voltage regulation, requiring innovative approaches to voltage management. Moreover, while the complexity of the power protection system might seem daunting, it can be effectively navigated with appropriate expertise and technology. For a thorough examination of the differences between DC and AC microgrids, refer to the detailed comparative performance analysis in Table 9.1 (Fig. 9.6).

Exercise 9.4

- Describe the DC/AC system operational flexibility to allow for the reconfiguration of existing facilities.
- Describe which DC/AC systems are vulnerable or complex to interface with power electronics devices.

Exercise 9.5
Describe a static transfer switch (STS) that connects it to the primary grid. What is the main purpose of the switch?

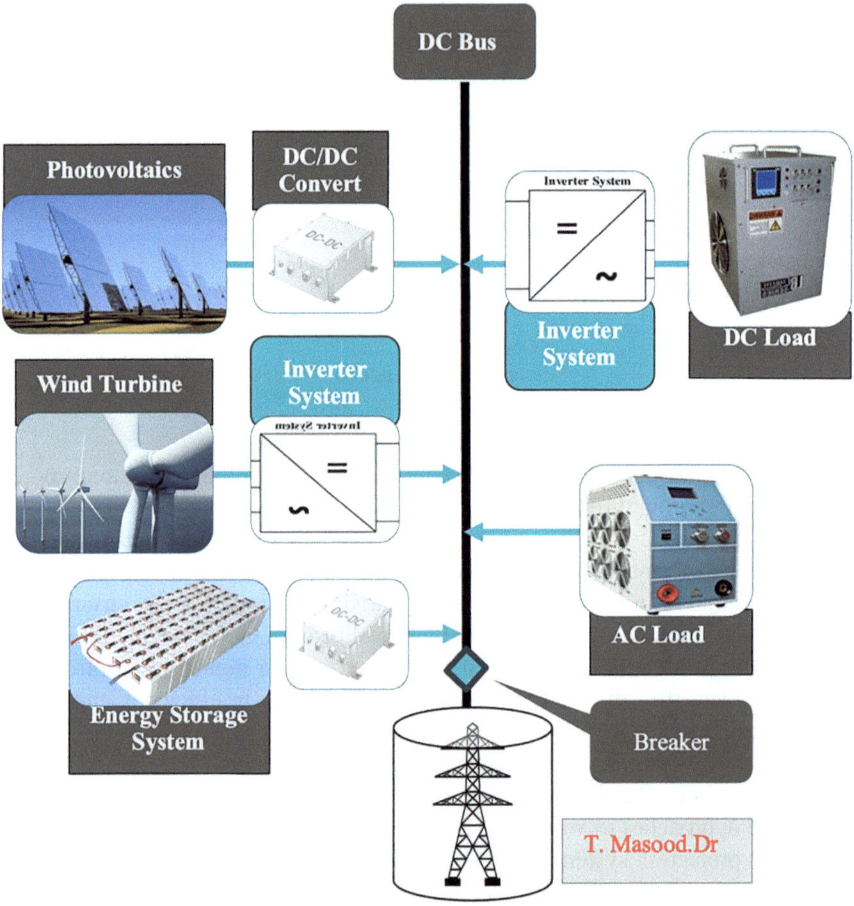

Fig. 9.6 (DC microgrid operations and control schematic) Diagram illustrating a renewable energy system connected to a DC bus. The system includes photovoltaics linked to a DC/DC converter, a wind turbine connected to an inverter system, and an energy storage system also linked to a DC/DC converter. These components feed into the DC bus, which supplies power to both DC and AC loads through an inverter system. A breaker is shown connected to the DC bus, indicating a safety mechanism

9.4 Hybrid Microgrid Operations and Control

A static transfer switch (STS) is an essential device that serves as a vital bridge to the primary electric grid, ensuring uninterrupted power connectivity. This sophisticated system incorporates a power electronic converter interface, which plays a pivotal role in precisely regulating the flow of electricity between various networks and the utility grid. This interface continuously adjusts the direction of power transfer in response to the dynamic and fluctuating balance between energy load demands and available power generation capabilities.

One of the standout features of this innovative interface is its ability to minimize conversion stages. This not only streamlines the entire energy management process but also significantly enhances the reliability and stability of the power system. By reducing the number of necessary interfacing devices, the STS simplifies the overall system architecture, making it easier to maintain and operate.

Perhaps most importantly, the implementation of this advanced technology drives down energy costs dramatically. By optimizing power management and efficiency, the static transfer switch proves to be an exceptionally smart and economically prudent choice for effectively managing and safeguarding power systems, ultimately contributing to a more sustainable energy future, as shown in Fig. 9.7 [8, 9].

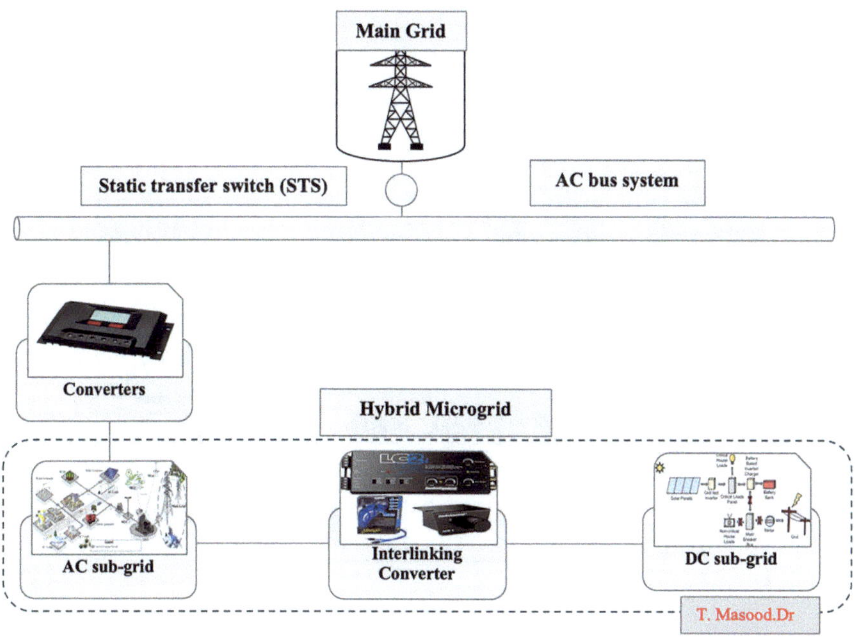

Fig. 9.7 (Hybrid microgrid operations and control schematic) Diagram illustrating a hybrid microgrid system connected to the main grid. The main grid is linked via a static transfer switch (STS) and an AC bus system. Below, converters connect to the hybrid microgrid, which includes an AC sub-grid and a DC sub-grid. An interlinking converter facilitates the connection between the AC and DC sub-grids

9.5 Microgrid Control

The objectives of microgrid control are essential for ensuring the efficient operation and reliability of the system. These objectives encompass several critical aspects, including the independent control of active and reactive power, the correction of voltage sags and system imbalances, and the ability to meet the dynamic load requirements of the grid. To achieve these goals, the microgrid control process is organized into a meticulously structured framework comprising four distinct control boundaries: control architecture, control method, control mode, and control level.

This structured approach is further refined into three specific levels of control: primary, secondary, and tertiary. Each level plays a pivotal role in facilitating a seamless and efficient management of the microgrid, enhancing the overall performance of the system, as illustrated in Fig. 9.8.

Tertiary control serves as a crucial component within this hierarchy, aimed at optimizing various operations and significantly improving efficiency. It operates at a global level, integrating feedback from both primary and secondary controls, and

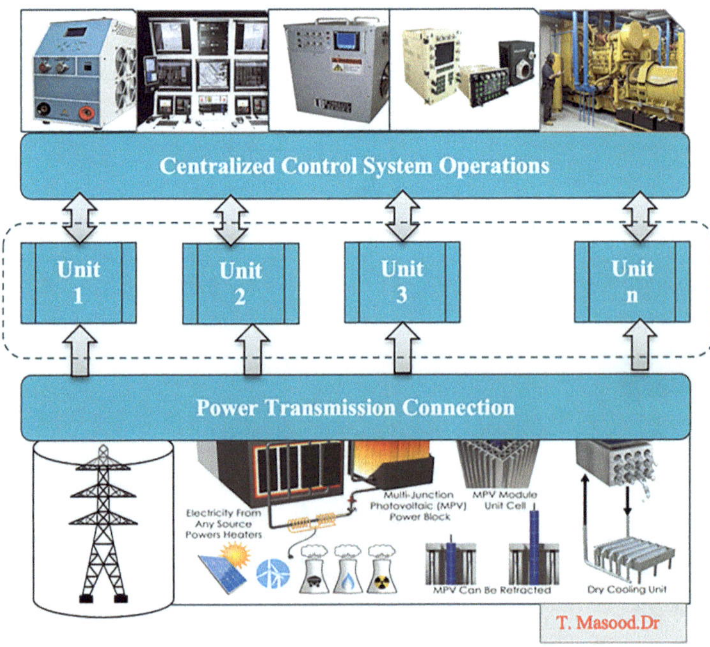

Fig. 9.8 (Centralized control) Diagram illustrating a centralized control system for power transmission. The top section shows various control panels and machinery labeled "Centralized Control System Operations." Four units labeled "Unit 1," "Unit 2," "Unit 3," and "Unit n" are connected, indicating a network. "Power Transmission Connection" features an electrical tower and a schematic of a multi-junction photovoltaic (MPV) system, including power blocks, MPV modules, and a dry cooling unit

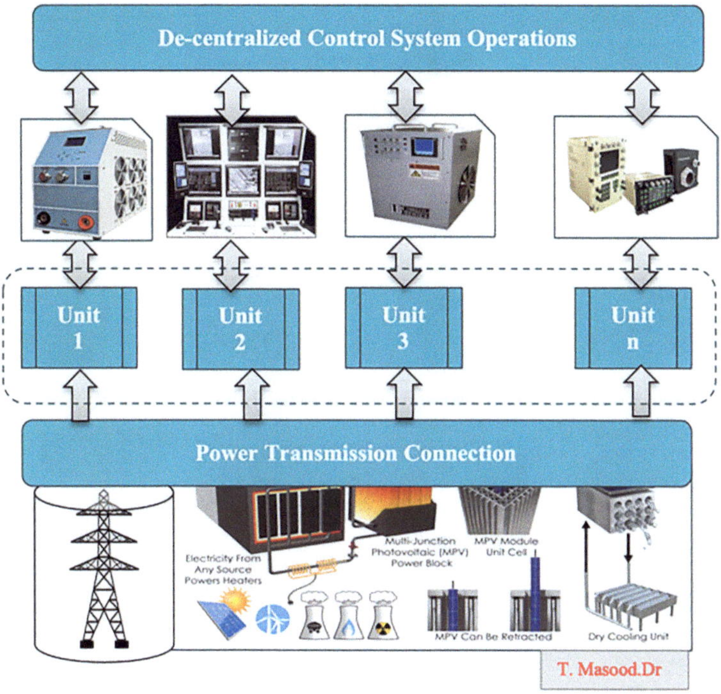

Fig. 9.9 (Decentralized control) Diagram illustrating a decentralized control system with multiple units. The top section shows various control devices labeled "De-centralized Control System Operations," connected to units labeled "Unit 1," "Unit 2," "Unit 3," and "Unit n." "Power Transmission Connection" features an electrical tower and components such as multi-junction photovoltaic power blocks, MPV modules, and a dry cooling unit. Text indicates electricity from any source powers heaters. The diagram emphasizes the integration of control systems with power transmission

effectively managing power flow in grid-tie mode. Secondary control is implemented to address and compensate for frequency deviations that arise from the actions of primary control and local level management. Meanwhile, primary control focuses on essential tasks such as maintaining voltage stability, preserving frequency stability, ensuring equitable power sharing among different sources, and managing protective devices to safeguard the system's integrity (Figs. 9.9 and 9.10) [10, 11].

9.5.1 Off-Grid and Islanded Microgrid

Islanded microgrid solutions are advanced energy systems that deliver reliable, grid-quality electricity to remote locations where conventional power infrastructure is absent or impractical. These microgrids are designed to operate independently, providing consistent energy access by utilizing a variety of local renewable resources, such as solar, wind, or biomass.

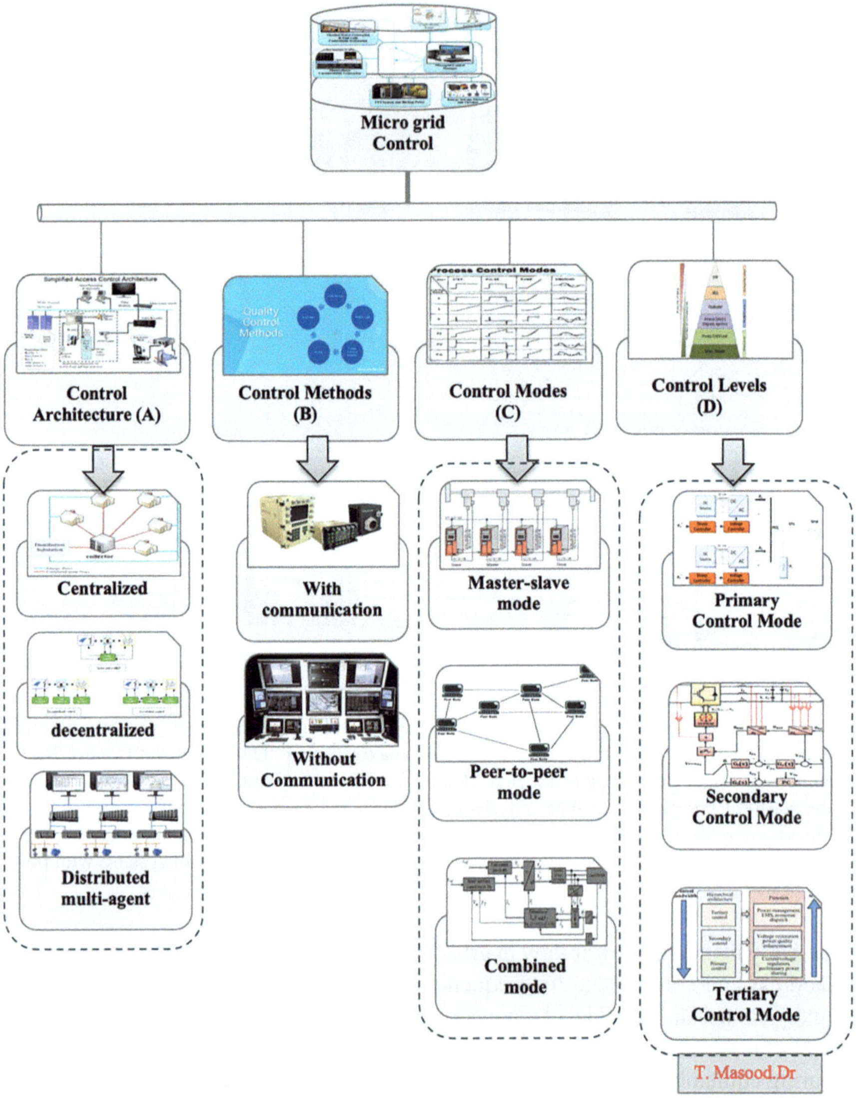

Fig. 9.10 (Microgrid operations and control schematic) - Diagram illustrating microgrid control systems, divided into four categories: Control Architecture, Control Methods, Control Modes, and Control Levels. Control Architecture includes centralized, decentralized, and distributed multi-agent systems. Control Methods are categorized as with or without communication. Control Modes feature master-slave, peer-to-peer, and combined modes. Control Levels consist of primary, secondary, and tertiary control modes

In addition to offering a lower cost of electricity compared to traditional grids, islanded microgrids significantly reduce environmental impact through their emphasis on clean energy sources. By minimizing reliance on fossil fuels, these systems

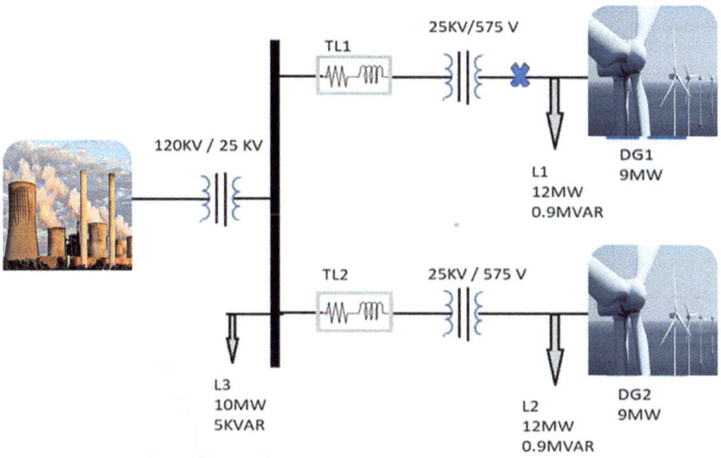

Fig. 9.11 (Off-grid and Islanded microgrid) Diagram illustrating an electrical power distribution system. On the left, a power plant generates electricity at 120 kV, stepping down to 25 kV. Two transmission lines, TL1 and TL2, connect to wind turbines on the right, each generating 9 MW at 25 kV/575 V. Line L1 outputs 12 MW and 0.9 MVAR, while L2 outputs the same. Line L3, connected to TL2, outputs 10 MW and 5 KVAR. Images of a power plant and wind turbines represent energy sources

contribute to a sustainable energy future while empowering communities with the tools they need to thrive. Overall, islanded microgrid solutions represent a transformative approach to energy access in underserved regions, fostering economic growth and environmental stewardship (Fig. 9.11) [12, 13].

9.5.2 Different Types of Islanding

A planned island represents a microgrid's strategic prowess to intentionally disconnect from the main power grid at a designated moment, allowing for precise control over its energy resources. This calculated measure ensures a continued supply of power during scheduled maintenance or other vital operations, reinforcing the microgrid's reliability and operational efficiency.

In contrast, an unplanned or dynamic islanding scenario arises when the microgrid swiftly identifies an issue within the main grid, such as a sudden power outage. This automatic response highlights the remarkable resilience and adaptability of the microgrid, enabling it to address unexpected challenges with agility and confidence. The dynamic islanding process is crucial; it assures that the microgrid can sustain stability and continue providing service, even amidst unforeseen disruptions [14].

Finally, a seamless island showcases the microgrid's exceptional ability to transition smoothly between planned and unplanned island modes without interrupting the power supply to its connected loads. This transition guarantees a consistent and

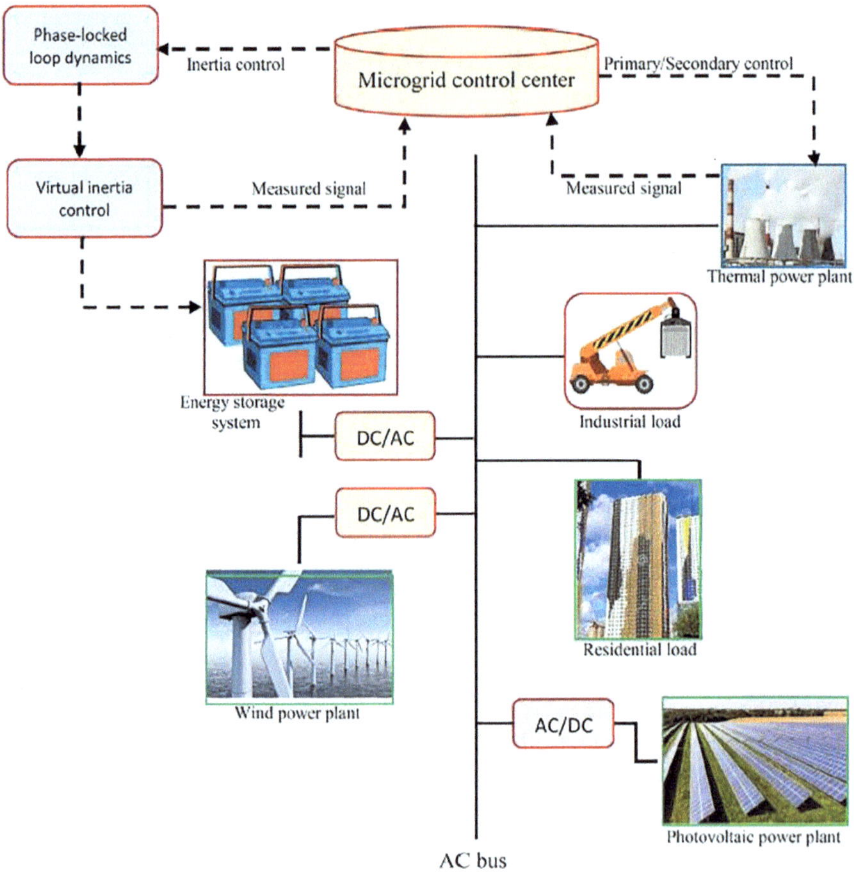

Fig. 9.12 (Different types of Islands) Diagram illustrating a microgrid control system. At the top, the "Microgrid control center" oversees various components. "Phase-locked loop dynamics" and "Virtual inertia control" feed into the system. Below, an "Energy storage system" connects through DC/AC converters. On the right, "Thermal power plant," "Industrial load," and "Residential load" are shown with images. "Wind power plant" and "Photovoltaic power plant" connect via AC/DC converters to the AC bus

uninterrupted experience for users, further emphasizing the microgrid's efficiency and reliability in delivering a stable energy flow. Embracing such advanced technology paves the way for a more dependable and sustainable energy future, fostering greater energy independence and resilience in our communities (Fig. 9.12).

9.5.3 Grid Following Inverters

Inverter-based technologies have long been confined to a grid-following control strategy, but the emergence of innovative alternative control strategies presents an opportunity to revolutionize grid integration. In grid-following configurations,

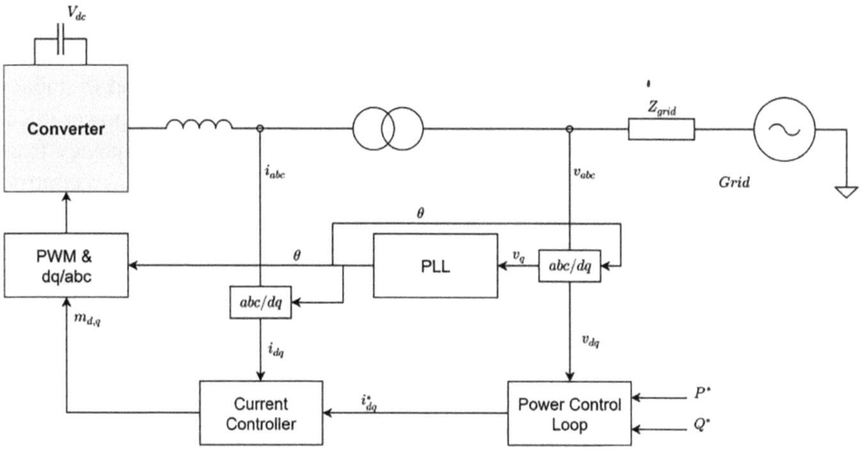

Fig. 9.13 (Grid following inverters) Diagram of a power control system for a converter connected to a grid. The system includes components such as a converter, PWM and dq/abc block, PLL, current controller, and power control loop. Arrows indicate the flow of signals and power. The grid is represented on the right, connected through an impedance

inverters receive vital voltage and frequency measurements from the grid, alongside active and reactive power setpoints dictated by the power plant controller (PPC).

However, grid-following inverters inherently lack the capacity to bolster grid strength. They operate as controllable current sources that synchronize with the grid, adapting to fluctuations in voltage and frequency. This limitation underscores the urgent need for alternative control strategies that can enhance grid resilience. The role of the power plant controller is crucial, as it monitors grid conditions but relies heavily on external setpoints from a supervisory system, which assesses the power system network. This reliance on higher-level control accentuates the existing limitations and amplifies the call for improved control solutions, as depicted in Fig. 9.13 [15, 16].

Moreover, the dependence on a higher supervisory control system introduces communication delays, preventing grid-following inverters from matching the rapid response of synchronous generators. They cannot replicate the fast-reacting nature essential for effective voltage and frequency regulation. Advancing to alternative control strategies is not just beneficial; it is imperative for the future stability and efficacy of our power systems.

Figure 9.13 demonstrates that the majority of inverter-based resources rely on a phase-locked loop (PLL) to effectively synchronize with the grid while functioning as constant current sources. This illustration presents a generic inverter schematic, clearly showing how essential the PLL is as a key component in the overall control system that links the inverter to the network.

The inverter's "grid-following" operation harnesses a sophisticated closed-loop control system to accurately derive the grid phase and frequency. This complexity underscores the advanced nature of the system you are engaged with, which is vital for ensuring optimal performance and reliability in modern energy solutions.

9.5.4 Grid-Forming Inverters

A grid-forming inverter plays a crucial role in ensuring that voltage and frequency setpoints are upheld in the grid. Its ability to adjust active and reactive power injection or absorption is directly tied to deviations in grid voltage and frequency from their ideal values. As a result, a grid-following inverter effectively acts as a controllable voltage source, enhancing grid stability.

Despite their expertise in forming and sustaining grid voltage and frequency, grid-forming inverters face significant hurdles in fully replicating the dynamic characteristics of synchronous generators during changes in the network. Sudden fluctuations in loading conditions or unexpected contingency events can drive the grid's voltage and frequency away from their nominal values, presenting a serious challenge for these inverters.

The rapid shifts in grid voltage and frequency often encountered in weak networks are alarming. In such scenarios, the absence of inertia means that a conventional grid-forming inverter can only provide a limited boost to network robustness. This limitation not only underscores the vulnerabilities in our current systems but also emphasizes the pressing need for innovative solutions in the realms of power systems and renewable energy. Addressing these challenges is essential for a more resilient and reliable energy future [17, 18].

Figure 9.14 shows that, based on virtual synchronous generator concepts, SG behaviors in P and Q are emulated using a voltage source (amplitude and frequency) instead of a current source. Auxiliary energy storages are used to support the functions.

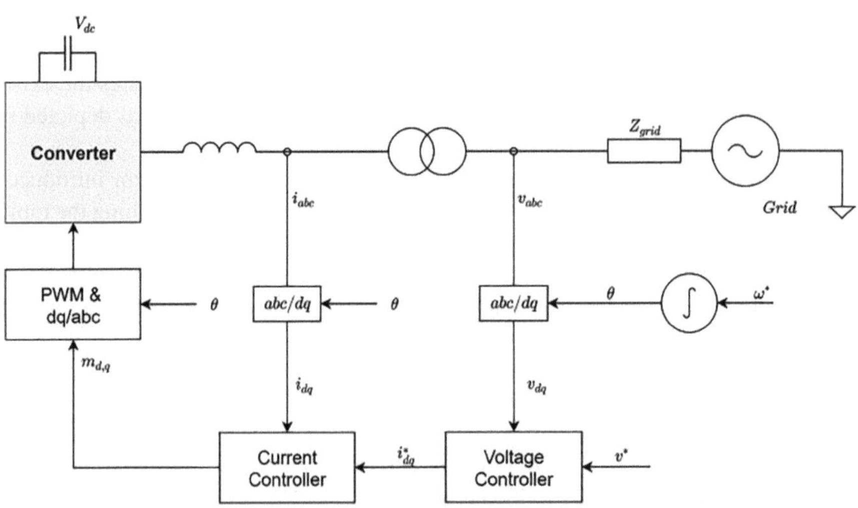

Fig. 9.14 (Grid-forming inverters) Diagram of a power conversion system showing the flow of electrical signals. The system includes a converter connected to a PWM and dq/abc block, which interfaces with current and voltage controllers. The current controller manages the current flow, while the voltage controller regulates voltage. The system outputs to a grid, with labeled connections indicating various electrical parameters such as voltage and current

9.6 Short-Circuit Study

Short-circuit studies play a vital role in identifying the worst-case fault currents under N-X conditions or other contingency scenarios. From these studies, we derive a short-circuit ratio (SCR) value that represents the strength of the point of interconnection (POI), making our work critical for system integrity.

Nominal power plant controller (PPC) parameters—such as active and reactive power rise times, ramp rates, and step-change magnitudes—are not randomly assigned. They are carefully calculated based on worst-case scenarios, emphasizing the importance of your expertise and diligence.

Adjusting PPC parameters is not a static procedure; it is a dynamic process that responds to SCR detection feedback from a nearby power meter at the POI, while adhering to bounded gain factor droop laws.

Regular monitoring and alarm-based reviews of detected SCR trends are not just advisable—they are essential. These practices enable us to promptly identify the need for updated site studies or adjustments to nominal PPC parameters and gain factors, thereby safeguarding the reliability of the entire system. Your proactive engagement in this process is crucial for optimal performance [19].

9.6.1 Advantages of Short Circuit Ratio Study

- Short-circuit studies derive an SCR value representing the "strength" of POI.
- PPC parameters are fixed according to the worst-case conditions.
- PPC parameters remain unchanged.

9.6.2 Traditional Short Circuit Ratio (SCR) Method

- SCR is defined as the ratio between short circuit apparent power (SCMVA) from a 3LG fault at the given location in the power system to the rating of the inverter-based resources at that location.

9.7 Strong and Weak Microgrid

A robust power grid, underscored by a high stability capacity ratio (SCR), exemplifies the strength and reliability of our energy systems. This kind of grid skillfully handles the fluctuations in both power generation and consumption, ensuring that the voltage levels and phase angles remain steady and within safe limits. This vital ability to uphold stable voltage, even amidst varying levels of demand from consumers and producers alike, guarantees a consistent and dependable energy supply for everyone involved.

Conversely, a fragile grid, marked by a low SCR, reveals a system that is considerably susceptible to the unpredictable changes in active and reactive power injection. Such weaknesses can lead to significant instability in the power system, compromising its reliability and leaving users vulnerable to outages. Therefore, making the necessary investments in a strong and resilient grid is crucial for securing a trustworthy and sustainable energy future for all communities [20].

Issue associated with Weak Grid—Fault Ride Through

1. HVRT.
2. Unintentional Islanding issues.
3. Control instability.

Bibliography

1. Jamil Abdo & Tariq Masood "New Technique to develop Intelligent, highly reliable & Efficient Unmanned Power Generation Modular" 22nd International Conference on Renewable Energies and Power Quality (ICREPQ'24) Bilbao (Spain), 26th to 28th June 2024. https://www.icrepq.com/icrepq24/392-24-abdo.pdf
2. Jamil Abdo & Tariq Masood "Operational Reliability- Production Best Practice Pragmatic Case Studies & Improvement" published in Proceedings of the 9th North American Conference on Industrial Engineering and Operations Management Washington D.C., USA, June 4-6, 2020
3. Tariq Masood & Jamil Abdo "Maintenance Strategy Production Best Practices—Pragmatic Case Studies and Operations Improvement" published in Proceedings of the 9th North American Conference on Industrial Engineering and Operations Management Washington D.C., USA, June 4-6, 2020
4. Jamil Abdo & Tariq Masood "Techno Economics Analysis to Reshape Global Clean and Green Energy" 22nd International Conference on Renewable Energies and Power Quality (ICREPQ'24) Bilbao (Spain), 26th to 28th June 2024. https://www.icrepq.com/icrepq24/415-24-abdo.pdf
5. M.Saeed Ullah, Tariq Masood, Umair Zahid, H. Abuzar Ahsan "Electrochemical Oxidation of Dissolved organic Dye in Aqueous Solution' Published in IMASCON-2019
6. Nasser Alemadi, Tariq Masood, R Aggarwal, Suhail A Qureshi, D Kothari "Novel selection approach of STATCOM device tailored and optimized engineering configuration to implement in between Qatar and Bahrain" published in Int. J. Autom. Power Eng. https://doi.org/10.14355/ijape.2014.0302.01
7. Tariq Masood & Jamil Abdo "New Technique to Enhance Hybrid Renewable Energy Resources' Operational Flexibility and Reliability" 22nd International Conference on Renewable Energies and Power Quality (ICREPQ'24) Bilbao (Spain), 26th to 28th June 2024. https://www.icrepq.com/icrepq24/396-24-masood.pdf
8. Tariq Masood and Jamil Abdo "Review: Uninterrupted Liquid Hydrogen (E-Methanol) Production framework to deliver Clean & Green Energy" Published 22nd International Conference on Renewable Energies and Power Quality (ICREPQ'24) Bilbao (Spain), 26th to 28th June 2024. https://www.icrepq.com/icrepq24/414-24-masood.pdf
9. Tariq Masood, Abdel Aty Edris, Suhail Aftab Qureshi, Muhammad Tajammal, Murtaza Hashmi, DP Kothari, Samer Kareem "SSSC tailored to optimization performance in Between United Arab Emirates and Oman on the GCC Power Grid" published in the 13th IET International conference on AC and DC power Transmission Conference on 14–16 February, 2017. http://ieeexplore.ieee.org/document/7934934/
10. Tariq Masood, Abdel Aty Edris, Suhail Aftab Qureshi, Muhammad Tajammal, Murtaza Hashmi, DP Kothari, Samer Kareem "Smart Grid Operations and Control Challenge by Implementing

STATCOM Tailored to Optimization Performance at the GCC Power Grid" published in the 13th IET International conference on AC and DC power Transmission Conference on 14–16 February, 2017. http://ieeexplore.ieee.org/document/7934988/
11. Tariq Masood, Muhammad Tajammal, Samer Kareem, Suhail Aftab Qureshi, Abdul Jabbar Khan, DP Kothari, Nair Jameel "Unified power Flow Controller Modeling and Analysis Technique on the GCC Power Grid Operations and Control" published IEEE Energy Con, 2018 June 7-9 Limassol Cyprus. https://ieeexplore.ieee.org/document/8398834/
12. Tariq Masood, Muhammad Tajammal, Samer Kareem, Suhail Aftab Qureshi, Murtaza Hashmi, Nasir Jameel "Unified power Flow Controller (UPFC) Modeling and Analysis Technique by using Matlab and EMTP" published (ICREQP, 18 Salamanca Spain 21–23 March, 2018). http://www.icrepq.com/icrepq18/250-18-masood.pdf
13. Tariq Masood, Muhammad Tajammal, Samer Kareem, Suhail Aftab Qureshi, Murtaza Hashmi, Naser Jameel "STATCOM Controller Performance Enhancement and Optimization Technique" published (ICREQP, 18 Salamanca Spain 21–23 March, 2018). http://www.icrepq.com/icrepq18/249-18-masood.pdf
14. Tariq Masood, Muhammad Tajammal, Samer Kareem, Suhail Aftab Qureshi, Murtaza Hashmi, DP Kothari, "Smart Grid Operational functions and Control Challenges by Implementing SSSC Tailored to Optimize performance in between Qatar and KSA on the GCC Electrical-power grid" published in the ASTESJ Journal Advances in Science, Technology and Engineering Systems Journal Vol. 2, No. 1, 30.09.2017. https://astesj.com/special-issue-4/
15. Tariq Masood, Raj K Aggarwal, Suhail Aftab Qureshi, RAJ Khan "STATCOM model against SVC control model performance analyses technique by MATLAB" published in International Conference on Renewable Energies and Power Quality (ICREPQ'10) https://icrepq.com/icrepq'10/655-Masood.pdf
16. Tariq Masood, N Al-Emadi, SA Qureshi, YY Almulla, AJ Khan, SK Shah "FACTS Control Devices (Statcom, SSSC and UPFC) Re-Configuration Techniques By PSIM/MATLAB" Published in: 2007 International Conference on Electrical Engineeringhttps://ieeexplore.ieee.org/abstract/document/4287295
17. Tariq Masood, Raj K Aggarwal, SA Qureshi, RAJ Khan "STATCOM and SVC control operations and optimization during network fault conditions" https://ieeexplore.ieee.org/document/5636868
18. Tariq Masood, RK Aggarwal, SA Qureshi, RAJ Khan "Statcom control reconfiguration technique for steady state and dynamic performance optimization during network fault conditions" published in International Conference on Renewable Energies and Power Quality (ICREPQ'10) Granada (Spain) https://www.icrepq.com/icrepq'10/656-Masood.pdf
19. Tariq Masood, Samer Karim Shah, Muhammad Tajammal, "Unified power flow controller modeling and analysis technique on the GCC power grid" published in 2018 IEEE International Energy Conference (ENERGYCON) https://ieeexplore.ieee.org/abstract/document/8398834
20. Tariq Masood, Suhail A Qureshi, RK Aggarwal, DP Kothari "Multiple FACTS devices integration through distributed control system and its application on the GCC interconnection" published in 2013 7th IEEE GCC Conference and Exhibition (GCC) https://ieeexplore.ieee.org/abstract/document/6705750

Power Transmission Line Characteristics 10

Contents

10.1 Introduction ... 242
10.2 The Series Inductance of a Two-Conductor Line ... 243
10.3 Mutual Inductance and Electromagnetic Interference ... 245
10.4 The Self-Inductance of a Coaxial Cable ... 247
10.5 The Smart Shunt Capacitance of a Two-Conductor Line ... 248
10.6 The Concentric Cable ... 250
10.7 The Simple Line Method ... 252
10.8 The Line Performance Chart ... 257
Self-Assessment Questions ... 262
Bibliography ... 264

Abstract

When you have completed this chapter, you will be able to: Determine the capacitance and inductance of lines and cables in simple regular arrangement. Calculate the voltage induced between parallel arrangements of conductors. Setup a line representation using four-terminal network to calculate the conditions on lines. Use characteristic impedance to calculate changes in conditions on a transmission line. Use π and T models of transmission lines to make simple calculation of voltage, current, load flow and power factor conditions on a line. Construct a simple chart to show how the operating condition of a line may be changed by changes in terminal voltages. Use the simple chart, or mathematical analysis, to determine changes in power and var flow caused by changes in reactive loads.

Supplementary Information The online version contains supplementary material available at https://doi.org/10.1007/978-3-031-94312-6_10.

T. Masood et al., *Protection of Electrical Power Transmission Systems*, Power Systems, https://doi.org/10.1007/978-3-031-94312-6_10

Objectives

When you have completed this chapter, you should be able to:

- Determine the capacitance and inductance of lines and cables in a simple regular arrangement.
- Calculate the voltage induced between parallel arrangements of conductors.
- Set up a line representation using a four-terminal network to calculate the conditions on lines.
- Use characteristic impedance to calculate changes in conditions on a transmission line.
- Use π and T models of transmission lines to make simple calculations of voltage, current, load flow, and power factor conditions on a line.
- Construct a simple chart to show how the operating condition of a line may be changed by changes in terminal voltages.
- Use the simple chart or mathematical analysis to determine changes in power and var. flow caused by changes in reactive loads.

10.1 Introduction

The simplest view of a transmission line is that it is a length of conductor used to connect a load to a source or to interconnect two power supply systems. As with any conductor, it has inductance, capacitance, and resistance. The simple view would suggest that the resistance is the most important factor because the I^2R loss and its associated heating will place the only limit on power transfer.

This limit exists, but for high voltage, long transmission lines, other factors are just as important. We have already seen that there is a theoretical limit on the power that can be transferred between a synchronous source $\boldsymbol{E}$ and synchronous load $\boldsymbol{V}$, and that the limit is determined by the load angle between the two voltages and by the transfer impedance between them. The inductance is a critical component of this transfer impedance.

Capacitance is also significant on long, high-voltage power transmission lines. Reactive current has to flow along the line to charge up the capacitance of the line. This will affect the voltage distribution along the line. An underground cable is far more capacitive than an overhead line and will need proportionately more capacitive current to charge it. The design and operations of the long transmission lines are not therefore a simple issue, and lines are a very significant part of the transmission and distribution network [1].

The inductance, capacitance, and resistance of a transmission line are distributed parameters. That is, each infinitesimal chapter of a line has its own inductance, capacitance, and resistance. An accurate mathematical model of this is complex, but we only need such accuracy when considering high frequency and transients' effect on the power transmission line, or when dealing with exceptionally long power transmission lines. Fortunately, we can establish the main principles of power line operations by using good approximate models that are easy to set up and deal with.

In this chapter, we will be examining how the physical dimensions of a line determine its inductance and capacitance, and we will use these derived values to establish simple models to describe the operations of the power transmission lines. The SMART models will be based on π and T network approximation, which is justified for the line length of up to 150, 500, 1000, and 1500 km. These models are then used to construct a simple line chart from which we can determine the practical power transfer capabilities of the line.

10.2 The Series Inductance of a Two-Conductor Line

Figure 10.1 represents a chapter through two long, cylindrical, parallel conductors A and B. The effect of earth is ignored, and the current I is assumed to go up conductor A and return via conductor B.

The magnetic field set up by a long, cylindrical conductor has a circular symmetry about the axis of the conductor, and at a radius x away from the axis, the magnetizing force H is given by:

$$H = \frac{1}{2\pi x} \tag{10.1}$$

The magnetic flux density is given by $B = \mu_0 H$, where μ_0 is the permeability of free space and has a value $\mu_0 = 4\pi \times 10^{-7} Hm^{-1}$; thus, the magnetic density at a radius x from the axis of conductor A due to the current I in A is:

$$B_x = \mu_0 H = \frac{\mu_0 I}{2\pi x} \tag{10.2}$$

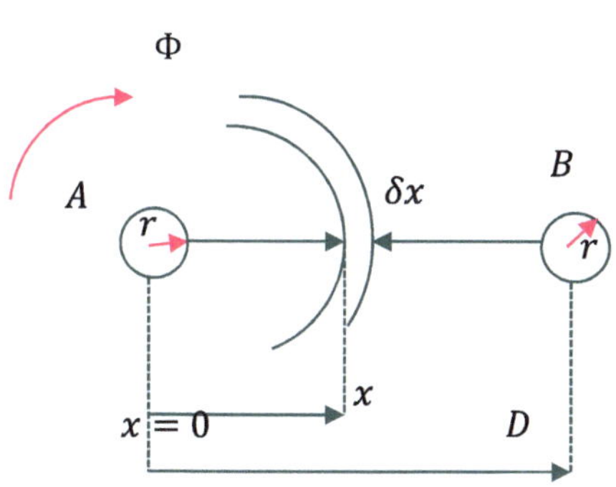

Fig. 10.1 Represents a chapter through two long, cylindrical, parallel conductors. Diagram illustrating a fluid flow system with two circular sections labeled A and B. Both sections have arrows labeled "r" indicating radial directions. A curved arrow labeled "Φ" points from A. A horizontal line connects A and B, with a segment labeled "δx" between them. The distance from A to the start of δx is marked "x = 0," and the distance from A to B is labeled "D"

Imagine now a thin cylindrical shell of radius x, length l, and wall thickness δx, centered on the axis of A. The magnetic flux through this shell will be:

$$\delta\Phi = B_x \times l \times \delta x \tag{10.3}$$

The total magnetic flux existing between the surface of conductor A and the surface of conductor B will be the sum of a series of small fluxes $\delta\Phi$ as the variable x increases from $x = r$ *to* $x = D - r$, i.e.,

$$\Phi = \int_r^{D-r} \frac{\mu_0 I}{2\pi x} l\, dx = \frac{\mu_0 I}{2\pi x} l\, In\left(\frac{D-r}{r}\right) \tag{10.4}$$

This flux links the conducting loop formed by the conductors A and B. We also have a flux created by the returning current in B [2]. This will add to the flux created by the current I in A, and simple symmetry suggests that it will have the same magnitude. So, the total flux linking the two conductors is:

$$\Phi_T = \frac{\mu_0 I}{\pi} l\, In\left(\frac{D-r}{r}\right) \tag{10.5}$$

Inductance is defined as flux-linkages per ampere, so we can define the self-inductance of a two-conductor line as L, where:

$$L = \frac{\Phi_T}{I} = \frac{\mu_0}{\pi} l\, In\left(\frac{D-r}{r}\right) \text{Henries}(\text{H}) \tag{10.6}$$

or, writing this as a per-unit length value

$$L = \frac{\Phi_T}{I} = \frac{\mu_0}{\pi} l\, In\left(\frac{D-r}{r}\right) \text{Henries}(\text{H})$$

The calculation for a three-conductor line is a little more involved and becomes difficult if the line is not symmetrically placed or if the conductors are transposed. In a transposed line, the conductors are crossed over every so often, so that the "a" conductor moves to the position of the "b" conductor, the "b" conductor moves to the position of the "c" conductor, and the "c" conductor moves to the position of the "a" conductor. This tends to happen only on very long lines.

The effects of earth wires and of the ground also need to be taken into account for an exact solution, particularly when transient conditions are being analyzed. You can imagine how difficult this can be when you consider that the resistivity of the ground will vary with ground and weather conditions along the length of the line.

However, we have developed a simple equation for the inductance of a simple line, which shows the way that inductance varies with line separation. We also have a straightforward method of analysis, which we can use to calculate the electromagnetic interference between adjacent lines [3].

10.3 Mutual Inductance and Electromagnetic Interference

Figure 10.2 represents a two-conductor power line, *AB*, and a parallel two-conductor communication circuit, *CD*. We will define the distances separating the conductors as $d_{AB,}d_{AC}$, and so on and assume that the radius of all conductors is small with respect to the distances between conductors [4].

Following the procedure of the previous chapter, we can define the flux passing between conductors *C* and *D* due to current *I* in conductor *A* at a distance *x* from the axis of *A* as:

$$\delta\Phi = \frac{\mu_0 I}{2\pi x} l\delta x \tag{10.7}$$

Total flux linking the circuit CD due to the current in A is:

$$\Phi_A = \int_{x=d_{AC}}^{x=d_{AD}} \frac{\mu_0 I}{2\pi x} l\, dx = \frac{\mu_0 I}{2\pi} l\, In\left(\frac{d_{AD}}{d_{AC}}\right) \tag{10.8}$$

The flux linking the circuit CD due to the current I in conductor B is:

$$\Phi_B = \frac{\mu_0 I}{2\pi} l\, In\left(\frac{d_{BD}}{d_{BC}}\right) \tag{10.9}$$

However, Φ_B is opposing Φ_A so that the total effective flux linkage with CD is:

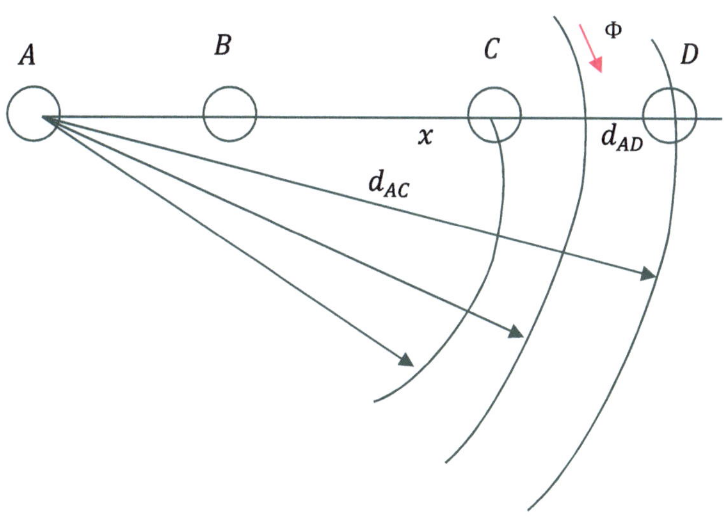

Fig. 10.2 Represents a two-conductor power line. Diagram illustrating wavefronts and rays. Three circular wavefronts emanate from a point source on the left, with arrows indicating the direction of wave propagation. The wavefronts are perpendicular to the rays, which are straight lines extending outward. A pink arrow labeled with the Greek letter Phi (Φ) indicates the angle of incidence

$$\Phi_T = \frac{\mu_0 I}{2\pi x} l \left[In\left(\frac{d_{BD}}{d_{BC}}\right) - In\left(\frac{d_{AD}}{d_{AC}}\right)\right] \tag{10.10}$$

The mutual inductance, M, between the two circuits is defined as flux linkages per ampere, so that:

$$\text{M} = \frac{\mu_0 I}{2\pi x} l \left[In\left(\frac{d_{BD}}{d_{BC}}\frac{d_{AD}}{d_{AC}}\right)\right] H \tag{10.11}$$

or

$$\text{M} = \frac{\mu_0 I}{2\pi x} l \left[In\left(\frac{d_{BD}}{d_{BC}}\frac{d_{AD}}{d_{AC}}\right)\right] Hm^{-1}$$

Calculating the induced e.m.f., *e*, is now straightforward because:

$$e = M\frac{dI}{dt} \tag{10.12}$$

and, if

$$I = isin\omega t$$

Then:

$$e = Mi\frac{dI}{dt}sin\omega t = Mi\omega sin\omega t\left(\text{per metre length}\right)$$

Exercise 10.1

Four long, straight conductors lie in the same horizontal plane with 1 cm between adjacent conductors. Supply a circuit with 500 mA r.m.s current at 12.5 kHz. Determine the e.m.f per meter run induced around the loop formed by the right-hand conductors and the mutual inductance between the two loops per unit length.

Assume a steady current in the left-hand conductors of Fig. 10.2. The flux linking the loop CD as a result of current *I* in A is:

$$\Phi_A = \int_{x=d_{AC}}^{x=d_{AD}} \frac{\mu_0 I}{2\pi x} l\,dx = \frac{\mu_0 I}{2\pi} l\,In\left(\frac{3}{2}\right) \tag{10.13}$$

The flux linking CD as a result of current *I* in B is:

$$\Phi_B = \frac{\mu_0 I}{2\pi} l\,In2$$

The total flux linking the loop CD in an anticlockwise direction about A is:

$$\Phi_T = \frac{\mu_0 I}{2\pi} l\,In\left(\frac{4}{3}\right)$$

The mutual inductance per unit length, M, is the flux linkage per ampere, which is calculated as $0.58 \times 10^{-7} Hm^{-1}$ [5].

The induced e.m.f per meter run of CD is:

$$e = M\frac{dI}{dt} = M \times 0.5 \times 2\pi f \times \cos 2\pi ft$$

The r.m.s value is calculated as 2.3 mVm^{-1}.

This example of mutual coupling should seem obvious, and it clearly has relevance to situations where communication circuits run parallel to power lines. But, mutual coupling of this sort also exists between the various phases of a multi-phase conductor system; between adjacent circuits, where, for example, two three-phase supplies run in parallel on one set of overhead towers; and between phase conductors and the earth wire. The series impedance of long transmission conductors is therefore quite a complex thing. It depends not only on the self-impedance of the conductor but also on the effects of currents in adjacent conductors and, possibly, the ground.

10.4 The Self-Inductance of a Coaxial Cable

In this case, we assume that the current I is supplied by the central conductor and that it returns via a second conductor or, in the case of low-level signals, along the coaxial sheath [6]. Either way, the conducting sheath of a coaxial cable protects the inner conductor from the effects of externally generated electromagnetic fields, as shown in Fig. 10.3. The only magnetic flux within the cable is produced by the current I in the main conductor. Following our earlier example, we have:

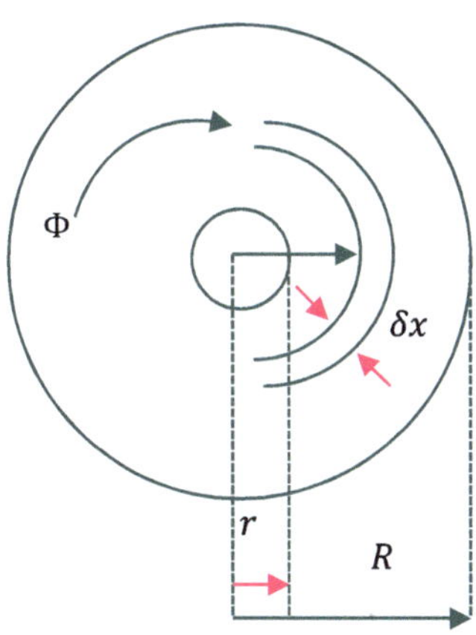

Fig. 10.3 Externally generated electromagnetic fields. A diagram of a spiral with arrows indicating direction and measurements. The spiral is labeled with the Greek letter Phi (Φ). Two radial lines are marked with distances: a smaller radius labeled "r" and a larger radius labeled "R." A segment of the spiral is highlighted with a distance labeled "δx." Arrows show the direction of the spiral's curve and the radial measurements

$$\delta\Phi = \frac{\mu_0 I}{2\pi} l\,\delta x \tag{10.14}$$

$$\Phi_T = \frac{\mu_0 I}{2\pi} l\, In\left(\frac{R}{r}\right) \tag{10.15}$$

$$\mathrm{L} = \frac{\Phi_T}{I} = \frac{\mu_0}{2\pi} l\, In\left(\frac{R}{r}\right) \tag{10.16}$$

$$L = \frac{\mu_0}{2\pi} l\, In\left(\frac{R}{r}\right) Hm^{-1} \tag{10.17}$$

If the outer sheath is an effective screen and is earthed correctly at each end, then there should be no mutual coupling between the conductor and any other circuit.

High-voltage cables are usually of a coaxial construction with solid metal screening around each phase conductor. Each phase conductor is laid separately. Medium-voltage, three-phase cables may have all three-phase conductors bound within one cable structure, but each phase conductor will be individually screened, which will minimize mutual coupling and hence mutual inductance. Low-voltage wiring will not have screening about the conductors, and mutual electromagnetic coupling will be maximized. This will be a major problem when there are high-frequency fields or where transient conditions occur. Wiring in any control or data network that may be susceptible to unwanted interference should always be screened [7].

10.5 The Smart Shunt Capacitance of a Two-Conductor Line

Figure 10.4 represents a chapter through two long, cylindrical conductors. The effect of earth is ignored, and the change in displacement is balanced with +Q coulombs on conductor A and –Q coulombs on conductor B.

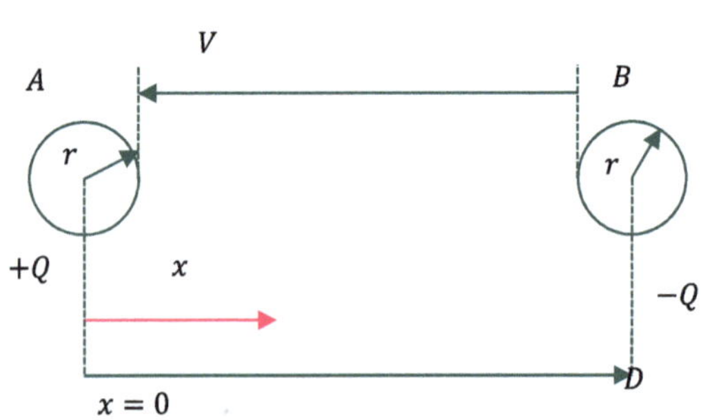

Fig. 10.4 Represents a chapter through two long, cylindrical conductors. Diagram illustrating two charged spheres, labeled A and B, with radii r. Sphere A has a positive charge +Q, and sphere B has a negative charge –Q. The distance between the spheres is marked as x, with a reference point at x = 0. An arrow labeled V points from B to A, indicating a potential difference

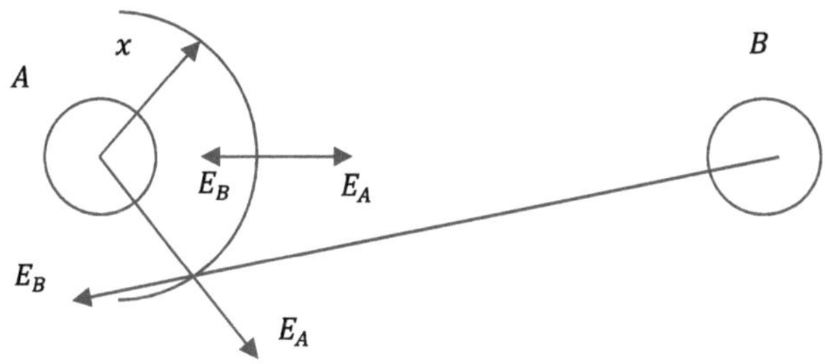

Fig. 10.5 Electric field strength has two components. Diagram illustrating two circles labeled A and B. Circle A has two arrows labeled \(E_A\) and \(E_B\) pointing outward, with \(E_A\) directed horizontally and \(E_B\) directed diagonally downward. An arc labeled \(x\) extends from the center of circle A. A line connects circle A to circle B

Electric field strength E is defined as acting positively along the line of increasing x. We can then define the potential of a point at $x = a$ with respect to a point $x = b$ as:

$$V_{ab} = -\int_b^a E dx.$$

At a point a general distance x away from the axis of conductor A, as defined in Fig. 10.5, the electric field strength has two components: one due to the charge on A and one due to the charge on B. We assume that $Q_A = -Q_B$ so that the total charge on the system is zero [8].

Theoretically, it does not matter which path we take when we determine the potential difference A and B, but for ease of calculation, we choose the one that gives the easiest solution. In this case, the choice is the shortest line between the two axes. At the point x along this line, the two components of E are:

E *at* x due to charge on A:

$$E_A = \frac{Q}{2\pi\varepsilon_0 xl}$$

And E *at* x due to charge at B:

$$E_B = \frac{-Q}{2\pi\varepsilon_0 (d-x)l}$$

If you are unsure about this, then you need to refer to a basic test on electrical theory and look up Gauss law.

The constant ε_0 is the permittivity of free space and has a value $10^{-9}/36\pi$ Farads per meter (Fm^{-1}). If the material between the conductors is an insulator other than free space (or air), then we have to multiply ε_0 by the value of relative permittivity for the material [9].

The resultant electrical field strength E_R is:

$$E_R = \frac{Q}{2\pi\varepsilon_0 xl} - \left[\frac{-Q}{2\pi\varepsilon_0 (d-x)}\right] \tag{10.18}$$

So that the voltage between A and B is:

$$V_{AB} = -\int_{d-r}^{r} \frac{Q}{2\pi\varepsilon_0 xl}\left[\frac{1}{x} + \frac{1}{d-x}\right]dx \tag{10.19}$$

$$V_{AB} = \frac{Q}{\pi\varepsilon_0 xl} In\left(\frac{d-r}{r}\right)$$

This gives the capacitance C as

$$C = \frac{Q}{V_{AB}} = \frac{\pi\varepsilon_0 xl}{In\left(\frac{d-r}{r}\right)} \text{Farads}(\text{F}) \tag{10.20}$$

Or alternatively, the capacitance per unit length is:

$$C = \frac{\pi\varepsilon_0}{In\left(\frac{d-r}{r}\right)} \text{Fm}^{-1}$$

Exercise 10.2
Two long, straight, cylindrical conductors run in parallel to form a "go and return" circuit. Each conductor has an outside diameter of 2 cm, and they are separated by a distance of 1 m between the centers.
Determine the capacitance between the conductor per meter run.

The answer is obtained by straight substitution and should be $C = 6.04\frac{pF}{m}$.

10.6 The Concentric Cable

Assume a charge Q on the inner conductor so that the electric field E acts radially outward from the center in the direction of increasing x (Fig. 10.6). The potential of the inner conductor with respect to the outer is then

$$V = \int_a^r Edx$$

But the electric field strength E *at* X *is*

$$E = \frac{Q}{\pi\varepsilon_0 xl}$$

so that

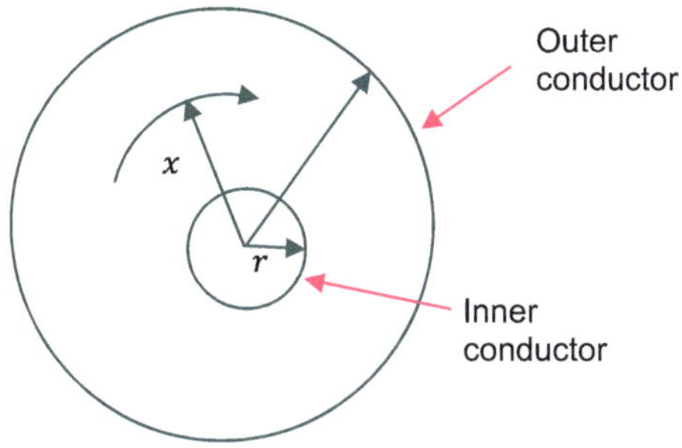

Fig. 10.6 Inner conductor electric field. Diagram illustrating two concentric circles labeled as inner and outer conductors. The inner circle is marked with radius "r," and the outer circle is marked with radius "x." Arrows indicate the direction from the center to the perimeters of both circles

$$V = \int_a^r \frac{Q}{2\pi\varepsilon_0 xl} dx = \frac{Q}{2\pi\varepsilon_0 xl} In\left(\frac{a}{r}\right)$$

which gives:

$$C = \frac{2\pi\varepsilon_0}{In\left(\frac{a}{r}\right)} \text{Farads per meter length}$$

> **Exercise 10.3**
> A concentric cable consists of an inner conductor of diameter 2 cm and an outer conductor of internal diameter 6 cm. Polythene tape is wrapped around the inner conductor to a thickness of 1 cm, and then a layer of nylon tape is laid on top of the polythene to completely fill the remaining space between the two conductors. Calculate the capacitance of the cable per meter run. Take the relative permittivity for polythene as 2.4 and the relative permittivity of nylon as 3.6.

A cross chapter through the cable is shown in Fig. 10.7. The voltage between the inner and outer conductors is given by:

$$V = \int_a^r E_1 dx - \int_a^r E_2 dx \tag{10.21}$$

$$E_1 \; at \; x \; for \; r_2 < x < r_3 = \frac{Q}{2\pi\varepsilon_0\varepsilon_1 xl} \tag{10.22}$$

$$E_2 \; at \; x \; for \; r_2 < x < r_3 = \frac{Q}{2\pi\varepsilon_0\varepsilon_2 xl} \tag{10.23}$$

so that

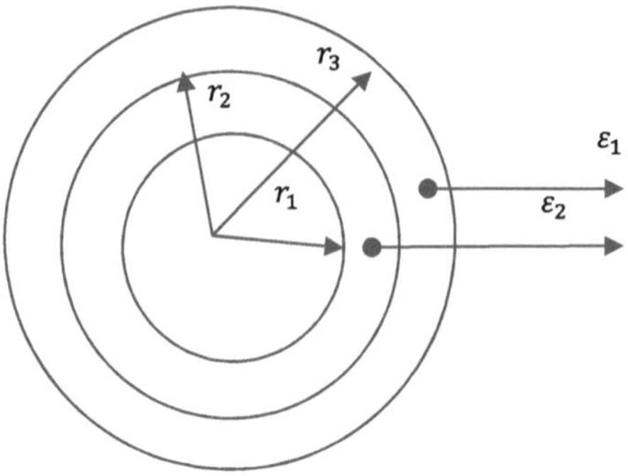

Fig. 10.7 Voltage between the inner and outer conductors. Diagram of three concentric circles with radii. Two arrows extend from the circles to the right, indicating different directions or forces

$$V = \frac{Q}{2\pi\varepsilon_0\varepsilon_1 xl}\left[\frac{1}{\varepsilon_1} In\left(\frac{r_3}{r_2}\right) + \frac{1}{\varepsilon_2} In\left(\frac{r_2}{r_1}\right)\right] \tag{10.24}$$

Substitution gives:

$$C = \frac{Q}{VI} = 140\frac{pF}{m} \tag{10.25}$$

10.7 The Simple Line Method

From what we have done so far, we can see that the simplest representation of a two-conductor line will show a series reactance $X_L = 2\pi fL$ per unit length and a shunt reactance:

$$Xc = \frac{1}{2\pi fC} \tag{10.26}$$

Per unit length, where L and C have the values already defined. If we allow for the possibility of losses along the line, then the series impedance $Z = (R + j\omega L)$ should be used, and it will be easier to work with a shunt admittance $Y = j\omega C$ rather than shunt reactance. For lines less than about 150 km long, we can choose one of two approximate representations: either the **Π** equivalent circuit or the ***T*** equivalent circuit. These are illustrated in Figs. 10.8 and 10.9, respectively [9].

In practice, there is little to choose between the two models, although for some applications, one may be better suited than the other.

Now, what we need to do is to develop the models a little bit further so that we end up with the standard four-terminal (or two-port) representation of a transmission line. This will enable us to represent the line by the equivalent circuit model shown in Fig. 10.10 and by the following two equations [10]. Unfortunately, this means a bit of tedious manipulation, but it is worth doing because it will show us

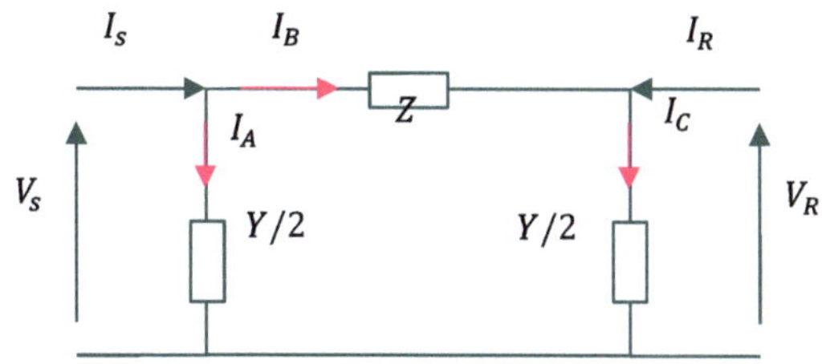

Fig. 10.8 *Π* equivalent circuit model. The diagram illustrates an electrical circuit with labeled components and currents. It features a central impedance labeled "Z" connected between two admittances labeled "Y/2" on either side. Currents are marked with arrows indicating direction. Voltages are shown as vertical arrows at the ends of the circuit

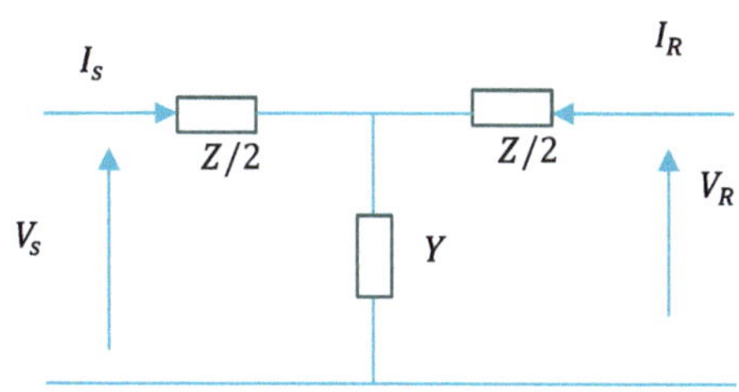

Fig. 10.9 $\mathbb{T}$ equivalent circuit model. The diagram illustrates an electrical network with two impedances labeled "Z/2" connected in series on a horizontal line. Between these impedances, a vertical line connects to an admittance labeled "Y." Arrows labeled "I_S" and "I_R" indicate current flow to the left and right, respectively. Voltage sources "V_S" and "V_R" are shown as upward arrows on the left and right sides

Fig. 10.10 Represent the equivalent circuit model. Diagram of a rectangular block with four labeled points: A, B, C, and D. An arrow labeled "Vs" points toward the left side of the block, and an arrow labeled "VR" points away from the right side. The diagram likely represents a system with input and output variables

exactly how the ***A B C D*** constants are related to the basic parameters of the overhead line (or even underground cable):

$$V_S = AV_R + BI_R$$

$$I_S = CV_R + DI_R$$

Exercise 10.4

Refer back to Fig. 8.8 and try to show that the following equations are valid:

$$V_s = \left[1 + \frac{ZY}{2}\right]V_R + ZI_R$$

$$I_s = \left[Y + \frac{ZY^2}{2}\right]V_R + \left[1 + \frac{ZY}{2}\right]V_R$$

From Fig. 8.8, we have:

$$V_s = ZI_B + \frac{I_C}{Y/2}$$

$$I_B = I_R + I_C$$

$$I_C = \frac{V_R Y}{2}$$

$$\therefore I_B = I_R + \frac{V_R Y}{2}$$

$$V_s = Z\left[I_R + \frac{V_R Y}{2}\right] + V_R$$

or

$$V_s = \left[1 + \frac{ZY}{2}\right]V_R + ZI_R.$$

We also have:

$$I_S = I_A + I_B$$

and

$$V_R = \left[\frac{I_A}{Y/2} - ZI_B\right]$$

or

$$I_A = \left(V_R + ZI_B\right)\frac{Y}{2}$$

$$\therefore I_S = \left(V_R + ZI_B\right)\frac{Y}{2} + I_B$$

or

$$I_S = \frac{V_R Y}{2} + I_B \left[\frac{ZY}{2} + 1 \right]$$

Substituting for I_B gives us:

$$I_S = \frac{V_R Y}{2} + \left[I_R + \frac{V_R Y}{2} \right] \left[\frac{ZY}{2} + 1 \right]$$

or

$$I_S = \left[Y + \frac{ZY^2}{4} \right] V_R + \left[\frac{ZY}{2} + 1 \right] I_R$$

We have shown that the **Π** circuit can be represented by the following two-port matrix equation:

$$\begin{bmatrix} V_S \\ I_S \end{bmatrix} = \begin{bmatrix} \frac{ZY}{2}+1 & Z \\ Y + \frac{ZY^2}{4} & \frac{ZY}{2}+1 \end{bmatrix} \begin{bmatrix} V_R \\ I_R \end{bmatrix}$$

$$\begin{bmatrix} V_S \\ I_S \end{bmatrix} = \begin{bmatrix} A & B \\ C & D \end{bmatrix} \begin{bmatrix} V_R \\ I_R \end{bmatrix}$$

The values ***A B C D*** can be read directly from the matrix:

Exercise 10.5

If you are still feeling strong, show that

$$AD - BC = 1.0$$

The T equivalent circuit gives rise to a similar matrix representation as follows:

$$\begin{bmatrix} V_S \\ I_S \end{bmatrix} = \begin{bmatrix} \frac{ZY}{2}+1 & Z + \frac{ZY^2}{4} \\ Y & \frac{ZY}{2}+1 \end{bmatrix} \begin{bmatrix} V_R \\ I_R \end{bmatrix}$$

Whichever model we use, we can see that the **B** constant is dominated by the series impedance Z, and the C constant is dominated by the shunt admittance Y [11]. The A and D constants are equal in both models and will have a value close to unity.

Note 1: The A B C D constants are complex numbers.

Note 2: The equations for V_s and I_s refer to phase quantities.

Exercise 10.6

An overhead line has a total series impedance:
$Z = (4 + j12)\Omega$ and a shunt admittance of

$$Y = \left(0 + j0.75 \times 10^{-2}\right) S.$$

Determine the A B C D constants for a Π representation of the line.

For the **Π** network

$$A = \frac{ZY}{2} + 1 = 1 + \frac{(4 + j12)\left(0 + j0.75 \times 10^{-2}\right)}{2} = 0.955 + j0.15$$

$$B = Z = 4 + j12$$

$$C = Y + \frac{ZY^2}{4} = -0.00056 + j0.733\,S$$

$$D = A.$$

Exercise 10.7

A balanced three-phase overhead line has constant:

$$A = 0.9\angle 10^\circ$$

and

$$B = 100.\angle 64^\circ$$

The receiving-end line voltage is held at 400 kV. Determine the line voltage at the sending end needed to deliver 900 MW at 0.9 power factor leading at the receiving end.

$$S = \frac{900}{0.9} = 1000\,MVA$$

$$I_R = \frac{1000 \times 10^6}{\sqrt{3} \times 400 \times 10^3} = 1443\angle 26^\circ$$

$$V_s = AV_R + BI_R$$

$$V_s = 0.9 \times \frac{400\angle 10^\circ}{\sqrt{3}} + 1443 \times 100\angle(26+64)^\circ V$$

$$V_s = 204.64 + j36.1 + j144.3 kV$$

$$V_s = 272.8\angle 41.4^\circ kV$$

$$V_{s(Line)} = 472.5\angle 41.4^\circ kV$$

10.8 The Line Performance Chart

As shown earlier

$$\begin{bmatrix} V_S \\ I_S \end{bmatrix} = \begin{bmatrix} A & B \\ C & D \end{bmatrix} \begin{bmatrix} V_R \\ I_R \end{bmatrix}$$

from which

$$V_S = AV_R + BI_R.$$

Taking the receiving-end voltage V_R as the reference phasor, we can draw the phasor diagram of Fig. 10.11 to represent this equation.

In this figure, we have split the receiving-end current I_R into two components: I_{RP} which is in phase with V_R and I_{RQ} which is at a right angle to V_R. I_{RQ} is the quadrature component of the load current, and in this case, it is taken as a leading component of the load current and in this case, it is taken as a leading component, i.e., the receiving-end power factor is assumed to be leading [12].

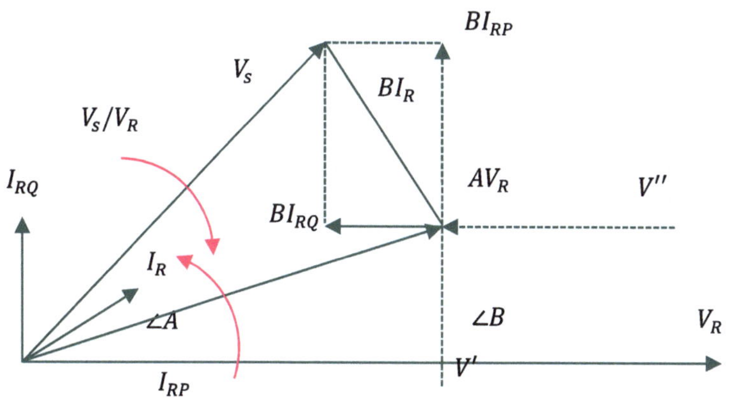

Fig. 10.11 Phasor diagram. A vector diagram illustrating electrical components in a circuit. The diagram includes vectors originating from a common point. Arrows indicate directions and magnitudes of voltages and currents. Angles \(\angle A\) and \(\angle B\) are marked with curved arrows. The diagram uses solid and dashed lines to differentiate between components

The active and reactive power at the receiving end is given by:

$$P = 3|V_R| I_{RP}$$

$$Q_R = -3|V_R| I_{RQ}$$

[Note the negative sign to represent the leading vars.]

Remember from the first chapter that lagging vars are associated with an inductive load in which the current lags the voltage, and by conversation, a positive sign is used. Leading vars are used for capacitive loads in which the current leads the voltage, and in this case, the sign for vars is negative. It can be said that an inductive load absorbs vars and a capacitive load generates vars [13].

From the phasor diagram of Fig. 8.11, it is apparent that, because the B parameter of the line is effectively a constant for any given application, the following relationship applies

$$|B| I_{RP} \propto P_R$$

$$|B| I_{RQ} \propto Q_R$$

We can use this to scale the phasor diagram in units of power. In particular, any voltages, V′ say, along the BI_{RP} axis and V″ along the BI_{RQ} axis, are related to the received power by:

$$\frac{P_R}{V'} = \frac{3|V_R|}{|B|}$$

$$\frac{Q_R}{V''} = \frac{3|V_R|}{|B|}$$

Exercise 10.8

A three-phase transmission line has a two-part B parameter of $50 \angle 70^\circ$ *Ω/phase*. The sending- and receiving-end voltages are held constant at 400 kV. If a line performance chart is now constructed, what will every 10 kV along the V′ and V″ axes actually represent in terms of real power (MW) and reactive power (Mvar).

$$|V_R| = |V_S| = \frac{400}{\sqrt{3}} = 231\,kV$$

Now, for every 10 kV along the BI_{RP} axis on the chart,

$$I_{RP} = \frac{10\times10^3}{|B|} = 200\,A$$

Every 10 kV along with axis therefore corresponds to:

$$3 \times 231 \times 10^3 \times 200 = 138.6\,MW$$

Along the BI_{RQ} axis, every 10 kV corresponds to 138.6 Mvar.

It can be seen therefore that we can scale the BI_{RP} and BI_{RQ} axes to represent real and reactive power at the receiving end of the line. This forms the basis of the line performance chart [14].

A positive value of P_R clearly shows that power is arriving at the receiving end, whereas a negative value of P_R means that power is being sent from the receiving end to the sending end of the line. A negative value of the Q_R means that the reactive power received at the receiving end of the line is leading, and conversely, when BI_{RQ} is positive, the reactive power received is lagging. These two cases correspond to a capacitive load and an inductive load, respectively.

We can now generalize the line chart as shown in Fig. 10.12, which shows the axes for real and reactive with the origin for power measurement at the tip of the AV_R phasor.

Several important features of the active and reactive power transfer characteristics of a transmission line are readily apparent from Fig. 10.12. In practice, we must arrange that the magnitude of the voltages at sending and receiving ends does not deviate significantly from the normal designed levels, since any significant deviation from normal will be reflected down through the network to the power distribution system and will result in a poor quality of supply from the point of view of voltage stability. Normal, if the receiving-end voltage is maintained at its nominal value, then the magnitude of the sending-end voltage should be such that:

$$0.95 < \left|\frac{V_S}{V_R}\right| < 1.05$$

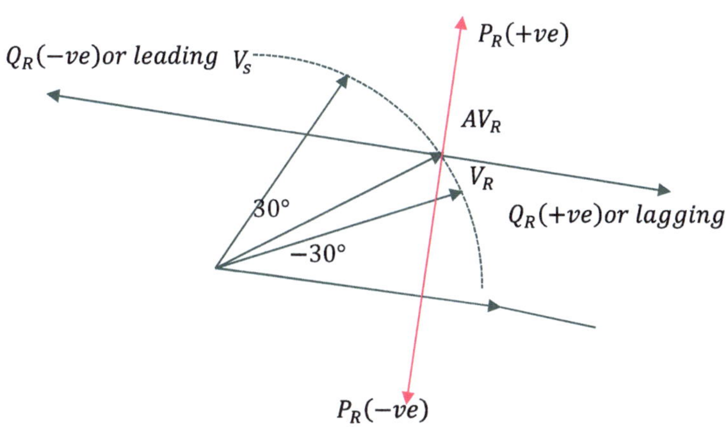

Fig. 10.12 Axes for real and reactive with the origin for power measurement. Diagram illustrating a phasor representation with vectors. The vectors form angles of 30° and – 30° with the horizontal axis. The diagram shows the relationship between active and reactive power components in an electrical system

There is also a requirement to limit the angle between the voltages at each end of the line. This is the load angle of the line, and for stability reasons, this is limited to within ±30.

It is thus clear that, in practice, one is not entirely free to specify any arbitrary power P_R or reactive power Q_R to be transmitted to the receiving end of a transmission line. In fact, it is very evident that, within the typical range of voltages possible, only certain reactive powers can actually be transmitted to the receiving end. In a practical system, these constraints are met by employing special equipment on the system, essentially to control the reactive and active power flows [15].

It should be noted that the reactive power associated with power systems is very closely bound up with the whole matter of voltage control, whereas the active power transmitted is largely dictated by the load angle of the line.

An overhead earth wire can help to improve the capacity of a line to transfer power. It can be shown that the earth wire reduces the effective series impedance of the line, which, of course, means a reduction in the B parameter.

If you think back to the work we did on synchronous source and synchronous loads, you may remember that the maximum power that could be transmitted between two voltages depends inversely on the transfer impedance. Thus, a reduction in B will mean an increase in the maximum possible power transfer [16].

Surge Impedance Loading

This is an interesting case where the load on the line is resistive, i.e., at unity power factor, and equal in value to the characteristic impedance of the line:

$$Z_0 = \sqrt{\left(Z/Y\right)} = \sqrt{\left(L/C\right)}$$

With this loading, there is no voltage drop along the line (assuming that the line is lossless) and there is no reactive power flow along the line.

Example

A balanced three-phase overhead line has constant $A = 0.9 \angle 10°$ and $B = 110 \angle 65°$.

The receiving- and sending-end line voltage is maintained at 325 kV and 385 kV, respectively, to deliver a three-phase load of 1500 and 0.9 power factor leading.

(a) Draw a line chart to represent this condition.
(b) Given that the real power has been reduced to 750 MW from 15,000 and that the terminal voltages remain unchanged, determine the reactive power now received end of the line.
(c) If the total VA at the receiving end were to be reduced to zero and the receiving-end voltage were to be maintained at 325 kV, what value would be sending-end voltage need to be?

We will adopt a slightly different approach to line charts here and construct it as a current diagram rather than a power diagram.

The basic equation is:

$$V_S = AV_R + BI_R$$

so

$$I_R = \frac{V_s}{B} - \frac{AV_R}{B}$$

Allowing for the phase angle, and taking V_R as the reference phasor, we have

$$I_R \angle \varnothing = \frac{|V_S| \angle \delta}{|B| \angle \beta} - \frac{|A| \angle \alpha}{|B| \angle \beta} \times |V_R| \angle 0$$

$$\frac{AV_R}{B} = 2.81 \times 10^3 \angle -55° A$$

$$\frac{V_s}{B} = 3.5 \times 10^3 \angle (\delta - 65)° A$$

$$I = \frac{500 \times 10^6}{500 \times 10^3} = 1.71 \times 10^3 \angle 25.84° A$$

If we choose a scale of 1 $cm = 0.6 \times 10^{-3}$ A, then we can draw the three phasors as in Fig. 10.13.

Line OS = 4.68 cm represents the phasor:

$$\left(\frac{AV_R}{B} \angle -55° A \right)$$

Line OT = 2.85 cm represents the phasor I.

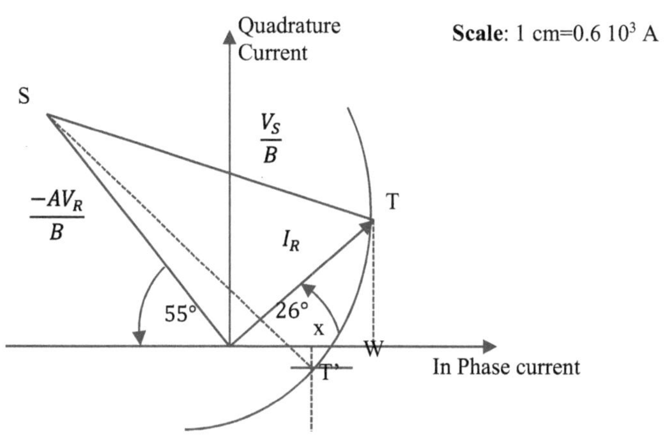

Fig. 10.13 Current phasor diagram. A phasor diagram illustrating electrical currents with labeled vectors and angles. The vertical axis represents quadrature current, and the horizontal axis represents in-phase current. The diagram includes vectors and angles of 55° and 26° marked between the vectors. Points S, T, W, and X are indicated, with a scale of 1 cm equal to 0.6×10^3 A

The line ST then represents the phasor

$$\frac{V_s}{B}.$$

The horizontal intercept *OW* represents the in-phase current.

We now reduce the real power to one-half the original value, so the new in-phase current is represented by *OX where OX* = *OW*/2.

The terminal voltage remains unchanged, so the length of phasors OS and OT cannot change. The angle of

$$\frac{AV_R}{B}$$

cannot change, so the only possible variable is the angle between the phasors, $\angle OST$. The point T can therefore move on a circle of radius ST centered on S. The new operating point must therefore be the point T', which defines the new current XT'.

Thus, $I_Q = XT' = 0.5\ cm = 0.5 \times 0.6 \times 10^3 = 300\ A$:

$$Q = 3 \times 300 \times 325 \times 10^3 = 292.5\,Mvar$$

The power factor is now lagging.

Finally, if we reduce the receiving-end VA to zero, then the current phasor must be zero, so that:

$$\frac{AV_R}{B} = \frac{V_s}{B}$$

or

$$V_s = AV_R = 308.75\angle 10°kV.$$

Self-Assessment Questions

? Self-assessment question 8.1

An overhead line has a total series impedance *Z of* $(5 + j13)\Omega$ and a shunt admittance of $(j0.75 \times 10^{-2})s$:

Determine:

(a) the approximate ***A, B, C, D*** constants of the line based upon **Π** and **T** representations; and
(b) the ***A, B, C, D*** constants of the line when its capacitance is neglected.

Determine also the magnitude and phase of the sending-end voltage relative to the receiving-end voltage when a load of 30 MW at 0.85 power factor lagging is supplied. The receiving-end line voltage is 132 kV. Compare the results obtained by using the approximate equivalent circuits of the line and comment on the comparison. The given parameters are consistent with a line length of approximately 30 km.

? Self-assessment question 8.2

Outline the basis of a line performance chart for an overhead line.

A balanced three-phase overhead line has constants $A = 0.95 \angle 10°$ and $B = 110 \angle 65°$ if the receiving-end phase voltage is held constant at 325 kV, determine graphically or otherwise:

(a) The sending-end phase voltage necessary to deliver a load of 1500 MW at 0.9 power factor leading at the receiving end.
(b) The maximum power that can be transmitted if the sending-end voltage is held constant at the value determined in part (a).

? Self-assessment question 8.3

(c) An overhead transmission line consists of two long, parallel conductors, each of radius r, separated by a distance D. The conductors form a go-and-return circuit.

Starting with the magnetic circuital law, show that the self-inductance of the line is given by:

$$L = \frac{\mu_0}{\pi} In\left(\frac{D-r}{r}\right) Hm^{-1}$$

where $\mu_0 = 4\pi \times 10^{-7} Hm^{-1}$ is the permittivity of the free space.

(d) A four-wire three-phase, 50-Hz overhead line ***ABCN*** is represented in Fig. 10.14 with separation as shown. A communication line DE runs parallel to the three-phase conductor ***A,B,and C.***

A third harmonic current of 100 A **r.m.s** flows in the transmission line. Calculate the voltage induced in the communication line per kilometer length.

$$D_{AB} = D_{BC} = D_{BN} = 1.2$$

$$D_{CD} = 1m$$

$$D_{DE} = 0.5m$$

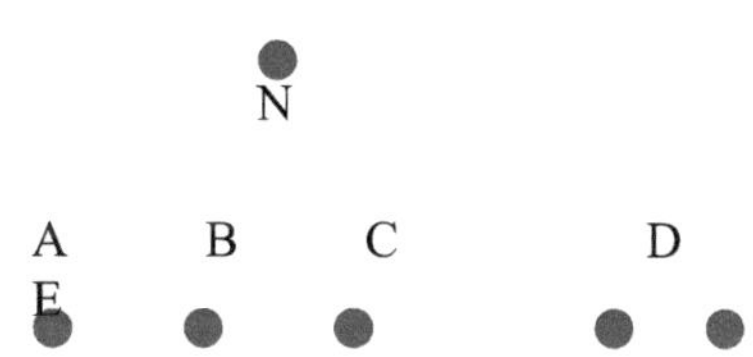

Fig. 10.14 A four-wire three-phase, 50-Hz overhead line ABCN is represented in Fig. 10.14 with separation as shown. A communication line DE runs parallel to the threephase conductor A, B, and C.

Self-assessment question 8.4

Conditions on a long balanced three-phase overhead transmission line are described by the following equation:

$$V_1 = AV_2 + BI_2$$

$$A = 0.95\angle 10°\text{and} = 110\angle 65°\Omega$$

The phase voltage at end 3 is held constant at 232 kV, and the following operating conditions are applied:

(a) The phase voltage at the sending-end V1 is constrained so that

$$0.95 < \left|\frac{V_1}{V_2}\right| < 1.05$$

(b) The phase angel, δ, between V_1 and V_2, is constrained so that

$$-30° < \delta < 30°.$$

Determine graphically or otherwise the limiting values of reactive power that can be transferred by the line to the receiving end under all allowed operating conditions.

Bibliography

1. L. Ndahepele; S. Chowdhurym "Impact of Distributed Generation on Traditional Protection in Distribution and Transmission Systems: A Review" Published in IEEE PES/IAS PowerAfrica in 2020
2. Cleberton Reiz; Jonatas Boas Leit "Impact Analysis of Distributed Generation on Protection Devices Coordination in Power Distribution Systems" 2021 IEEE PES Innovative Smart Grid Technologies Conference - Latin America (ISGT Latin America) in 2021
3. Jane Courtney; Aaron McDonnell "Impact on Distribution System Protection with the Integration of EG on the Distribution Network" published in 54th International Universities Power Engineering Conference (UPEC) in 2019
4. Lü Zhining; Ning Baifeng; Luo Weifeng; Liu Wei "Distribution networks ground fault protection method based on fuzzy measure fusion criterion"published in IEEE 8th International Conference on Advanced Power System Automation and Protection (APAP) in 2019
5. Luyao Pei; Cheng Xu; Jinsong Zhang; Bo Wang; Yining Ma "Artificial Intelligence Algorithm in Intelligent Power Supply and Distribution Network Design System published in IEEE 3rd International Conference on Electronic Technology, Communication and Information (ICETCI) in 2023
6. Seongil Kim; Byungchul Sung; Jeong-Hyeok Heo; Duck-Su Lee; Protection against Transient Overvoltages Caused by Intermittent Earth Faults in Unearthed Marine Power Distribution Network" published in IEEE International Conference on Power Systems Technology (POWERCON) in 2022
7. Yu Deng; Dingkang Liang; Jiahao Wang; Xinyi Song; Zehua Li; Zhetong Hong "Reliability Evaluation of Distribution System Based on Dynamic Bayesian Network" Published in 3rd Power System and Green Energy Conference (PSGEC) in 2023
8. Qin Lijun; Wang Ying; Zheng Shenrui; Guo Qing " Situation of study on distribution network protection containing distributed generation and application scheme of intelligent protection"

2011 4th International Conference on Electric Utility Deregulation and Restructuring and Power Technologies (DRPT)
9. Vassilis C. Nikolaidis; Evangelos Papanikolaou; Anastasia S. Safigianni "A Communication-Assisted Overcurrent Protection Scheme for Radial Distribution Systems With Distributed Generation" IEEE Transactions on Smart Grid
10. Yuwei Xiong; Wei Chen "Positive sequence current differential protection based on least square method for distribution network with distribution generation" 2021 IEEE 4th International Electrical and Energy Conference (CIEEC).
11. Zong Guo; Jiarui Li; Hongbo Lian; Shihua Zhao; Yang Chen; Renjie Dai; Xiuneng Xu; Yufeng Gu; Yangyang Liu "Optimization Method for Protection of Distribution Network with MW-PV Based on Multi-level Cooperative Protection Strategy" 2019 IEEE Sustainable Power and Energy Conference (iSPEC).
12. Rani Kumari; Bhukya K. Naick "Protection Coordination Impact on Reliability of Distribution Systems With Distributed Energy Resources" 2024 Third International Conference on Power, Control and Computing Technologies (ICPC2T).
13. Hana A. Halim; Melaty Amirruddin; Nur S. Noorpi; Nurhakimah M. Mukhtar "An Improved Protection Scheme for Smart Distribution Grid published in 2013 1st International Conference on Artificial Intelligence, Modelling and Simulation.
14. Jamil Abdo & Tariq Masood "New Technique to develop Intelligent, highly reliable & Efficient Unmanned Power Generation Modular" 22nd International Conference on Renewable Energies and Power Quality (ICREPQ'24) Bilbao (Spain), 26th to 28th June 2024. https://www.icrepq.com/icrepq24/392-24-abdo.pdf
15. Jamil Abdo & Tariq Masood "Operational Reliability- Production Best Practice Pragmatic Case Studies & Improvement" published in Proceedings of the 9th North American Conference on Industrial Engineering and Operations Management Washington D.C., USA, June 4–6, 2020
16. Tariq Masood & Jamil Abdo & "Maintenance Strategy Production Best Practices—Pragmatic Case Studies and Operations Improvement" published in Proceedings of the 9th North American Conference on Industrial Engineering and Operations Management Washington D.C., USA, June 4–6, 2020

Artificial Intelligence Industrial Application

11

Contents

11.1 Thinking Probabilistic 268
11.2 AI Machine Power and Disruption 270
11.3 Designing Reliable Systems 270
11.4 Embracing Uncertainty 273
11.5 AI System Prediction Technology 273
11.6 AI Prediction Machine Rules 274
11.7 AI Model Architecture 275
11.8 AI Models Tutorials 275
  11.8.1 AI Industrial Assessment Models 275
  11.8.2 Introduction to Machine Learning in Microgrid Operations 280
  11.8.3 Data Analytics in Microgrid Control 282
  11.8.4 Machine Learning Applications in Microgrid Control 283
  11.8.5 Case Studies and Practical Implementations 284
11.9 Challenges and Future Directions 286
11.10 Conclusion 288
Self-Assessment Questions 291
Bibliography 292

Abstract

At the end of this chapter, you will have been equipped with the knowledge of the general principles of AI components used for industrial applications. This includes probabilistic thinking, AI machine power and disruption, designing reliable systems, AI system reliable technology, AI machine prediction rules, machine learning in microgrid, **tutorial on simulating probability AI models**,

---

Supplementary Information The online version contains supplementary material available at https://doi.org/10.1007/978-3-031-94312-6_11.

T. Masood et al., *Protection of Electrical Power Transmission Systems*, Power Systems, https://doi.org/10.1007/978-3-031-94312-6_11

and the **challenges and future direction of AI in smart and microgrid operations and control.** This knowledge will empower you to apply these principles in your professional endeavors.

Objectives

By the end of this section, you will gain invaluable insights into the essential principles of AI components that are revolutionizing industrial applications. Discover how to embrace:

- Probabilistic thinking for informed decision-making.
- The power of AI machines and their potential to disrupt industries.
- Strategies for designing reliable systems that enhance efficiency.
- The importance of reliable technologies in AI systems.
- Effective AI machine prediction rules to improve accuracy.
- The role of machine learning (ML) in optimizing microgrids.
- A comprehensive tutorial on simulating AI models through probability.
- The challenges and future directions that will shape the industry.

This knowledge will empower you to leverage AI effectively and position yourself at the forefront of industrial innovation.

11.1 Thinking Probabilistic

Artificial intelligence (AI) has fundamentally transformed how we approach system analysis by introducing probabilistic thinking. In investigations of incidents like car accidents or industrial mishaps, it is essential to ask whether the Panelboard Operator or protection/control engineers witnessed the event. While we generally expect a clear "yes" or "no" response, AI provides more nuanced insights, indicating partial observations such as "sort of" or "a bit." For example, AI may identify a 0.01% likelihood of detecting a potential human presence preparing to act.

The integration of AI shifts systems from a deterministic framework to a probabilistic one. In many instances, existing systems are not fully equipped to handle this transition; thus, there is a critical opportunity to redesign systems to significantly enhance productivity.

To fully harness the potential of AI, many industrial applications must adopt new system-level solutions that integrate prediction, judgment, and regulatory functions. These elements are not just beneficial; they are indispensable for ensuring seamless operations during the shift from deterministic to probabilistic systems in the contemporary landscape of AI integration. Prediction and judgment are the cornerstones of effective decision-making. In a decision tree, prediction quantifies the likelihood of each potential outcome, while judgment evaluates the associated payoffs. Both components—prediction and judgment—are crucial for informed decision-making, as illustrated in Fig. 11.1.

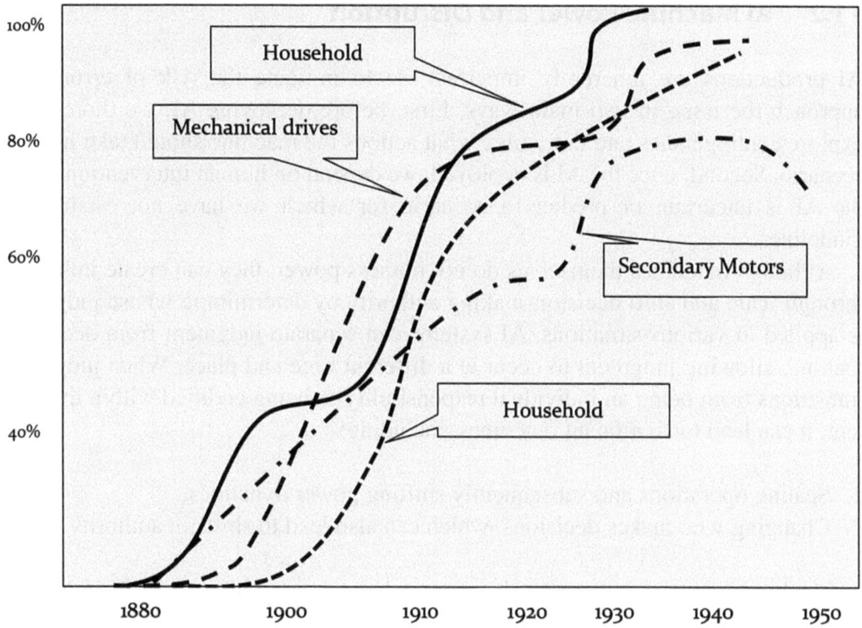

Fig. 11.1 (Modern productivity paradox) Line graph showing the adoption rates of household electricity, mechanical drives, and secondary motors from 1880 to 1950. The y-axis represents percentage adoption from 0% to 100%, and the x-axis shows years. Household electricity adoption rises sharply after 1900, reaching nearly 100% by 1950. Mechanical drives and secondary motors show slower adoption rates, with mechanical drives peaking around 80% and secondary motors around 60% by 1950

Despite the challenges inherent in implementing system-level innovations with AI, it is vital to initiate this process as soon as possible. The sooner we begin, the more AI learns, and subsequently, the more accurate and influential these systems will become. Feedback control loops are instrumental in shaping effective system designs. For AI to learn efficiently, it must have constant access to outcome data, which informs frequent feedback mechanisms.

Although machines do not make decisions independently, AI can effectively simulate decision-making processes. The judgments generated by AI guide actions, but we must acknowledge that its predictions are inherently imperfect. To mitigate risks associated with errors, we rigorously establish two lines of defense. First, we engage in thorough contingency planning before deploying AI, clearly defining the appropriate actions for various scenarios. Second, once deployed, human intervention is crucial when AI lacks high confidence in its predictions or encounters situations without predefined judgments. This "human-in-the-loop" approach underscores the essential role of human oversight in AI decision-making.

Exercise 11.1

Describe the power of AI industrial applications how to shift to newly designed system-level solutions

11.2 AI Machine Power and Disruption

AI productions are inherently imperfect, so to mitigate the risk of errors, we approach the issue in two main ways. First, before deploying AI, we thoroughly explore contingencies and determine what actions the machine should take in each scenario. Second, once the AI is deployed, we depend on human intervention when the AI is uncertain or predicts a scenario for which we have not established guidelines.

Although machines themselves do not possess power, they can create influence through scale and shift decision-making authority by determining whose judgment is applied in various situations. AI systems can separate judgment from decision-making, allowing judgment to occur at a different time and place. When judgment transitions from being an individual responsibility to being codified within the system, it can lead to significant outcomes, including:

1. Scaling operations and subsequently shifting power dynamics.
2. Changing who makes decisions, which can also lead to shifts in authority.

Machines can seem to make decisions when we are able to codify judgment effectively. The AI generates predictions and draws on human judgment to take appropriate actions. Despite the challenges associated with implementing system-level innovations with AI, there is a compelling reason to initiate these efforts sooner rather than later. AI provides a competitive advantage for first movers because it learns over time. The sooner it is deployed, the faster it can start learning; the more it learns, the better it becomes, thus making the new system more effective, as illustrated in Figs. 11.2 and 11.3.

Exercise 11.2

Describe the systems dedicated to AI which can decouple the judgment from decisions such that it can be provided at a different time and place if judgment shifts from individually deployed by people for each decision to instate codify into systems, then this can lead to number.

11.3 Designing Reliable Systems

Grasping the two main system designs and approaches is not merely important; it is essential for tackling the reduced reliability that can arise from the integration of AI-based decision-making. We cannot underestimate the weight of this responsibility.

- **Coordination and Modularity**

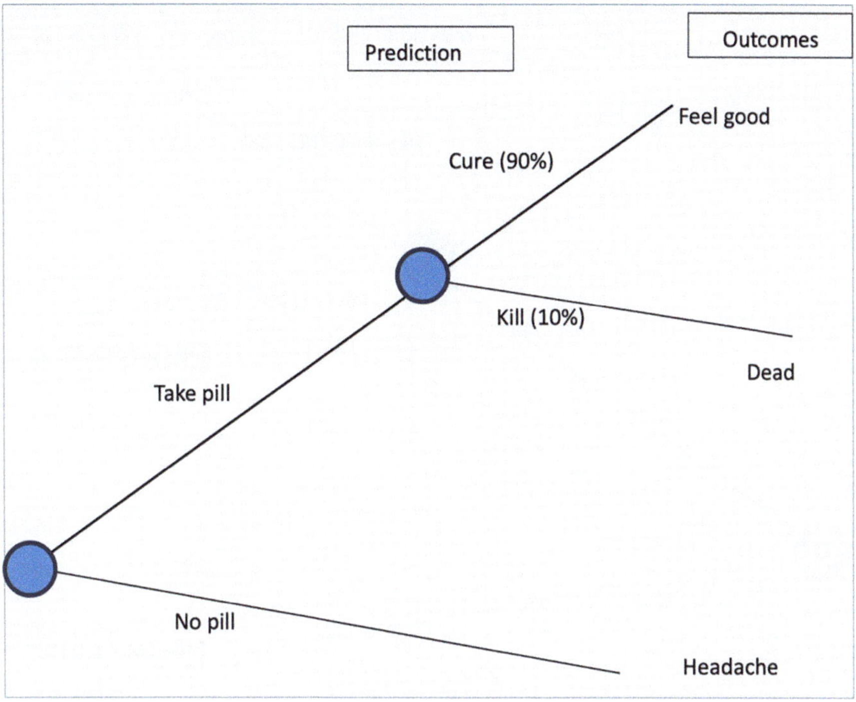

Fig. 11.2 (Pills decision tree) A decision tree diagram illustrating the outcomes of taking or not taking a pill. The "Take pill" branch leads to a prediction node with two outcomes: "Cure" with a 90% chance resulting in "Feel good," and "Kill" with a 10% chance resulting in "Dead." The "No pill" branch leads directly to "Headache." The diagram highlights the probabilities and potential consequences of each decision

Modularity plays a pivotal role in creating boundaries around AI, enabling enhanced decision-making while minimizing the drawbacks linked to misalignment between AI systems and other organizational components. Though modularity reduces coordination costs, it may sacrifice valuable synergies. Systems comprise an intricate web of decisions that interact with one another. For instance, with three binary choices, you end up with eight combinations; with ten interacting decisions, that number skyrockets to 1024 combinations, and with 20 decisions, it escalates to an astonishing 1,048,576 combinations. This illustrates how quickly systems with independent choices can become complex, underscoring the necessity for the tools and approaches we are discussing. Simulation emerges as a powerful asset in system design. By utilizing digital twins to mimic various combinations, we can harness AI to predict outcomes for each scenario, paving the way for an exciting future in system design.

Remember that decisions are never made in isolation. The results of individual decisions often ripple out, influencing subsequent choices and actions. This is why we opt for predetermined rules instead of real-time decisions at times: rules increase

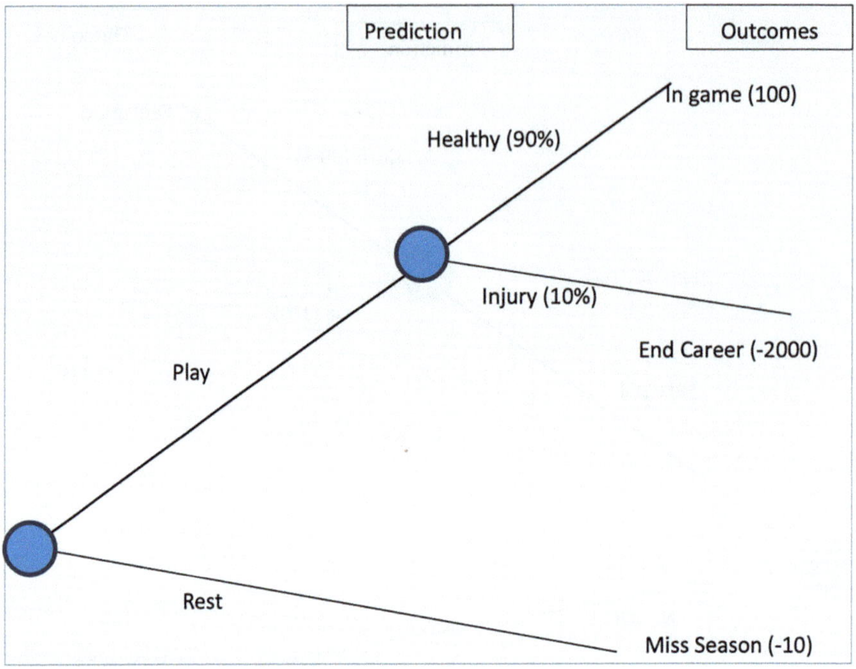

Fig. 11.3 (Play versus rest decision tree) A decision tree diagram illustrating predictions and outcomes for a player's decision to play or rest. The "Play" branch predicts a 90% chance of staying healthy, leading to an "In game" outcome with a value of 100, and a 10% chance of injury, leading to an "End Career" outcome with a value of −2000. The "Rest" branch leads to a "Miss Season" outcome with a value of −10

reliability. In this context, we might accept less-than-ideal localized choices in exchange for enhanced overall system reliability.

Reliability stands out as a critical characteristic of systems with interdependent decisions, highlighting the importance of your role in system design. Your expertise is invaluable to our success, and we want you to know how vital you are to this process. Your contributions are not simply important; they are fundamental to our collective achievements.

Exercise 11.3

Describe how to use digital twins to simulate different combinations and use AI to predict outcomes from each combination, offering a promising and exciting future for system design.

11.4 Embracing Uncertainty

Thinking in bets invites us to confront the inherent uncertainty that surrounds our decision-making processes, urging us to methodically evaluate the likelihood of various outcomes. When we identify an outcome that presents a considerable chance of occurrence, we can align our choices with that option, embracing a calculated risk. Conversely, when the probability appears low, we are prompted to explore alternative paths that may yield more favorable results. This mindset allows us to establish a robust decision-making framework characterized by clearly defined thresholds that embody our predictive assessments. The true power of this practical strategy becomes especially apparent when our forecasts turn out to be accurate, emphasizing the pivotal role that precise predictions play in steering our choices.

Now, consider the intricate and emotionally charged scenario of assessing refugees for entry into a host country—a decision that is imbued with layers of uncertainty and moral complexity. The concept of thinking in bets comes alive through the poignant personal experiences shared by refugee claimants. Their testimonies often reveal the delicate balance of risks and potential outcomes, illuminating a process that profoundly impacts individual lives while also influencing societal dynamics. By carefully weighing each personal account and reflecting on the varying levels of confidence involved in the decision-making process, we can cultivate a deeper understanding of the broader ramifications of our choices in such sensitive circumstances (Fig. 11.4).

11.5 AI System Prediction Technology

One of the most crucial aspects of AI production is the distinction between correlations and causation. While correlations can be sufficient for certain applications, when it comes to AI informing causal relationships, we must rely on randomized experiments to gather the necessary operational data from power utility plants. These experiments are the gold standard in statistics, enabling us to accurately determine cause-and-effect relationships.

Imagine the potential of the power utility operations recommendation engine in the production machine. What if it could significantly enhance its accuracy? Initially, it would excel at recommending various operational scenarios. But what if it reached a point where it could predict load with such precision that it could anticipate power utilities' needs for power quality and delivery? This could revolutionize the way we operate, shifting from a reactive to a proactive approach.

The AI that offers superior recommendations on the current platform is a game-changer. However, implementing this new model would necessitate a comprehensive redesign of our systems, particularly in how we manage the EMS (energy management system) supply and demand and its critical operating parameters. This

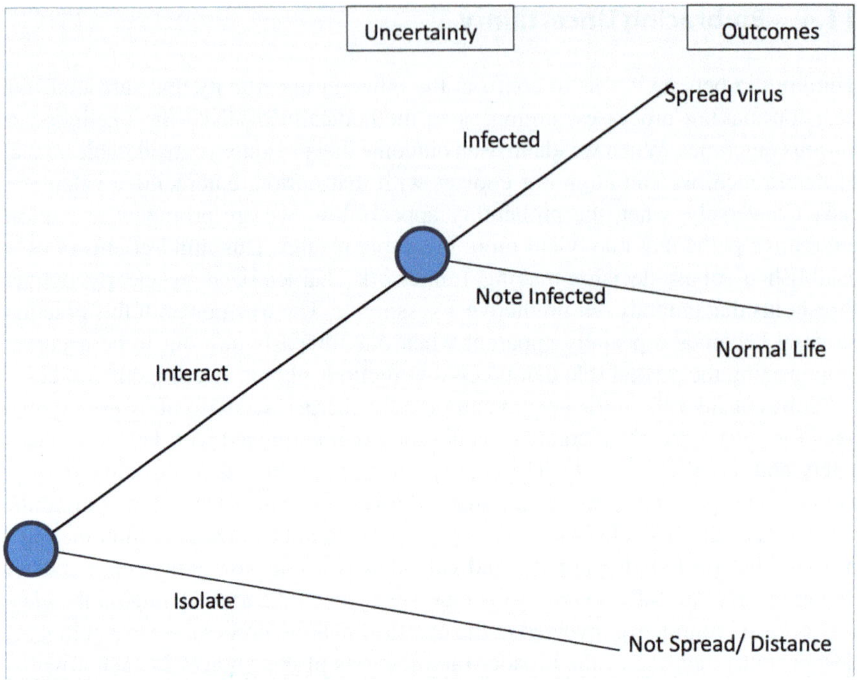

Fig. 11.4 (Isolate or interact decision tree) Flowchart illustrating decision paths related to virus spread. The starting point shows two options: "Interact" or "Isolate." Interacting leads to "Uncertainty" with outcomes of "Infected" resulting in "Spread virus" or "Not Infected" leading to "Normal Life." Isolating results in "Not Spread/Distance"

redesign is not just a suggestion, but an urgent necessity to keep pace with the rapidly evolving landscape of power utility operations.

Exercise 11.4
Describe the AI system and prediction technology.

11.6 AI Prediction Machine Rules

Turning rules into actionable decisions is essential for effectively leveraging predictive technologies. Yet, for such systems to thrive, they must be adaptable. The interconnection of rules often leads to complexities, making decision-making seem daunting within these frameworks.

Consider a compelling example related to COVID-19 where we developed a streamlined oil system involving 12 major companies. In this scenario, the CEO's leadership was instrumental in guiding senior teams on managing employee infectiousness through rapid antigen testing. This proactive approach not only allowed the 12 companies to maintain operations in a challenging environment but also prevented a potential deadlock within their existing system. The success of this initiative inspired over 2000 additional organizations to embrace a shift from rigid rules to flexible decisions, highlighting the critical role of strong leadership in navigating such transitions [1].

Exercise 11.5
Describe AI system prediction machine rules and prediction technology.

11.7 AI Model Architecture

Matrikon Process-NET is a powerful solution that efficiently gathers data from all production stations at Qatar Energy fields, including ABB-DCS, Honeywell-DCS, SCADA, SOLAR, Bentley Nevada, and Allen Bradley. By centralizing this data, it transforms information management, allowing users to present vital insights in various formats using Internet Explorer. This innovative platform not only streamlines operational activities but also harnesses the power of AI models to enhance decision-making and performance, as demonstrated in Fig. 11.5.

11.8 AI Models Tutorials

11.8.1 AI Industrial Assessment Models

Advanced Process AI: Machine/deep learning models are constituted by incorporating **(real-time and big historical data) with three layers of** supervised, unsupervised, and enforced forms of data. These layers represent different stages of data processing and model training. The supervised layer involves training the model on labeled data, the unsupervised layer deals with unstructured data to identify patterns, and the enforced layer ensures the model adheres to predefined rules and regulations.

Machine/deep learning is unconventional programming to develop different logic (through linear, non-linear performance equations, and neural networks). To resolve complex engineering problems, equipment design issues, and operational bottlenecks, trends, and patterns to predict current and potential downsides and deliver as per business needs in compliance with health, safety, and environment with a degree of precision, they developed and customized real-time actionable production dashboards and equipment performance AI models to optimize, predict, and plan seamless operations and to maximize production and uptime of the equipment to comply with product quality, health, safety, and environmental regulations, in line with consent-to-operate (CTO) government agency set legislation.

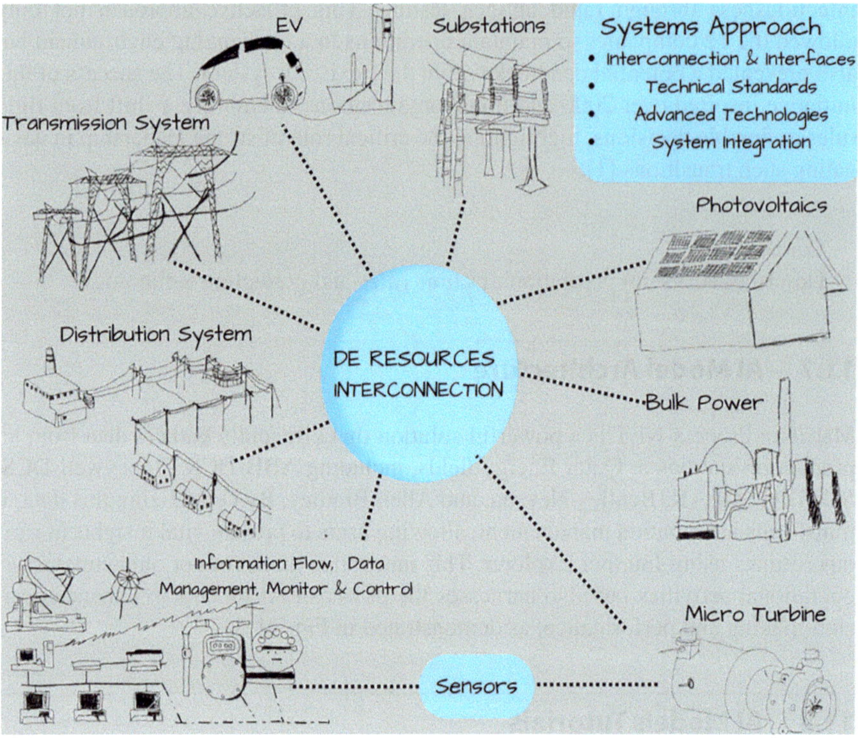

Fig. 11.5 (AI-Virtual schematic to collect and compile daily operations data) Diagram illustrating the interconnection of distributed energy resources. The central blue circle labeled "DE Resources Interconnection" connects to various components: transmission system, distribution system, substations, photovoltaics, bulk power, microturbine, and sensors. An electric vehicle (EV) is shown near the substations

The following are the data collection points; this is an essential fuel for AI models:

1. Total tags: 50,000 (as per attached Big Data CSV file).
2. Plants: K.Main, North and South, F.Main, North, South, Jaleha Main, Diyab.
3. 24 Injection manifolds
4. 600 Gas lift injection wells
5. 700 Production wells
6. 500 Power water injection wells
7. 150 Geothermal wells/renewable energy.

- Intelligent actionable production dashboard (achieve zero trip incidents, to maximize the uptime of the different plant operations) (Figs. 11.6 and 11.7).
- Intelligent Actionable Daily Production Reports to Surveillance and monitor daily operational activities (achieve set production target, with minimum operator intervention).

Fig. 11.6 (Actionable maintenance and operational reliability dashboard) The image displays a business performance dashboard with various metrics and a safety pyramid. Key performance indicators include production at 45,469 BOD, free cash flow at 119,247 QAR millions, national oil and gas demand at −5%, projects at 156 QAR millions, Qatarization at 6%, and health and safety with a lost time injury rate (LTIR) of 3. The safety pyramid for Q3 2015 shows no fatalities, 1 restricted work case, 1 medical treatment, 9 first aid cases, 5520 unsafe observations, and 72,630 total observations. A summary compares end Q2 2015 and end Q3 2015, highlighting a decrease in total observations from 334,869 to 72,630

- Intelligent flare surveillance models (to monitor and measure each molecule of hydrocarbon released during normal or abnormal operations without any measuring device).
- Intelligent actionable CPM (controller performance monitoring, enhance equipment performance to achieve zero trip incidents) (Figs. 11.8 and 11.9).

Advanced process AI leverages machine and deep learning models that integrate real-time and extensive historical data through three critical layers: supervised, unsupervised, and enforced data processing. Each layer plays a vital role in data processing and model training:

- **Supervised Layer**: This layer is dedicated to training the model using labeled data, ensuring precision in outputs.
- **Unsupervised Layer**: It tackles unstructured data to uncover critical patterns that drive insights.
- **Enforced Layer**: This layer guarantees that the model adheres strictly to established rules and regulations.

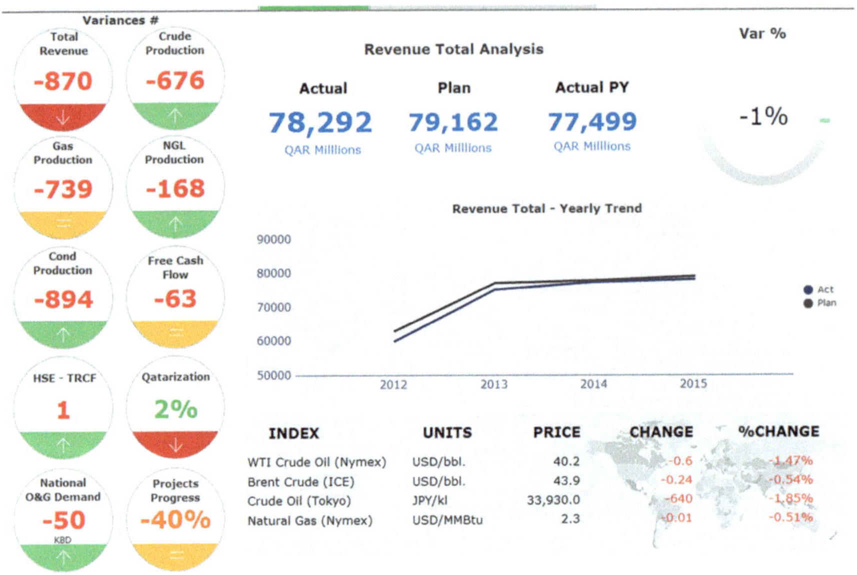

Fig. 11.7 (Actionable production dashboard-AI generated) The image is a financial dashboard displaying revenue analysis and various metrics. At the top, variances are shown with colored circles indicating performance: Total Revenue (−870), Crude Production (−676), Gas Production (−739), NGL Production (−168), Cond Production (−894), Free Cash Flow (−63), HSE–TRCF (1), and Qatarization (2%)

Machine and deep learning represent groundbreaking programming techniques that develop robust logical frameworks, including both linear and non-linear performance equations, as well as neural networks. These techniques are essential for overcoming complex engineering challenges, addressing equipment design issues, and eliminating operational bottlenecks. By identifying trends and patterns, we predict current and potential problems, delivering solutions that decisively meet business needs while ensuring full compliance with health, safety, and environmental standards.

Our commitment to operational excellence has led to the development of customized real-time actionable production dashboards and equipment performance AI models. These tools are designed to optimize, predict, and plan operations, maximizing production and equipment uptime while consistently maintaining product quality in alignment with government legislation (CTO—Consent to Operate). Data Collection Points: This information is indispensable for AI models and includes:

- Total tags: 50,000 (as per attached Big Data CSV file).
- Plants: K.Main, North and South, F.Main, North, South, Jaleha Main, Diyab.
- 24 Injection manifolds
- 600 Gas lift injection wells
- 700 Production wells

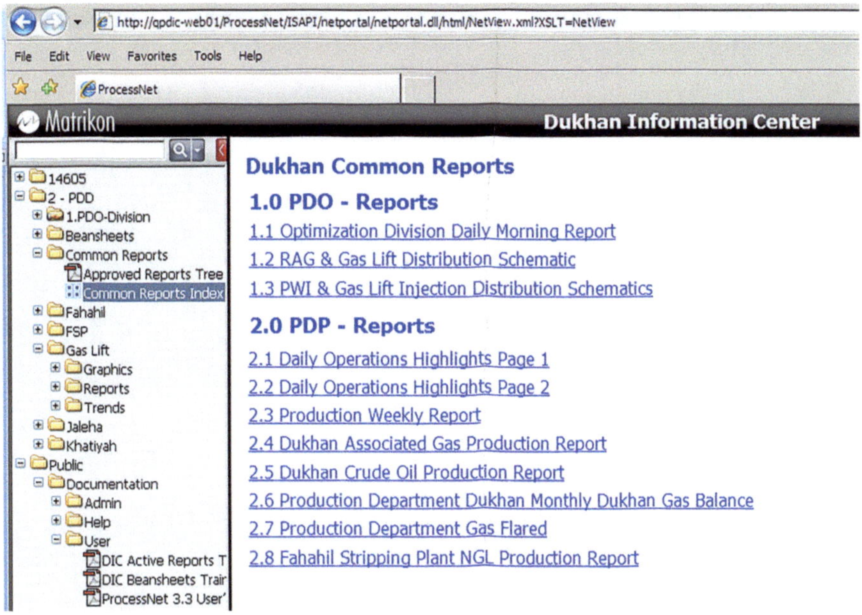

Fig. 11.8 (Actionable daily production reports senior management level-AI-based) A computer screen displaying the "Dukhan Information Center" interface with a navigation tree on the left and a list of reports on the right. Featuring links to various reports like "Optimization Division Daily Morning Report" and "Production Weekly Report"

- 500 Water injection wells
- 150 Geothermal wells/renewable energy sources.

Key Components

1. Intelligent actionable production dashboard: We are committed to achieving zero trip incidents and maximizing uptime across various plant operations.
2. Intelligent actionable daily production reports: These reports empower surveillance and monitoring of daily operational activities, ensuring that we meet production targets with minimal operator intervention.
3. Intelligent flare surveillance models: Our models accurately monitor and measure hydrocarbon releases during both normal and abnormal operations, eliminating the need for physical measuring devices.
4. Intelligent actionable CPM (Controller Performance Monitoring): This tool is pivotal in enhancing equipment performance, driving us toward zero trip incidents.

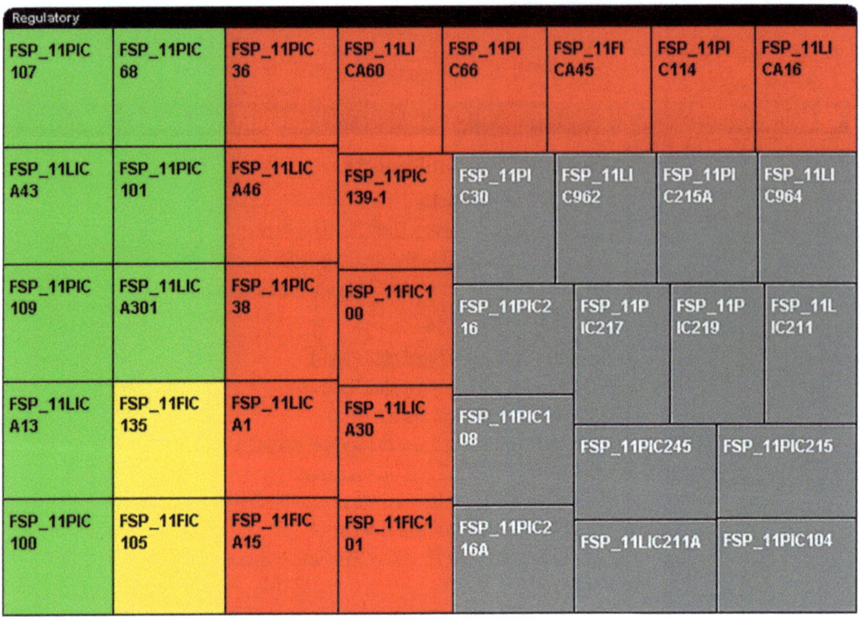

Fig. 11.9 (Equipment performance enhancement models—AI-based) A grid chart titled "Regulatory" with colored sections indicating different statuses. Green sections include codes such as FSP_11PIC 107 and FSP_11PIC 60, suggesting a positive status. Yellow sections, such as FSP_11PIC 109 and FSP_11PIC 100, indicate caution. Red sections, including FSP_11PIC 36 and FSP_11PIC A1, denote critical status. Gray sections, such as FSP_11PIC30 and FSP_11PIC215A, represent inactive or neutral status

11.8.2 Introduction to Machine Learning in Microgrid Operations

Machine learning (ML) represents a significant advancement in the field of microgrid operations, offering a range of techniques that enhance decision-making, automation, and optimization within complex energy systems. ML algorithms enable microgrids to handle dynamic environments where energy production and consumption are continuously fluctuating.

- **Supervised Learning:** Supervised learning, which uses labeled datasets to train models, is pivotal in predicting energy consumption, identifying equipment failures, and optimizing grid operations. For instance, regression models can predict future energy demand based on historical consumption data and external factors like weather conditions. Classification algorithms, such as decision trees and support vector machines, can categorize different types of grid disturbances, enabling faster and more accurate responses.
- **Unsupervised Learning:** Unsupervised learning does not rely on labeled data and is essential for uncovering hidden patterns in energy data. Techniques like clustering can group similar consumption patterns, helping to identify normal and abnormal behaviors in the grid. This is particularly useful for anomaly detec-

tion, where deviations from the norm can indicate potential issues such as equipment malfunctions or security breaches.

- **Reinforcement Learning:** Reinforcement learning is especially relevant for adaptive control systems in microgrids. In this approach, the system learns optimal strategies through trial and error, constantly improving its decisions based on real-time feedback. For example, reinforcement learning can be used to optimize the dispatch of energy storage systems, ensuring that energy is stored during periods of low demand and released during peak times to balance the grid.
- **Challenges and Benefits:** While ML offers significant advantages, such as improved accuracy in predictions and automation of complex processes, there are challenges to its implementation. These include the need for large and high-quality datasets, the computational resources required for training complex models, and the potential risks of overfitting, where models perform well on training data but fail to generalize to new data. Additionally, the interpretability of ML models is crucial, particularly in critical applications like energy management, where decisions must be transparent and justifiable.

In summary, ML is a powerful tool that enhances the efficiency, reliability, and resilience of microgrid operations. By integrating ML into energy systems, microgrids can become more adaptable and capable of handling the complexities of modern energy distribution and consumption.

Machine Learning (ML) marks a transformative leap forward in the realm of microgrid operations, offering an array of sophisticated techniques that fundamentally enhance decision-making, automation, and optimization in intricate energy systems. ML algorithms equip microgrids with the ability to adeptly navigate dynamic environments, where fluctuations in energy production and consumption occur continuously [2].

Supervised Learning At the core of ML, supervised learning serves as a crucial methodology that employs labeled datasets to train predictive models. This technique plays a vital role in forecasting energy consumption patterns, diagnosing equipment failures, and optimizing the overall operations of microgrids. For instance, regression models leverage historical energy consumption data along with various external factors, such as weather conditions, to accurately project future energy demand. Furthermore, classification algorithms such as decision trees and support vector machines categorize different types of grid disturbances, enabling utilities to respond to challenges with remarkable speed and precision.

Unsupervised Learning Unsupervised learning unveils hidden patterns within energy data, functioning without the need for labeled inputs. This method is instrumental in analyzing consumption behaviors, where techniques such as clustering group similar patterns together. By identifying both standard and anomalous behaviors within the grid, utilities can effectively pinpoint deviations that may indicate potential issues, including equipment malfunctions or security threats. This capability is crucial for maintaining the integrity and reliability of microgrid operations.

Reinforcement Learning Reinforcement learning represents a dynamic approach utilized in adaptive control systems for microgrids. This technique allows systems to discover optimal strategies through a process of trial and error, fostering continuous improvement in decision-making based on real-time feedback. For example, reinforcement learning can be employed to refine the dispatch of energy storage systems. This ensures that energy is efficiently stored during periods of low demand and judiciously released during peak consumption times, thereby maintaining a balanced and stable grid.

Challenges and Benefits While the advantages of ML are significant—ranging from improved accuracy in predictions to the automation of complex processes—there are also notable challenges to consider. The successful implementation of ML often hinges on access to large, high-quality datasets and the computational resources required to train intricate models. Moreover, there exists the risk of overfitting, where models may perform exceptionally on training data while failing to generalize effectively to new, unseen data. Additionally, the transparency of ML models is essential, particularly in critical applications such as energy management, where decision-making must be justifiable and comprehensible.

In summary, ML is not merely an enhancement; it is a cornerstone technology that greatly amplifies the efficiency, reliability, and resilience of microgrid operations. By weaving ML into the fabric of energy systems, microgrids can become more agile and adept at addressing the intricate challenges of modern energy distribution and consumption. These advancements pave the way for smarter, more responsive, and sustainable energy solutions.

Exercise 11.6
Describe the challenges and benefits of the AI system

11.8.3 Data Analytics in Microgrid Control

Data analytics plays a crucial role in microgrid control by enabling the analysis of vast amounts of data generated by energy production, consumption, and grid performance. This data-driven approach allows grid operators to make informed decisions that optimize energy distribution, enhance grid stability, and improve overall efficiency.

- **Energy Production Data:** Data from renewable energy sources, such as solar panels and wind turbines, is often highly variable. Analyzing this data allows operators to predict production levels based on factors such as weather forecasts and historical trends. This predictive capability is essential for planning energy distribution and ensuring that supply meets demand.

- **Energy Consumption Data:** Detailed consumption data provides insights into how energy is used by different sectors within the microgrid, such as residential, commercial, and industrial users. By analyzing this data, operators can identify patterns and trends, optimize load distribution, and implement demand response strategies to balance the grid during peak usage times.
- **Grid Performance Data:** Monitoring grid performance is critical for maintaining stability and preventing outages. Data on voltage levels, frequency, and the health of grid components can be continuously analyzed to detect potential issues before they escalate. Real-time analytics platforms can process this data as it is generated, enabling immediate responses to any deviations from normal operation.
- **Predictive Analytics:** Predictive analytics, using techniques such as time series analysis and regression models, allows operators to forecast future energy needs and adjust supply strategies accordingly. For example, machine learning models trained on historical data can predict spikes in demand, enabling the grid to prepare for increased energy consumption and avoid overloading.
- **Real-Time Data Analysis:** Real-time analytics is essential for maintaining grid stability, especially in microgrids with high levels of renewable energy integration. By analyzing data as it is generated, operators can quickly adjust energy flows, activate backup systems, and respond to unexpected changes in production or consumption. This capability is crucial for preventing disruptions and ensuring a reliable energy supply.
- **Big Data Technologies:** The vast amounts of data generated by microgrids require robust storage, processing, and visualization tools. Big data technologies such as Hadoop and Spark enable the efficient handling of large datasets, while advanced visualization tools help operators make sense of complex data. By leveraging these technologies, microgrids can optimize their operations and improve decision-making. Data analytics transforms the way microgrids are managed, providing the insights needed to optimize performance, enhance reliability, and reduce costs. However, implementing these technologies requires overcoming challenges related to data quality, integration, and privacy. As microgrids continue to evolve, the role of data analytics will become increasingly important in ensuring their success [3].

11.8.4 Machine Learning Applications in Microgrid Control

Machine learning (ML) offers practical solutions to many of the challenges faced in microgrid operations. By applying ML techniques, microgrids can improve their efficiency, reliability, and adaptability, leading to better overall performance.

- **Predictive Maintenance of Microgrid Components:** Predictive maintenance uses ML models to analyze data from sensors monitoring the condition of grid components such as transformers, inverters, and batteries. By identifying patterns that precede equipment failures, these models can predict when maintenance is needed, allowing operators to address issues before they lead to costly outages. For example, neural networks can be trained to recognize early warning signs of component wear and tear, enabling preemptive maintenance that reduces downtime and extends the lifespan of critical infrastructure.

- **Load Forecasting Using Machine Learning Models:** Load forecasting is crucial for balancing supply and demand within a microgrid. ML models, such as decision trees, gradient boosting, and artificial neural networks (ANNs), can predict future energy demand based on historical data, weather forecasts, and other relevant variables. Accurate load forecasting enables grid operators to plan energy distribution more effectively, reducing the need for expensive peak power plants and minimizing energy waste. In practice, these models can be deployed in real-time systems that continuously update predictions as new data becomes available, ensuring that the grid can respond quickly to changes in demand.
- **Optimization of Energy Resources With Machine Learning:** Optimizing the use of energy resources is essential for maximizing the efficiency of a microgrid. ML algorithms, such as genetic algorithms, particle swarm optimization, and reinforcement learning, can develop strategies for balancing the use of renewable energy sources, traditional generation, and energy storage systems. For example, reinforcement learning can be used to determine the optimal times to charge and discharge energy storage systems, ensuring that energy is stored when it is abundant and released when it is needed most. These optimization techniques help microgrids reduce costs, increase the use of renewable energy, and improve overall system performance.
- **Anomaly Detection in Microgrids:** Anomaly detection is critical for identifying irregularities that could indicate potential issues within the grid, such as equipment failures or cyber threats. ML techniques such as clustering algorithms, principal component analysis (PCA), and autoencoders can detect anomalies in energy consumption patterns, grid frequency, and voltage levels. By identifying these anomalies early, operators can take corrective action before they lead to significant problems. For example, an anomaly detection system might alert operators to unusual voltage fluctuations, allowing them to investigate and resolve the issue before it causes an outage. These applications demonstrate the practical impact of ML on microgrid operations, showing how these technologies can be used to enhance efficiency, reduce costs, and improve the reliability of energy systems. By integrating ML into their operations, microgrids can become more resilient and better equipped to handle the challenges of modern energy distribution.

11.8.5 Case Studies and Practical Implementations

To illustrate the profound real-world impact of machine learning (ML and data analytics on the operation of microgrids, this section presents a series of compelling case studies and practical implementations. These examples vividly demonstrate how these innovative technologies tackle complex challenges and significantly enhance the performance of microgrids.

Case Study 1: Predictive Maintenance in a Renewable Energy Microgrid
This case study delves into a cutting-edge microgrid that seamlessly integrates solar and wind power sources. To ensure the longevity and reliability of its critical infrastructure, the microgrid has adopted state-of-the-art predictive maintenance practices. Advanced sensors are strategically installed on transformers, inverters, and battery systems to continuously monitor key indicators such as temperature, vibration, and electrical currents. Leveraging sophisticated machine learning models—particularly neural networks—this data is rigorously analyzed to forecast potential equipment failures. By proactively addressing these predicted issues before they escalate into significant breakdowns, the microgrid not only reduces unplanned downtimes but also significantly lowers maintenance costs and extends the operational lifespan of its infrastructure. This case study provides compelling quantitative evidence, showcasing the dramatic decrease in unplanned outages and the substantial savings achieved via predictive maintenance strategies.

Case Study 2: Load Forecasting in an Urban Microgrid
In this illuminating case study, we explore an urban microgrid that caters to a diverse array of consumers, including residential homes, bustling commercial enterprises, and dynamic industrial facilities. The microgrid employs sophisticated ML-based load forecasting models to accurately predict energy demand on both an hourly and daily basis. The models are meticulously trained using historical consumption data alongside weather forecasts and other pertinent factors, resulting in highly precise energy demand predictions. By honing its ability to forecast demand accurately, the microgrid can adeptly optimize its energy generation and distribution strategies. This optimization minimizes reliance on costly peak power plants while simultaneously curbing energy waste. The case study emphasizes the transformative improvements in grid efficiency and the impressive cost savings that have been realized through the implementation of advanced load forecasting techniques.

Case Study 3: Optimization of Energy Resources in a Hybrid Microgrid
This case study reveals the innovative workings of a hybrid microgrid that adeptly combines renewable energy sources—such as solar and wind—with traditional power generation and energy storage systems. Through the application of intelligent ML optimization algorithms, including genetic algorithms and reinforcement learning, the microgrid develops sophisticated strategies for balancing energy supply and demand in real time. The optimization process is not simply reactive; it involves carefully determining the optimal timings for charging and discharging energy storage systems, as well as making strategic decisions about how much energy to draw from renewable versus non-renewable sources. The results of this strategic optimization reveal a compelling narrative of energy savings, a marked reduction in dependence on fossil fuels, and a notable enhancement of overall system efficiency.

Practical Implementation: Anomaly Detection System in a Smart Microgrid
In a further demonstration of technological prowess, this section presents an example of a smart microgrid that has successfully implemented an advanced anomaly detection system utilizing cutting-edge ML techniques. This system tirelessly monitors a range of critical parameters, including voltage levels, frequency fluctuations, and energy consumption patterns, to identify anomalies that may signal potential issues, such as equipment malfunctions or security breaches. Upon detecting an anomaly, the system promptly alerts operators, enabling them to investigate and rectify the situation before it escalates into a more serious problem. This forward-thinking implementation has significantly bolstered grid reliability and security, resulting in a noticeable reduction in outages and a much more rapid response to potential threats.

These detailed case studies and practical implementations collectively illustrate the substantial benefits of integrating ML and data analytics into microgrid operations. By harnessing these forward-looking technologies, microgrids are not only achieving greater efficiency, reliability, and resilience but are also making invaluable contributions to the overall stability and sustainability of modern energy systems.

11.9 Challenges and Future Directions

While machine learning (ML) and data analytics present transformative opportunities for enhancing the efficiency and reliability of microgrid operations, their successful implementation is accompanied by a host of challenges that must be navigated. This section delves into the key obstacles that need to be addressed and outlines promising future directions for research and development in this dynamic field [4].

Challenges

1. **Data Quality and Availability**: The efficacy of ML models is intricately tied to the quality and accessibility of data. In the context of microgrids, data is amassed from a wide array of sources, such as sensors embedded in infrastructure, smart meters tracking energy consumption, and various external databases. It is imperative to ensure that this data is not only accurate and complete but also current to train robust and reliable ML models. Unfortunately, the microgrid environment often presents hurdles such as data gaps, noise interference, and inconsistencies that can significantly compromise the integrity of the information. Moreover, effectively integrating and standardizing data from diverse sources poses a complex and labor-intensive challenge, requiring significant expertise and resources.
2. **Computational Resources**: The deployment of advanced ML models, especially in applications that require real-time analysis, demands considerable computational power. Processing vast datasets and executing sophisticated algorithms necessitate high-performance computing infrastructure, which may be limited in

some microgrid setups. Additionally, the ability to implement ML models in real-time environments hinges upon rapid and efficient data processing capabilities, underscoring the need for substantial investments in advanced computational resources that can handle these demanding tasks.

3. **Algorithmic Bias and Interpretability**: One of the inherent risks associated with ML models is the potential for bias, which can yield unfair or less-than-optimal outcomes. This concern is particularly acute in critical sectors like energy management, where biased models could inadvertently lead to an inequitable distribution of energy resources or flawed predictions regarding energy demand and supply. Furthermore, many ML models, particularly those utilizing deep learning techniques, often operate as black boxes, making it challenging to discern how decisions are made. Ensuring that these models are transparent and their workings are understandable is essential for fostering user trust and ensuring that decisions made by the models can be adequately justified.
4. **Integration With Existing Systems**: Seamlessly incorporating ML and data analytics into the pre-existing infrastructure of microgrids can prove to be a daunting task, particularly for systems that were established before these technologies became mainstream. Retrofitting legacy systems to accommodate ML applications may demand extensive modifications to both hardware and software, as well as the formulation of new protocols for effective data collection and analysis. Furthermore, stakeholders may exhibit resistance to adopting unfamiliar technologies, particularly if these advancements necessitate significant alterations to their established workflows, thereby posing a barrier to innovation.

Future Directions

1. **Advanced ML Algorithms**: The future trajectory of research is expected to emphasize the creation of more advanced ML algorithms tailored to handle the intricate complexities inherent in microgrid operations. Innovative algorithm development will focus on enhancing resilience to variations in data quality, enabling real-time processing capabilities, and fostering adaptability to the dynamic conditions of the grid. Additionally, there is a rising demand for algorithms that can effectively integrate with other emerging technologies, such as the Internet of Things (IoT) and blockchain, facilitating the creation of more cohesive and intelligent energy systems.
2. **Edge Computing and Distributed Analytics**: A promising future for ML in microgrids may revolve around adopting distributed computing strategies. In such frameworks, data analysis and decision-making occur closer to the data sources, often referred to as the "edge" of the network. By leveraging edge computing, organizations can significantly reduce latency and enhance the responsiveness of ML models, making them more effective in real-time applications. Distributed analytics can further ease the computational load, leading to more efficient resource utilization and diminishing the dependency on centralized processing facilities.

3. **Integration With Emerging Technologies**: The convergence of ML with other cutting-edge technologies, such as IoT for real-time data gathering and blockchain for secure energy trading, represents a rich area for exploration. Such integrations will pave the way for the development of energy systems that are not only more secure and efficient but also exceptionally flexible, allowing microgrids to operate seamlessly within broader decentralized networks. For instance, IoT devices can continuously monitor energy consumption and production metrics, facilitating optimal grid management. Simultaneously, blockchain technology can ensure secure and transparent energy transactions, while robust cybersecurity measures can safeguard against potential threats to grid integrity.
4. **Policy and Regulatory Considerations**: The integration of ML and Data Analytics within microgrid operations necessitates a careful examination of the associated policy and regulatory frameworks. Addressing crucial issues such as data privacy, navigating ethical concerns surrounding artificial intelligence, and establishing standards for the application of ML within critical infrastructure are all paramount for future developments. Policymakers and regulatory bodies will need to engage in close collaboration with industry stakeholders to forge frameworks that encourage innovation while ensuring the safety and reliability of energy systems.
5. **Collaboration and Innovation**: The enhancement of ML and data analytics within microgrids will rely heavily on sustained collaboration among academia, industry, and governmental entities. Cross-disciplinary partnerships will be vital for cultivating the next generation of energy management technologies and for addressing the various technical, social, and regulatory challenges that emerge in this fast-evolving landscape. By fostering an environment of cooperation and innovation, stakeholders can drive meaningful advancements that reshape the future of energy management. While machine learning (ML) and data analytics offer significant advantages for microgrid operations, their implementation is not without challenges. This section explores the obstacles that must be overcome and the future directions for research and development in this field.

11.10 Conclusion

The integration of machine learning (ML) and data analytics into microgrid operations marks a groundbreaking shift in the management of modern energy systems. These technologies present robust solutions for optimizing grid performance, enhancing reliability, and building resilience against increasingly complex and dynamic challenges.

Recap of Key Points In this chapter, we delved into the transformative role of ML and Data Analytics across several facets of microgrid operations, including predictive maintenance, load forecasting, energy resource optimization, and anomaly detection. Each application underscores the immense potential of these technologies to meet pressing challenges in energy management, addressing issues such as the variability of renewable energy sources and the critical need for more efficient and reliable grid operations.

Evolving Role of Machine Learning in Microgrid Operation and Control As ML technology continues to advance, its significance in microgrid operation and control will only increase. ML models are capable of continuously learning and adapting to shifting conditions, enabling microgrids to respond adeptly to fluctuations in supply and demand. This capability is vital in the realm of renewable energy, where production can be highly unpredictable. By analyzing vast data streams in real time, ML empowers autonomous grid management systems that can optimize energy distribution independently, reducing reliance on constant human oversight. Furthermore, the synergy between ML and other cutting-edge technologies, including the Internet of Things (IoT), blockchain, and robust cybersecurity measures, is set to revolutionize our energy networks. In this evolving landscape, microgrids will not only function as autonomous units but also as essential elements of a decentralized energy system that is both resilient and efficient. By embracing these advancements, we can pave the way for a more sustainable and reliable energy future.

Outlook for Research and Development

As we look to the future, it is imperative to identify several critical areas that require focused research and development to fully unlock the incredible potential of machine learning (ML) and data analytics in microgrids:

Advanced ML Algorithms There is a pressing need for the creation of advanced ML algorithms specifically designed to navigate the intricate and variable nature of microgrid operations. These advanced algorithms should excel in real-time data processing, exhibiting the ability to swiftly adapt to evolving conditions while maintaining robustness against fluctuations in data quality. The complexity of microgrids—comprising diverse energy sources, fluctuating demand, and variable environmental factors—demands sophisticated solutions to ensure optimal performance.

Edge Computing and Distributed Analytics The trajectory of ML in microgrids points toward innovative distributed computing approaches, wherein data analysis and decision-making occur in proximity to the data sources themselves. This paradigm shift is crucial as it significantly reduces latency—ensuring timely data processing—and enhances responsiveness to immediate operational challenges. By placing computing resources closer to the point of data generation, we can achieve a more dynamic and effective application of ML models in real-time scenarios.

Integration With Emerging Technologies The integration of machine learning with other groundbreaking technologies—such as the Internet of Things (IoT), blockchain, and advanced cybersecurity solutions—holds remarkable promise for the future landscape of energy systems. These synergistic relationships will forge more secure, efficient, and adaptable microgrids, facilitating seamless interactions within vast decentralized networks. For instance, IoT devices can provide real-time data from various energy sources, while blockchain technology can enhance transaction security and reliability, thus fostering a more integrated energy ecosystem.

Policy and Regulatory Considerations With the growing integration of ML and data analytics in microgrid operations, the establishment of new policy and regulatory frameworks is essential. These frameworks must prioritize the safe and responsible deployment of cutting-edge technologies. Policymakers and regulators must actively engage with industry stakeholders to formulate standards that not only support innovation but also uphold safety and reliability in energy systems. This collaborative effort is critical to building a regulatory environment that encourages the adoption of advanced technologies while addressing potential risks.

Collaboration and Innovation Advancing ML and data analytics in microgrids necessitates a robust framework for collaboration among academia, industry, and government entities. By cultivating interdisciplinary partnerships, we can ensure that these technologies are developed and rolled out in ways that maximize their potential benefits while minimizing associated risks. Such collaborations can drive innovation by combining diverse expertise, resources, and insights, which are essential for addressing the multifaceted challenges of modern energy systems.

Implications for the Protection of Power Distribution Systems The advancements in ML and data analytics have profound implications for enhancing the protection and resilience of power distribution systems. These technologies can significantly augment our predictive capabilities, enabling us to foresee and prevent faults, streamline energy flow, and swiftly identify and address security threats in real-time. As microgrids gain prominence and interconnectivity increases, their efficient and secure operation becomes vital to the overall stability of the power grid. By seamlessly integrating ML and data analytics into microgrid operations, we can engineer energy systems that not only deliver superior efficiency and reliability but also stand strong against future challenges. This integration is not merely beneficial; it is essential for the ongoing evolution of microgrids and power distribution systems, ultimately paving the way for a sustainable and reliable energy future.

Call to Action In closing, this chapter emphasizes a compelling call to action for heightened innovation and research in the application of ML and data analytics for microgrid operations. As the energy sector undergoes transformative shifts, the demand for revolutionary technologies and approaches to tackle the complexities and challenges of contemporary energy systems is more urgent than ever. Researchers, industry professionals, and policymakers are called upon to embrace the myriad possibilities that ML and data analytics offer. They should actively push the boundaries of what is possible and contribute meaningfully to the ongoing evolution of microgrids and power distribution systems.

The vast potential of ML and data analytics to transform microgrid operation and control cannot be overstated. By diligently advancing and implementing these technologies, we can construct energy systems that are not only markedly more efficient and reliable but also remarkably secure and resilient, poised to meet the demands of tomorrow's energy landscape.

Self-Assessment Questions

? Self-assessment question 14.1

A. Describe the evolving role of machine learning in microgrid operation and control and implications for the protection of power distribution systems.

? Self-assessment question 14.2

1. Describe the advanced ML algorithms, edge computing, and distributed analytics.
2. Integration with emerging technologies, policy and regulatory considerations, collaboration, and innovation.

? Self-assessment question 14.3

A. Describe the predictive maintenance of microgrid components and load forecasting using machine learning models.
B. Optimization of energy resources with machine learning, and anomaly detection in microgrids.

Bibliography

1. Gergely Hollósi; Dániel Ficzere; Attila Frankó; Máté Bancsics "Enabling Efficient Knowledge Sharing for Industrial AI" IEEE Network Operations and Management Symposium 2024.
2. M. B. Sergeeva; V. V. Voskobovich; A. M. Kukharenko "Data Processing in Industrial Internet of Things (IIoT) Applications : Industrial Agility"Wave Electronics and its Application in Information and Telecommunication Systems (WECONF) 2022
3. Parikshit Vasisht; Sudhakar Ranjan; Rahul J. Jadhav; "Artificial Intelligence as a Driver of Industrial Competitiveness" International Conference on Smart and Sustainable Technologies in Energy and Power Sectors (SSTEPS) 2022.
4. Martin Hollender; Chaojun Xu; Ruomu Tan "Engineering Challenges in Industrial AI" IEEE/ACM 3rd International Conference on AI Engineering—Software Engineering for AI (CAIN) 2024

Cybersecurity and Information Program's Power System

12

Contents

12.1 Introduction........ 294
12.2 Cybersecurity Operations........ 295
12.3 Cybersecurity Impact and Solution........ 296
12.3.1 Validate Process Mapping........ 298
12.3.2 Data Collection Framework........ 298
12.3.3 Data Collection: Introducing System Mapping........ 299
12.4 How Can You Help?........ 300
12.5 Cybersecurity Tailored Models........ 303
12.5.1 Project Plan Work for (IT Penetration Testing, Vulnerability & Risk Assessment Framework)........ 303
Self-Assessment Questions........ 309
Bibliography........ 309

Abstract

By the conclusion of this chapter, you will have a solid grasp of effective control and mitigation strategies essential for cybersecurity in industrial applications, particularly in the energy sector. You will gain insights into core cybersecurity principles and the operational challenges that arise within Electrical Power Systems, along with their implications and practical solutions. More importantly, you will have the opportunity to validate your understanding through cybersecurity process mapping and data collection techniques, boosting your confidence significantly. Finally, you will learn how to construct two distinct levels of project scope: first, for IT penetration testing and assessment; and second, specifically tailored to Electrical Power Systems and distributed electrical power generation. Embrace this knowledge to enhance your cybersecurity proficiency and secure our energy infrastructure effectively!

Supplementary Information The online version contains supplementary material available at https://doi.org/10.1007/978-3-031-94312-6_12.

T. Masood et al., *Protection of Electrical Power Transmission Systems*, Power Systems, https://doi.org/10.1007/978-3-031-94312-6_12

Objectives

At the conclusion of this section, you will gain valuable insights into the essential principles of the control and mitigation plan for cybersecurity components tailored for industrial applications, particularly within the energy sector.

The key areas of focus include:

- Cybersecurity introduction.
- Cybersecurity operations.
- Cybersecurity impacts and solutions.
- Validation of cybersecurity process mapping.
- Data collection and mapping.
- Project scope for IT penetration and assessment.

The scoping exercise stands as a pivotal element in determining the effectiveness of cybersecurity measures for a company's critical infrastructure. This endeavor transcends mere task completion; it represents a vital initiative to guarantee the precision and thoroughness of operational and control data. In light of the escalating demand for electric power and the aging infrastructure prevalent in many countries, concerns grow regarding suppliers' capacity to meet energy needs and the associated risks of greenhouse gas emissions. This complex landscape, coupled with limited capital investment in new transmission systems, underscores the urgent need for the development of innovative, cost-effective power systems. Additionally, the rapid advancements in renewable energy sources (RES)—particularly photovoltaic (PV) technologies—are driving the swift integration of PV systems into both residential and commercial settings. It is imperative to rigorously validate that all quality assurance (QA) process steps align accurately with the reference model. Furthermore, every industrial control system (ICS) must be meticulously mapped to the quality assurance framework, ensuring that no errors are tolerated.

With over 25 years of experience as an Operations Advisor at Qatar Energy, I can confidently vouch for the successful implementation of this groundbreaking approach at Qatar Energy. Embracing these principles is not just beneficial; it is essential for securing a sustainable and resilient energy future [1].

12.1 Introduction

The repercussions of security breaches in the smart grid are significant and can profoundly impact the entire infrastructure, including national economies and individual lives. The urgency of addressing these risks is heightened by the alarming rise in malicious electronic intrusions aimed at the grid's nodes and consumer devices. Reports of such attacks have emerged in several countries, including Finland, Turkey, and Ukraine. Consequently, it is imperative that we implement robust and comprehensive security solutions as integral components of demand side management (DSM) and demand response (DR) strategies.

To tackle these critical challenges and mitigate their potential impacts on power distribution infrastructure, we propose a cutting-edge convex optimization system. This system will establish secure DSM and DR strategies that not only minimize energy costs for consumers and power utilities but also safeguard against overloads in distribution transformers. It will adeptly accommodate the fluctuating power demands from sources such as electric vehicles (EVs) and plug-in electric vehicles (PEVs), thereby advancing the existing smart grid architecture through the effective utilization of smart meters. Distribution transformers are pivotal assets within the power grid and are frequently exposed to overload conditions that can lead to premature failures. Currently, grid operators lack real-time access to the loading conditions of these transformers, obscuring their aging status. During peak power consumption periods, excessively aged and thermally overloaded transformers are at a heightened risk of failure, which disrupts the electric power supply. Moreover, shifts in customer load behavior, driven by market incentives, can exacerbate the overload situation for these transformers.

Effective network modeling depends on the insightful analysis of smart meter data, providing essential visibility into the distribution network. Smart meters are a crucial element of advanced metering infrastructure (AMI) and function as gateways to the smart grid. AMI encompasses energy consumption meters with sophisticated two-way communication capabilities, empowering local utility servers to analyze data received from customers. This system ensures that both utilities and customers can visualize, process, and manage their energy needs with precision. By leveraging smart meter readings, we can accurately estimate the actual loading of distribution transformers, resulting in enhanced operational efficiency and strategic planning for distribution systems [2].

Exercise 12.1
Describe the DSM and DR strategies to address the cybersecurity issues.

12.2 Cybersecurity Operations

The growing integration of sophisticated cyber infrastructures into electric distribution networks—such as power electronic transformers, phasor measurement units (PMUs), and remote terminal units (RTUs)—has created a pressing concern regarding cyber vulnerabilities and the potential for malicious cyber-attacks. As we look to the future of smart grid power distribution systems, it is imperative that we focus on two pivotal objectives that can enhance both security and functionality.

First, we must strive to elevate the cybersecurity framework governing our distribution grids. This can be achieved by developing an innovative, adaptable, and scalable Moving Target Defense (MTD) framework, specifically designed to thwart the increasing threat posed by False Data Injection (FDI) attacks. Such an approach will not only enhance the robustness of our networks but also ensure that our systems remain resilient against sophisticated and evolving threats. Second, we can significantly improve the balance and efficiency of the distribution system by utilizing state-of-the-art three-to-single phase power electronic transformers. These transformers will facilitate single-phase lateral connections tapped onto three-phase main feeders, thereby optimizing power distribution and minimizing losses within our grid. However, we must remain vigilant about the vulnerabilities that can be exploited by attackers:

Active Interception Cybercriminals can engage in malicious activities that compromise the integrity of communication between smart meters and utility companies. They may drop or alter critical messages, leading to the transmission of false data. Such large-scale interference can have catastrophic consequences, as the smart grid could end up responding to demands that do not actually exist, resulting in improper shedding decisions and overall system failure.

Passive Interception On a more insidious level, hackers can intercept communications to gain access to sensitive personal information. For instance, they might determine the specific rooms or devices in active use, opening the door to more severe and targeted attacks that exploit this knowledge. In response to these escalating threats, we advocate for a comprehensive two-layer security mechanism tailored for the smart grid. The first layer will integrate a secure-by-design prevention mechanism, leveraging the capabilities of blockchain technology to ensure a high level of security and protect user privacy against potential breaches. The second layer comprises a vigilant detection mechanism that continuously monitors data exchanges between utility servers and smart meters, searching for any unusual behaviors or anomalies that may indicate an intrusion. By implementing these advanced strategies, we can fortify our smart grid, ensuring that it remains a safe and reliable energy distribution network in the face of evolving cyber threats [3].

12.3 Cybersecurity Impact and Solution

This section decisively tackles significant challenges and presents effective approaches to address the issue: We propose a Hidden Moving Target Defense (HMTD) strategy designed to enhance stealth against attackers. Existing Moving Target Defense (MTD) strategies often fail, as alert attackers can detect MTD activation before launching a False Data Injection (FDI) attack. Our HMTD strategy overcomes this limitation by remaining undetectable even in the presence of measurement noise, thereby significantly boosting its effectiveness in real-world scenarios.

To ensure rapid assessment of the stealthiness of MTD strategies, we will develop analytical feasibility conditions that leverage the structural characteristics of attackers' bad data detection problems. These conditions will establish a solid theoretical foundation for understanding the stealth capabilities of MTD strategies, enabling precise detection and robust implementation of countermeasures.

Furthermore, we will model the interaction between the system operator and attackers as a repeated Bayesian Stackelberg Game, employing an approximate dynamic programming approach to derive solutions. This robust framework will effectively predict attackers' strategies and optimize the system's defense mechanisms by integrating both historical behavior and dynamic responses.

Exercise 12.2
Describe the HMTD, MTD, and FDI strategies to determine impact and solutions.

We are committed to developing a robust HMTD-based reconfiguration model for unbalanced radial distribution networks. To accomplish this, we will implement an Alternating Direction Method of Multipliers (ADMM)-based semidefinite programming approach, ensuring efficient and scalable optimization that significantly enhances network security and resilience. In addition, we will establish a High-Performance Computing (HPC) and Power Hardware-in-the-Loop (PHIL)-enabled testbed for rigorous proof-of-concept validation and demonstration. Our customized test system at Kansas State University will effectively simulate and validate the designed models in real-world power grid scenarios.

Furthermore, we will design a groundbreaking topology for single-phase power distribution that guarantees balance across the three-phase power system. This innovative topology will enhance system efficiency and stability while effectively tackling the challenges presented by unbalanced loads in modern distribution networks. We will develop a sophisticated three-to-single-phase power electronic transformer topology that enables remote connection control, voltage regulation, power factor correction, metering, and protection for any single-phase lateral tapped onto a three-phase power distribution feeder. A comprehensive study will examine the impact of unbalanced enhancement alongside the proposed MTD strategy, using a targeted electrical power distribution system as a case study.

Innovation This section will detail our applicable and scalable MTD framework to combat False Data Injection (FDI) attacks and enhance phase balancing in three-phase electric distribution networks through three-to-single-phase power electronic transformers. We will execute thorough theoretical analysis, model development, algorithmic design, numerical validation, testbed development, and real-time simulation. Our work will push the boundaries of MTD knowledge in the power systems field, equipping distribution system operators with critical insights on secure operations of the distribution network [4].

Expected Impacts We will initiate outreach and educational activities that significantly raise public awareness of cyber-attacks and deepen understanding of cybersecurity in the smart grid among stakeholders, ratepayers, regulators, utility executives, and market participants. Our initiatives will promote widespread acceptance of transformative new technologies, paving the way for a new age of grid cybersecurity that empowers the next generation of power grids.

Exercise 12.3
Describe the innovation and expected impact on the vulnerability of the power operations network.

12.3.1 Validate Process Mapping

In this section, we explore the critical integration of Demand Side Management (DSM) and Demand Response (DR) technologies to ensure the efficient delivery of power while addressing the diverse needs of both commercial and domestic consumers within large electrical power systems. A key advancement in distribution networks is the active participation of customers in DSM and DR programs, which not only empowers consumers but also optimizes energy usage. Most existing DSM and DR strategies focus on minimizing customer power consumption based on established load priority policies during peak demand periods. Moreover, the smart grid is revolutionizing the landscape of energy distribution by acting as a sophisticated communication network that integrates smart meters, control devices, computers, automation technologies, and other innovative solutions. Effective communication among all entities in the smart grid is essential, creating a robust, complex network that must address the inherent security vulnerabilities of cyber systems.

Alarmingly, as of January 2017, a report from the U.S. Department of Energy highlighted the ongoing risks associated with cyber-attacks on the U.S. electric grid. This underscores the urgent need for enhanced security measures to protect our energy infrastructure, ensuring stability and reliability for all consumers.

Exercise 12.4
Describe the mapping framework to integrate the DSM and DR program.

12.3.2 Data Collection Framework

My previous organization successfully identified 423 systems, up from the initial 371. This increase not only highlights our commitment to thorough research but also demonstrates our ability to adapt to new information, as we discovered new systems while reevaluating those that are no longer relevant (Fig. 12.1).

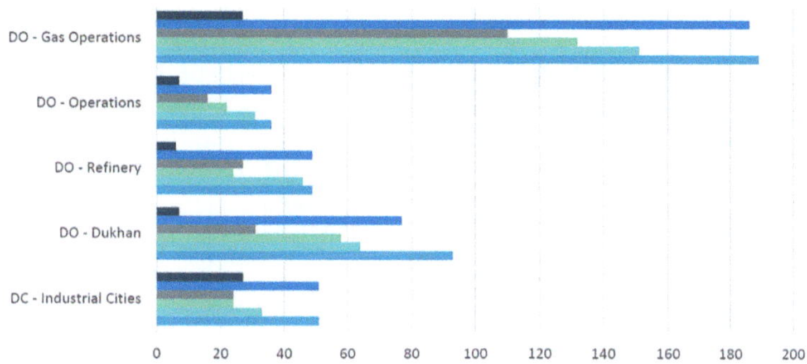

	DC - Industrial Cities	DO - Dukhan	DO - Refinery	DO - Operations	DO - Gas Operations
Number of New Systems	27	7	6	7	27
Number of Systems assigned to Unit	51	77	49	36	186
Number of Systems with Network Diagram	24	31	27	16	110
Number of Systems with Process Overview	24	58	24	22	132
Number of Systems with Questionnaire	33	64	46	31	151
Total Number of Systems	51	93	49	36	189

Fig. 12.1 Data collection framework. A horizontal bar chart and table compare system metrics across five departments: DC—Industrial Cities, DO—Dukhan, DO—Refinery, DO—Operations, and DO—Gas Operations. The chart shows five categories: New Systems, Systems Assigned to Unit, Systems with Network Diagram, Systems with Process Overview, and Systems with Questionnaire

As we prepare to advance to the next phase of the program, it is essential for the CIS Department to gain clarity on two important aspects:

- Which critical operations of our company are encompassed by the data we have received?
- What is the fundamental error in concentrating exclusively on the specified systems?

Understanding these points will help us make informed decisions moving forward.

12.3.3 Data Collection: Introducing System Mapping

A system map is a detailed representation of my energy company's Industrial Control System (ICS), depicting the intricate connections between various elements of industrial automation and control systems, and their integration with essential business processes. This comprehensive mapping system visually conveys the quality and accessibility of information related to the ICS, highlighting its significance in identifying vulnerabilities. It effectively emphasizes the vital role of prioritizing

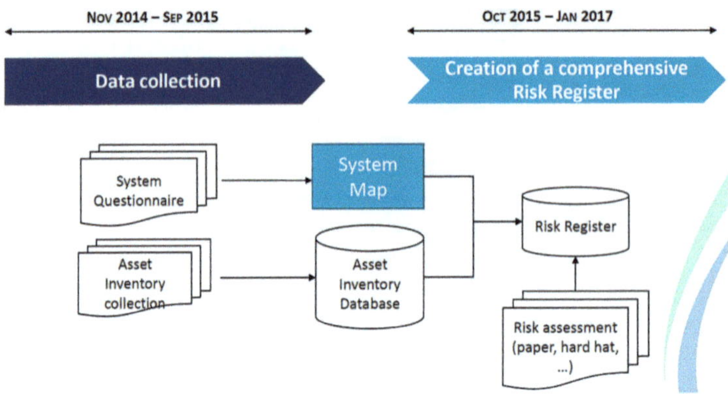

Fig. 12.2 Data collection and mapping framework. Flowchart illustrating the process of creating a comprehensive risk register from November 2014 to January 2017. The first phase, "Data collection" from November 2014 to September 2015, involves gathering information through a system questionnaire and asset inventory collection, feeding into a system map and asset inventory database

risk assessments to safeguard operational integrity and ensure the efficient functioning of our services (Figs. 12.2 and 12.3).

> **Exercise 12.5**
> Describe the system mapping structure and its operational decision tree also describe the validation process mapping framework stepwise with its elements.

12.4 How Can You Help?

The quality of the scoping exercise is crucial for the effectiveness of our cybersecurity measures to protect the company's critical infrastructure. Therefore, your thorough review of the prepared information is essential to validating its accuracy and completeness.

To ensure that we achieve the highest standards, we kindly ask you to:

- Confirm that all steps of the quality evaluation (QE) process are accurately aligned with the reference model.
- Verify that all Industrial Control Systems are correctly mapped to Qatar Petroleum's (QP's) processes.
- Provide insights regarding any "**Orphan Systems**."

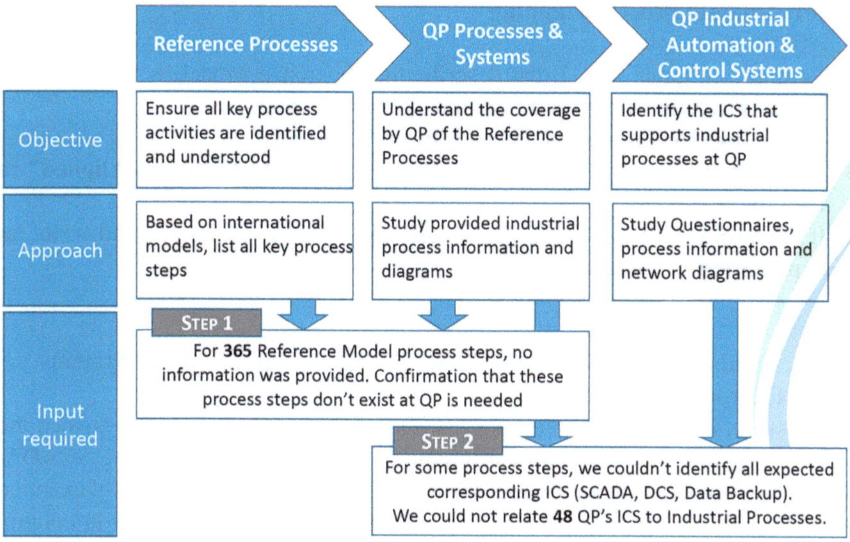

Fig. 12.3 Data collection and map structure framework. The image is a flowchart detailing the objectives, approaches, and input requirements for three categories: Reference Processes, QP Processes and Systems, and QP Industrial Automation and Control Systems. Under "Objective," Reference Processes aim to ensure that all key process activities are identified and understood

Step 1: Validate Process Mapping
Please compare columns E to J, which outline the Reference Model, with column K, representing the equivalent systems within Qatar Petroleum.

- Document any relevant comments in column V.
- If the information aligns perfectly, please note "Reference and QP Systems Aligned" in column W.
- If there are discrepancies, clearly state this in column W and include a detailed explanation in column V.

Your expertise in this review process will greatly enhance our cybersecurity framework and ensure robust protection of our critical assets. Thank you for your collaboration!

The quality of the scoping exercise will have a direct impact on the effectiveness of cybersecurity measures to secure the selected company's critical infrastructure. Therefore, we would highly appreciate your review of the prepared information to validate its exactness and completeness.

More Precisely:

- Validate that all QE process steps are correctly mapped to the reference model.
- Validate that all Industrial Control Systems are correctly mapped to QP's processes.
- Provide information about the "**Orphan System**".

Step 1: Validate Process Mapping

1. Compare columns E to J that describe the Reference Model, with column K, which is the equivalent system within Qatar Petroleum.
2. **Write any relevant comments in column V.**
3. If the information is aligned, state "**Reference and QP Systems Aligned" in column W.**
4. If the information is not aligned, please **state this in column W and write an explanation in column V** (Table 12.1).

Step 2: Validate the Mapping of ICS/SCADA/Distributed Control Systems (DCS) to Processes

In this stage, it is essential to ensure that there is a precise alignment between the Qatar Petroleum systems identified in column K and the corresponding SCADA/ICS systems outlined in columns R through T. Take the time to carefully examine each entry and determine if all relevant information has been included in column V. If you find that everything is in order and properly matched, indicate this in column W by writing, "QP Systems and SCADA/ICS Systems Aligned." However, if you identify any inconsistencies or gaps in the information, please document these discrepancies in column W and provide a detailed explanation in column V. Your meticulous attention to detail is vital for maintaining accuracy throughout this important validation process! (Table 12.2)

Table 12.1 Validating process map structure framework. A table titled "QP System Mapping" with columns: Location, Site, Domain, Operation, System, and QP System Mapping. The data includes entries for "DKN" and "JDGS" locations. The "DYB" site, highlighted in yellow, is associated with "Onshore Fields" and "Primary Process" operations, covering systems like "Gathering system manifolds"

LOCATION	SITE	DOMAIN	OPERATION	SYSTEM	QP SYSTEM MAPPING
DKN	DYB	Onshore Fields	Primary Process	Gathering system manifolds	Diyab Manifold - Gathering system manifolds
DKN	DYB	Onshore Fields	Primary Process	Lift gas manifold	Diyab Manifold - Lift gas manifold
DKN	DYB	Onshore Fields	Primary Process	Chemical Injection System (corrosion inhibitor/demulsifier)	Diyab Manifold - Chemical Injection System (corrosion inhibitor/demulsifier)
DKN	DYB	Onshore Fields	Utility	Drains system	Diyab Manifold - Drains system
DKN	DYB	Onshore Fields	Primary Process	Launchers for export pipelines	Diyab Manifold - Launchers for export pipelines
DKN	DYB	Onshore Fields	Primary Process	Receiver for liftgas pipeline	Diyab Manifold - Receiver for liftgas pipeline
DKN	DYB	Onshore Fields	Utility	Instrument air packages for LG manifolds	Diyab Manifold - Instrument air packages for LG manifolds
DKN	JDGS	Degassing	Primary Process	Oil Gathering and Separation	Jaleha Degassing Station - Oil Gathering and Separation
DKN	JDGS	Degassing	Primary Process	Oil Storage and Pumping	Jaleha Degassing Station - Oil Storage and Pumping

Table 12.2 Map process structure framework. A table listing systems at the Dukhan Gas Recycling Plant. Columns include Site, Unit, System Label, System Name, and System Full Name. The site is consistently Dukhan, and the unit is a Gas Recycling Plant. System labels include RGD-DCS, RGD-FGS, and RGD-ESD, among others. System names range from Distributed Control System to Recycle Compressor Packages

Site	Unit	System Label	System Name	System full Name
Dukhan	Gas Recycling Plant	RGD-DCS	RGD - Distributed Control System (DCS)	RGD - Distributed Control System (DCS)
Dukhan	Gas Recycling Plant	RGD-FGS	RGD - Fire & Gas Monitoring System (FGS)	RGD - Fire & Gas Monitoring System (FGS)
Dukhan	Gas Recycling Plant	RGD-ESD	RGD - Emergency Shutdown System (ESD)	RGD - Emergency Shutdown System (ESD)
Dukhan	Gas Recycling Plant	RGD-AUX	Aux. Ctrl Sys	Aux. Ctrl Sys - Gas Recycling Plant
Dukhan	Gas Recycling Plant	RGD-TBN	Turbine Ctrl Sys.	Turbine Ctrl Sys. - Gas Recycling Plant
Dukhan	Gas Recycling Plant	RGD-ATS	Antisurge Ctrl. Sys.	Antisurge Ctrl. Sys. - Gas Recycling Plant
Dukhan	Gas Recycling Plant	RG-DCS-SRV	DCS-E-server	DCS-E-server - Gas Recycling Plant
Dukhan	Gas Recycling Plant	RGD-PLC-MIAC	RGD - Main Instrument Air Compressor PLC Package	RGD - Main Instrument Air Compressor Package
Dukhan	Gas Recycling Plant	RGD-PLC-HOH	RGD - Hot Oil Heater Package	RGD - Hot Oil Heater Package
Dukhan	Gas Recycling Plant	RGD-PLC-HWC	RGD - Hydraulic wellheads control package	RGD - Hydraulic wellheads control Package
Dukhan	Gas Recycling Plant	RGD-PLC-RNCP	RGD - Reinjection Compressor Packages	RGD - Reinjection Compressor Packages
Dukhan	Gas Recycling Plant	RGD-PLC-RECP	RGD - Recycle Compressor Packages	RGD - Recycle Compressor Packages

Observation

1. The "ICS LIST" tab in the spreadsheet showcases a detailed Systems List table, which is essential for tracking various systems within our operations.
2. Each system is assigned a unique identifier known as the "System Questionnaire ID." This internal reference plays a vital role in systematically organizing and identifying all documentation that we receive from QP.
3. The format of this identifier consists of the characters "D0XXX," ensuring consistency and clarity. You can find this identifier prominently displayed in the first three columns of the QP Systems List table, providing a straightforward method for locating and referencing crucial system information (Table 12.3).

12.5 Cybersecurity Tailored Models

12.5.1 Project Plan Work for (IT Penetration Testing, Vulnerability & Risk Assessment Framework)

Operational Summary

Project plan for IT penetration testing, vulnerability, and risk assessment at a combined cycle gas turbine (CCGT) power plant. This project plan details a structured and thorough approach for conducting an exhaustive IT penetration test,

Table 12.3 Map process structure framework. A table titled "QP System Mapping" with columns for Operational ICS, Safety ICS, Auxiliary ICS, and Overall Consequences Rating. Rows list systems such as "Diyab Manifold—Lift gas manifold" and "Diyab Manifold—Chemical Injection System," both rated as "Very Low" in consequences

QP system mapping	Operational ICS	Safety ICS	Auxiliary ICS	Overall consequences rating
Diyab Manifold—Lift gas manifold	Emerson DeltaV ICSS	Emerson DeltaV ICSS	STM network (for DIC) MPLS network DIC Data Collection Node	**Very Low**
Diyab Manifold—Chemical Injection System (corrosion inhibitor/demulsifier)	Emerson DeltaV ICSS	Emerson DeltaV ICSS	STM network (for DIC) MPLS network DIC Data Collection Node	Very Low
Diyab Manifold—Drains system	Emerson DeltaV ICSS	Emerson DeltaV ICSS	STM network (for DIC) MPLS network DIC Data Collection Node	#N/A
Diyab Manifold—Launchers for export pipelines	Emerson DeltaV ICSS	Emerson DeltaV ICSS	STM network (for DIC) MPLS network DIC Data Collection Node	#N/A
Diyab Manifold—Receiver for lift gas pipeline	Emerson DeltaV ICSS	Emerson DeltaV ICSS	STM network (for DIC) MPLS network DIC Data Collection Node	#N/A

vulnerability assessment, and risk assessment specifically tailored for a combined cycle gas turbine (CCGT) power plant. This facility is characterized by its multiple energy blocks, diverse operational facilities, and vital IT infrastructure components, including distributed control systems (DCS) and complex communication systems. Our primary objective is to meticulously evaluate the security posture of the IT infrastructure, to systematically identify vulnerabilities, assess the risks associated with these vulnerabilities, and ultimately provide well-informed, actionable recommendations aimed at enhancing the overall security framework.

Objectives

Identify Vulnerabilities: We will systematically uncover potential security weaknesses within the IT infrastructure. Our exploration will cover all layers of the system, ensuring that no stone is left unturned. **Assess Risks**: We will conduct a comprehensive evaluation of the risks linked to the identified vulnerabilities, analyzing the potential consequences on operational integrity. **Evaluate Security Posture**: A thorough examination will be conducted to determine the effectiveness and robustness of existing security controls, helping us understand their strengths and weaknesses. **Recommend Improvements**: We will provide specific, actionable recommendations designed to enhance security and mitigate potential risks, ensuring a more resilient IT infrastructure.

Scope of Work

IT Infrastructure to Be Assessed
- **Firewalls:** The first line of defense against external threats, we will evaluate the configurations and rules in place.
- **Switches and Routers**: These critical components will be reviewed for security vulnerabilities and potential misconfigurations.
- **Servers (Physical and Virtual):** Both physical and virtual servers will be meticulously assessed to identify vulnerabilities in operating systems, applications, and configurations.
- **Workstations:** We will examine user workstations for security threats and compliance with security standards.
- **Web Applications**: A thorough analysis will be performed to discover potential vulnerabilities in the web applications that are essential for business operations.
- **Wireless Network**: We will evaluate the security measures protecting the wireless network and assess possible points of exploitation.
- **Remote Access Systems**: Security for remote access will be scrutinized to ensure secure connections for remote employees.
- **Surveillance Network and Tools**: We will analyze surveillance systems to ensure their integrity and security, considering their role in operational monitoring.

DCS Systems to Be Considered
Siemens SPPA-T3000: We will assess this system comprehensively to identify any vulnerabilities that could affect performance or security.

GE/Alstom DCS: A thorough evaluation of this DCS will also be conducted to ensure its robustness against potential threats.

Methodology

Preparation and Planning
- **Define Scope:** We will engage with stakeholders to frame and confirm the scope and objectives of the assessment, ensuring alignment with business goals.
- **Resource Allocation**: Our team will be assembled, consisting of skilled professionals with expertise in IT security, and resources will be allocated efficiently for optimal performance.

- **Initial Assessment**: We will perform a preliminary gathering of information regarding both the IT infrastructure and DCS systems to inform the subsequent steps.

Reconnaissance

- **Network Discovery**: Detailed mapping of the network topology will be performed to identify all active devices, forming a comprehensive view of the network landscape.
- **Information Gathering**: We will gather a wealth of information on system configurations, software versions, and existing security measures, setting the stage for deeper analysis.

Vulnerability Scanning

- **Automated Scanning**: Utilizing advanced vulnerability scanning tools, we will systematically identify known vulnerabilities across the network and systems.
- **Manual Testing**: Our team will conduct thorough manual testing to discover complex vulnerabilities that automated tools may miss, employing strategies that reflect real-world attack scenarios.

Penetration Testing

- **Exploit Testing**: We will attempt to exploit identified vulnerabilities to gauge their potential impact and understand the extent to which they could be leveraged against the organization.
- **Access Control Testing**: The effectiveness of current access controls and permissions will be rigorously evaluated, ensuring that only authorized users have access to critical systems.

Risk Assessment

- **Impact Analysis**: We will assess the potential business impact of each vulnerability identified, taking into account operational priorities and the nature of the vulnerabilities.
- **Likelihood Assessment**: Each identified vulnerability will be analyzed to determine the likelihood of exploitation, providing a clear risk profile for decision-making.

Reporting

- **Executive Summary**: We will craft a high-level overview of our findings and recommendations, tailored for upper management and key decision-makers.
- **Detailed Findings**: A comprehensive presentation of our detailed findings will be provided, including categorization of severity levels and associated risk assessments for full transparency.

- **Recommendations:** We will lay out specific and targeted recommendations for mitigating the identified risks, providing clear pathways to enhance security.

Tools Used

- **Vulnerability Scanners**: We will deploy leading tools such as Nessus and OpenVAS, which are crucial for identifying vulnerabilities at scale.
- **Penetration Testing Tools**: Our toolkit will include Metasploit, Burp Suite, and Nmap, which are essential for conducting thorough penetration testing.
- **Network Analysis Tools**: We will utilize analysis software such as Wireshark and TCP Dump to monitor traffic and assess the security posture.
- **Web Application Testing Tools:** Tools such as OWASP ZAP and Burp Suite will be employed to rigorously test web applications for vulnerabilities.
- **Risk Assessment Tools**: Risk assessment and analysis will be conducted using tools such as Risk Watch and FAIR, ensuring a comprehensive understanding of risk.

Assessment and Testing Environment

Environment Setup

- **Test Environment:** A controlled environment will be established, replicating the production environment to ensure that test results are applicable and reliable.
- **Data Protection:** Stringent measures will be taken to safeguard sensitive data during the assessment, ensuring compliance with all applicable regulations.

Testing Phases

- **Pre-Assessment**: We will confirm the readiness and defined scope of the testing environment, ensuring that all systems and stakeholders are prepared for the engagement.
- **Active Testing:** Our team will carry out testing adhering strictly to the outlined methodology, ensuring thoroughness and consistency.
- **Post-Assessment:** We will validate our results and ensure that no disruption occurs within normal operations, providing a seamless transition back to routine activities.

Finding Overview

Severity Levels

- **Critical:** Immediate corrective action will be mandated; these vulnerabilities pose a severe risk to operational integrity.
- **High:** Significant vulnerabilities with the potential for substantial negative impact will be highlighted, requiring prompt attention.

- **Medium:** Vulnerabilities of moderate impact will be noted, necessitating remediation but not requiring immediate action.
- **Low:** Minor vulnerabilities that present minimal risk will be documented for future reference.

Risk Assessment Matrix

Severity Versus Impact: We will categorize risks in relation to their severity and the potential impact on business operations, utilizing a clear matrix format for easy reference.

Security Controls and Levels

- **Current Controls**: A thorough review of existing security controls and their effectiveness will be conducted, emphasizing a holistic understanding of security posture.
- **Control Recommendations**: We will propose robust enhancements or additional controls aimed at effectively addressing any identified vulnerabilities.

Training Needs

- **Staff Training:** Identifying training needs for IT staff will be a priority based on our assessment findings, ensuring ongoing professional development.
- **Security Awareness**: Comprehensive security awareness training will be developed and delivered for all employees, fostering a culture of security throughout the organization.
- **Technical Training**: Dedicated technical training will be provided for system administrators, focusing on new security measures and best practices for optimal performance.

Deliverables

- **Comprehensive Report**: A meticulously detailed report will be produced, encompassing all findings, risk assessments, and recommendations, serving as a valuable resource for ongoing security efforts.
- **Executive Summary**: A concise, high-level summary suitable for senior management will be crafted, ensuring clarity on key findings and recommendations.
- **Presentation:** We will prepare and deliver a structured presentation of our findings and recommendations to key stakeholders, articulating the insights gained and the proposed next steps. This descriptive rewrite aims to provide a thorough understanding of the project scope, methodology, and deliverables, ensuring clarity and depth in articulating the approach to assessing and enhancing the security posture of the CCGT power plant's IT infrastructure.

Time Line (Table 12.4)

Table 12.4 IT Penetration Assessment and Framework. A table outlines a project timeline with phases, durations, and unspecified start and end dates. The phases include Preparation and Planning (2 weeks), Reconnaissance (1 week), Vulnerability Scanning (2 weeks), Penetration Testing (3 weeks), Risk Assessment (1 week), Reporting (2 weeks), and Review and Finalization (1 week). All start and end dates are marked as TBD (to be determined)

Phase	Duration	Start Date	End Date
Preparation and Planning	2 weeks	TBD	TBD
Reconnaissance	1 week	TBD	TBD
Vulnerability Scanning	2 weeks	TBD	TBD
Penetration Testing	3 weeks	TBD	TBD
Risk Assessment	1 week	TBD	TBD
Reporting	2 weeks	TBD	TBD
Review and Finalization	1 week	TBD	TBD

Self-Assessment Questions

? Self-assessment question 12.1

Describe the system map and its operational model of the selected Energy company's ICS, providing the spectrum.

? Self-assessment question 12.2

Develop a project plan work for (its penetration testing, vulnerability, and risk assessment framework) for power utility operations.

Bibliography

1. Yaofu Cao; Xiaomeng Li; Junwen Liu; "Research on interacted response technology of cyber security protection devices based on deep reinforcement learning oriented to new generation of power system" 2nd International Conference on Electrical Engineering and Control Science (IC2ECS) 2022
2. Dabin Sun; Bowei Wang "Research on the Design of the Implementation Plan of Network Security Level Protection of Information Security" 7th International Symposium on Mechatronics and Industrial Informatics (ISMII) 2021
3. Dong Liu; Yang Liu; Zhicheng Liu "Analysis and Reflection on the Situation of Industrial Information Security Ransomware Attacks" 8th International Conference on Data Science in Cyberspace (DSC) 2023
4. Xinli Xiong; Qian Yao; Qiankun Ren "Mission-Oriented Security Framework: An Approach to Embrace Cyber Resilience in Design and Action" 7th International Conference on Cryptography, Security and Privacy (CSP) 2023

Self-Assessment Question's Answers: Chapter Wise

Chapter 2: Transducers

SAQ 2.1

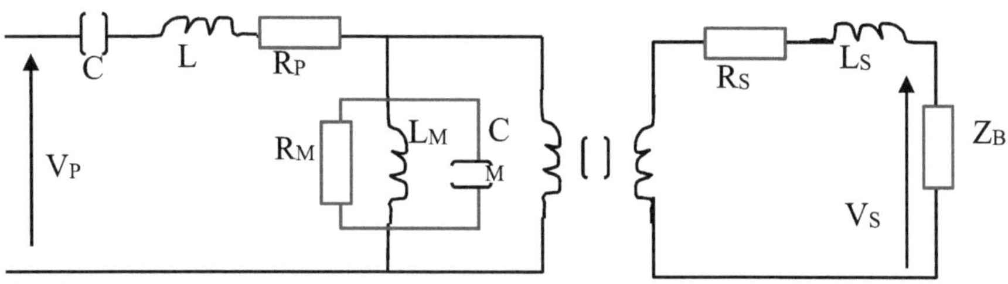

SAQ 2.1a figure

The transient response of CVT is much inferior to that of a conventional VT. The circuit above shows that CVT is a series resonant circuit. For example, under conditions of sudden voltage (such as due to a fault) on the primary, the interaction of series capacitance C and magnetizing inductance L_M can cause oscillation at, typically, 10 kHz. Also, under similar conditions, oscillations at around 300 Hz can greatly influence the transient response of a CVT. It can dominate the CVT behavior, allowing the same CVT with two different burdens to exhibit different transient. In particular, a reactive burden constitutes an additional tuned circuit, further a reactive burden constitutes an additional tuned circuit, further degrading the CVT transient response.

T. Masood et al., *Protection of Electrical Power Transmission Systems*, Power Systems, https://doi.org/10.1007/978-3-031-94312-6

SAQ 2.1 b

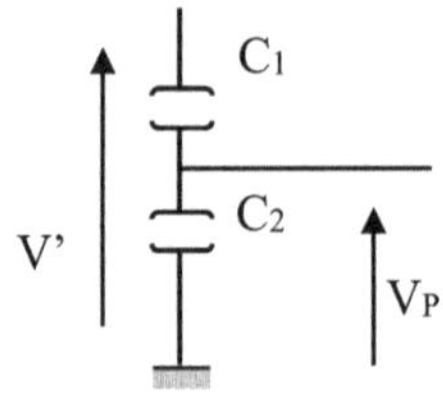

SAQ 2.1b figure

$$V_P = \frac{V_P}{C_1 + C_2} V'_p$$

where
C_1 = high-voltage capacitor
C_2 = low-voltage capacitor
Now

$$V'_P = \frac{400}{\sqrt{3}} kV = 231 kV \; and \; V_P = 10 kV$$

Therefore

$$10 = \frac{231 C_1}{C_1 + C_2}$$

This gives C2 as:

$$C_2 = \frac{221 C_1}{10} = \frac{221 \times 1000 \times 10^{-12}}{10} F$$

Or

$$C_2 = 22.1 nF$$

(ii) In the equivalent circuit

$$C = C_1 + C_2 = (22.1 + 1) nF$$

Or

$$C = 23.1 nF$$

Now the tuning inductance L is chosen to tune out C at power frequency, that is, 50 Hz.

We thus have:

$$\omega_0 L = \frac{1}{\omega_0 C}$$

or

$$L = \frac{1}{\omega_0^2 C} = \frac{10^9}{314.2^2 \times 23.1} = 438H$$

(iii)

$$L_M = 4kH$$

$$C = 23.1nF$$

We can say that

$$\omega L_M = \frac{1}{\omega C}$$

$$\omega^2 = \frac{1}{L_M C} = \frac{10^9}{23.1 \times 4 \times 10^9}$$

This gives

$$\omega = 104\,rad$$

or

$$f = 16.5\,Hz$$

The resonating frequency is 1/3rd of the nominal system frequency. This condition can occur when energizing a CVT and is called the ferroresonance phenomenon. A resistive burden Z_B reduces the likelihood of such a condition, and it is also common to build in special anti-ferroresonance devices.

SAQ 2.2 a

The primary factors of concern in the design of a CT are the quality of the CT core and the cross-sectional area of the core. A very high quality core material is required to keep the magnetizing (or exciting) impedance at a high level, thus minimizing the exciting current (the latter is mainly responsible for the errors in a CT). In a CT, the core has to support large flux swings particularly when the primary current is offset under transient conditions. Thus it is important to ensure that the core cross-sectional area is sufficiently large to support a large flux. SAQ 2.2 figure shows the equivalent circuit of a CT.

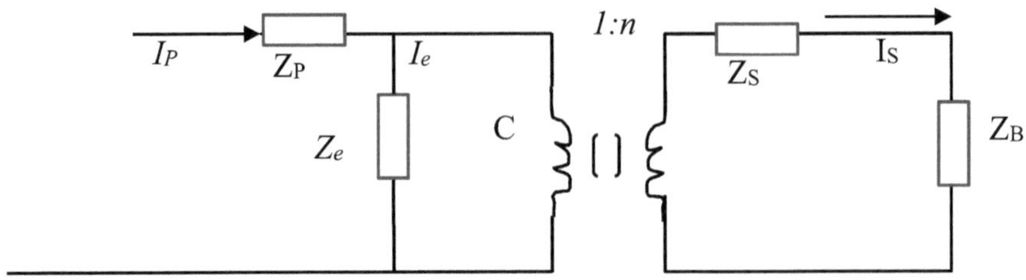

SAQ 2.2 figure

From the above, it can be seen that:

$$I_s\left(\text{referred to the primary}\right) = I_P - I_e$$

where I_e is the exciting current causing the errors in the secondary. Since I_e is a complex quantity, it is apparent that it causes both a phase and an amplitude errors in the secondary current. The ratio error can be partially compensated by reducing the secondary winding by one or two turns. The phase error can be reduced by making the burden Z_B moderately inductive so that I_S and I_e are roughly in phase.

SAQ 2.2 b

The primary fault current

$$i_P = I_P\left[\sin\left(\omega_0 t - \frac{\pi}{2}\right) + e^{\frac{-Rt}{L}}\right]$$

where R_b is the burden resistance referred to the primary.

Therefore, the flux developed across the CT core is

$$\Phi = \frac{1}{N_P}\int_{t_1}^{t_2} V_P dt$$

Integrating for each component in turn gives

The steady-state peak flux:

$$\Phi_A = \frac{I_P R_b}{N_P}\int_{\pi/\omega 0}^{3\pi/2\omega 0}\sin\left(\omega_0 t - \frac{\pi}{2}\right)dt = \frac{I_P R_b}{N_P \omega_0} \tag{A1}$$

The transient flux is given by:

$$\Phi_B = \frac{I_P R_b}{N_P}\int_0^{\infty} e^{\frac{-Rt}{L}} dt = \frac{I_P R_b}{N_P}\times\frac{L}{R} \tag{A2}$$

From A1 and A2, we can see that:

$$\Phi_B = \frac{\omega_0 L}{R}\Phi_A = \frac{X}{R}\Phi_A$$

The CT core has to support both fluxes, so that the total flux developed is given as:

$$\Phi_T = \Phi_A\left[1+\frac{X}{R}\right]$$

SAQ 2.2c

Let the flux associated with full load current be

$$\Phi_{fl}$$

Then the saturation flux

$$\Phi_{sat} = 100\Phi_{fl}$$

The steady state-flux associated with the fault current is given as

$$\Phi_A = 10\Phi_{fl}$$

Now from (b) above, the total flux developed across the CT core is given as:

$$\Phi_T = \Phi_A\left[1+\frac{X}{R}\right]$$

Therefore

$$10\Phi_{fl}\left[1+\frac{X}{R}\right] = 100\Phi_{fl}$$

or

$$\frac{X}{R} = 9$$

Chapter 3: Overcurrent Relays

SAQ 3.1

Source Impedance

"Source capacity varies between 500 and 100 MVA."
33 kV system

$$Z_s = \frac{V^2}{Z}$$

$$Z_{s500} = \frac{33\times 33}{500} = 2.178\,\Omega$$

$$Z_{s1000} = \frac{33 \times 33}{1000} = 1.089\,\Omega$$

Relay location	In-zone fault impedance minimum	In-zone fault impedance maximum	In-zone fault current maximum	In-zone fault current minimum	Maximum load current	Ratio
A	1.089	2.928	17 495	6507	1050A	1200:5
B	1.839	4.078	10 360	4627	750A	1000:5
C	2.989	6.578	6374	2896	350A	500:5
D	5.489	8.078	3471	2359	150A	250:5

	Relay location	**Maximum relay setting	Maximum relay settings	Chosen relay setting	Chosen PSM%
Note 1	A	2336	2100	2100	2100
Note 2	B	1448	1500	1500	150
	C	1180	700	700	150
	D	Note 3	300	375	150

**chosen to provide back-up protection for the following feeder

Note 1: A setting of 200% would have given a primary equivalent settings of 2400 A. the settings of 175% is in range

Note 2: A compromise is required, but when the available settings are considered, 150% is the only choice.

Note 3: The minimum fault current expected in the load circuit is unknown. A sensitive setting is therefore preferred provide that it is greater than $2(I_{LOAD})_{MAX}$

TMS Settings for Relay C

Grading between relays C and D is required for the maximum fault current experienced by relay D. For this current

$$t_D = \frac{13.5T}{\left[\frac{I_f}{PI_N}\right] - 1}$$

$$t_D = \frac{13.5 \times 0.1}{\left[\frac{3471}{375}\right] - 1} = 164\,ms$$

For the grading margin of 400 ms, t_c operating time is therefore

$$t_c = 164 + 400 = 564\,ms$$

The TMS settings for relay C t_c is therefore derived from:

$$t_C = \frac{13.5T_c}{\left[\frac{I_f}{PI_N}\right] - 1}$$

that is,

$$0.564 = \frac{13.5T_c}{\left[\frac{3471}{375}\right] - 1}$$

$$T_c = 0.152$$

This setting is not available. The alternatives are 0.1 and 0.2. A setting of 0.1 would dramatically erode the grading margin whereas as setting of 0.2 would provide an enhanced grading margin. The choice of T_c is:

$$\therefore T_c = 0.2$$

Similarly for relay B its setting is chosen from the grading between relay B and C at the maximum fault current experienced by relay C.

$$\therefore t_c = \frac{13.5T_c}{\left[\frac{I_f}{PI_N}\right] - 1}$$

$$t_c = \frac{13.5 \times 0.2}{\left[\frac{6374}{750}\right] - 1} = 360\,ms$$

$$t_c = 360 + 400 = 760\,ms$$

T_B is obtained from

$$t_B = \frac{13.5T_B}{\left[\frac{I_f}{PI_N}\right] - 1}$$

$$0.760 = \frac{13.5T_B}{\left[\frac{6374}{750}\right] - 1}$$

$$T_B = 0.183$$

As before the available settings are 0.1 or 0.2, the chosen setting is:

$$T_B = 0.2$$

Similarly for relay A, its setting is chosen form the grading between relays A and B at the maximum current experienced by relay B.

$$t_B = \frac{13.5T_B}{\left[\frac{I_f}{PI_N}\right] - 1}$$

$$t_B = \frac{13.5 \times 0.2}{\left[\frac{10360}{1500}\right] - 1} = 457$$

$$t_A = 457 + 400 = 857\,ms$$

T_A is obtained from

$$t_A = \frac{13.5T_A}{\left[\frac{I_f}{PI_N}\right] - 1}$$

$$0.857 = \frac{13.5T_A}{\left[\frac{10360}{2100}\right] - 1}$$

$$\therefore T_A = 0.250$$

The available settings are 0.2 or 0.3. In order to maintain grading margin, the chosen setting is:

$$T_A = 0$$

Chapter 4: Distance Protection

SAQ 4.1

The classic quadrilateral characteristic, as shown in Figure 4.14 can be produced using four linear characteristics as shown in SAQ 4.1 figure as where a trip output is given all four characteristics are in their operate state.

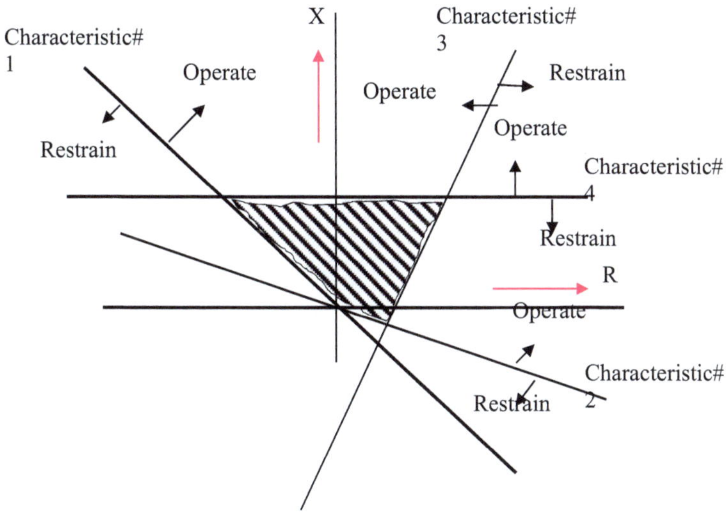

SAQ 4.1 figure a

Characteristics 1: A directional characteristic required to constraint the quadrilateral in the fourth quadrant

Characteristics 2: Also, a directional characteristic providing the main directional response of the quadrilateral

Characteristics 3: A linear characteristic, similar to a load blinder, providing constraint on the resistive reach and reducing problems with load encroachment.

Characteristics 4: A linear characteristic, similar to a reactive characteristic, providing the reach control for the protected line with some droop to accommodate load export and import.

The trip output is provided when all four comparators have operated. An antirace time delay is introduced to prevent am-functioning which could arise from the different response times of the comparators providing the characteristics shown in SAQ 4.1 figure b.

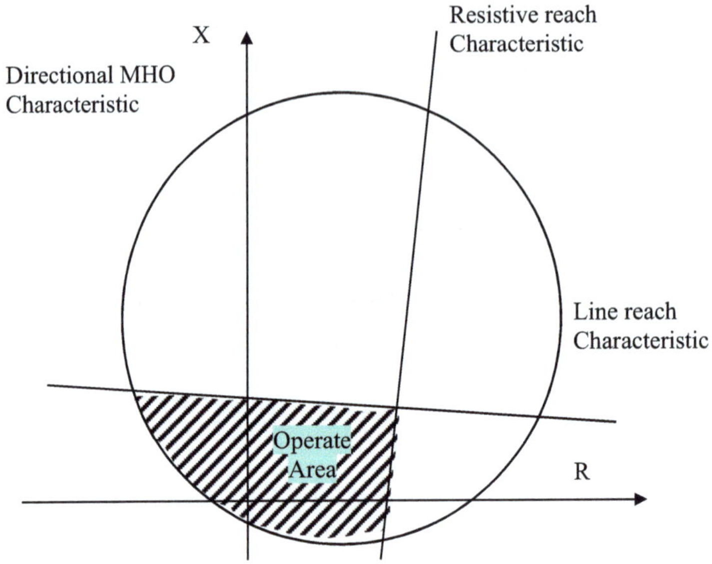

SAQ 4.1 figure b

In this implementation, the MHO characteristics provides the features previously provided by characteristics 1 and 2. An antirace time delay is still required for the final trip command to prevent malfunctioning which could arise from the different response times of the comparators.

SAQ 4.2

The positive sequence impedance of the line AB, Z_{L1} is given by:

$$Z = \ell_{AB} \times Z_1$$

$$Z_{L1} = 150(0.0383 + j0.320)\Omega$$

$$Z_{L1} = 5.745 + j48\,\Omega$$

$$Z'_{L1} = (5.745 + j48)\frac{2000 \times 110}{400000}\Omega$$

$$Z'_{L1} = 3.16 + j26.4\,\Omega$$

The zone 1 reach is set to 80% of this value

$$\therefore Z_{R1} = 0.8(3.16 + j26.4)\Omega$$

$$Z_{R1} = (2.528 + j21.12)\Omega$$

$$Z_{R1} = 2\angle 83.2°\Omega$$

The zone 1 setting would therefore be at

$$Z_{R1} = 21.3\Omega$$

The angle is generally set lower than the line angle, and hence the suggested angle setting is

$$\theta_{R1} = 75°$$

The zone 2 setting is also based on the positive sequence impedance of the protected line AB and

$$Z_{R2} = \frac{120}{100} \times Z_{L1}$$

$$Z_{R2} = \frac{120}{100} \times (3.16 + j26.4)\Omega$$

$$Z_{R2} = 31.9\angle 83.2°\Omega$$

The zone 2 setting would therefore be at:

$$Z_{R2} = 31.9\Omega$$

Again, the angle is generally set lower than the line angle, and hence the suggested angle setting is:

$$\theta_{R2} = 75°$$

The forward reach for the zone 3 characteristic, Z_{R3F} is set by 150% of the positive sequence impedance of lines Ab and BC as seen by the relay. Therefore:

$$Z_{R3F} = \frac{150}{100}(\ell_{AB} + \ell_{BC})Z_i \frac{CT}{VT}$$

$$Z_{R3F} = \frac{150}{100}(150 + 50)(0.0383 + j0.32)\left(\frac{2000}{1} \frac{110}{400000}\right)\Omega$$

$$Z_{R3F} = 6.32 + j52.8\Omega$$

$$Z_{R3F} = 53.18\angle 83.2°\Omega$$

The forward reach setting for zone 3 is therefore

$$Z_{R3F} = 53.2\Omega$$

With an angle of:

$$\theta_{R3} = 75°$$

Using the argument already presented for Z_1 and Z_2

The reverse reach for the zone 3 characteristic, Z_{R3R} is set to 10% of the positive sequence impedance of the line AB as seen by the relay, therefore

$$Z_{R3R} = \frac{10}{100}(\ell_{AB})Z_i\frac{CT}{VT} = \frac{10}{100}\times 150(0.0383 + j0.32)\left(\frac{2000}{1}\frac{110}{400000}\right)\Omega$$
$$= 0.316 + j2.64\,\Omega = 2.66\angle 83.2°\Omega$$

The reverse reach setting for the zone 3 is therefore

$$Z_{R3R} = 2.66\,\Omega$$

With an angle of

$$\theta_{R3R} = 75°$$

The residual compensation setting K, is derived from:

$$K = \frac{(Z_0 - Z_1)}{3Z_1}$$

$$K = \frac{(0.1474 + j0.837) - (0.0383 + j0.32)}{3(0.0383 + j0.32)}$$

$$K = \frac{0.1091 + j0.517}{0.1149 + j0.96}$$

$$K = \frac{(0.5088 - j0.0453)}{0.9669}$$

$$K = 0.5283\angle -5°\Omega$$

The residual compensation setting is therefore:

$$K = 0.5283$$

SAQ4.3

To provide MAIN A and MAIN B protection schemes, the two scheme should be able to operate independently and each provide instantaneous protection for the protected feeder. Using this approach, a failure of MIAN A protection should not compromise the operation of the MAIN B protection and vice versa.

The simplest approach to the MIAN A protection is to use the scheme zone 1 characteristics directly. As shown in SAQ 4.3 figure a the relay at A will provide instantaneous protection for it zone 1 and the relay of B will provide instantanous protection for it zone 1.

Since the main B protection has to be independent of the MAIN A protection, it must therefore relay on the zone 2 and or zone 3 characteristics.

Combining the operation of the zone 2 characteristic with communication channel provides the **PERMISSIVE OVERREACHING SCHEME**. This scheme provides instantaneous tripping and satisfied the needs of a MAIN B scheme. This scheme is illustrated in SAQ4.3 figure b. The zone 3 characteristics can be used to complement the operation of the "permissive overreaching scheme, and also provide overall back-up protection.

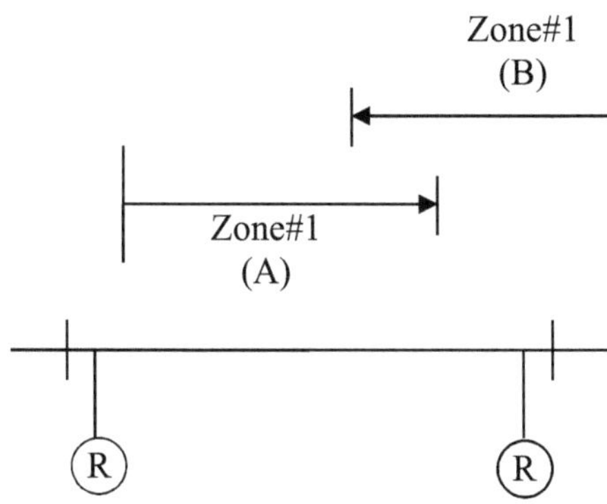

SAQ4.3 figure a: Main A protection

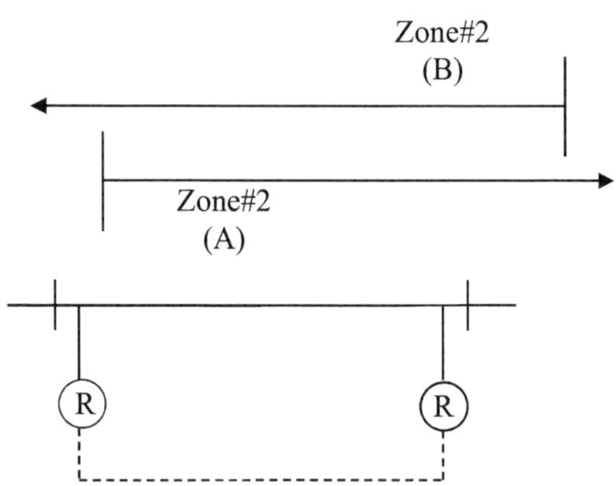

SAQ4.3 figure b: Main B protection

This type of protection has been widely used for transmission lines operating at all voltages. It is particularly attractive when using single zone, fault scheme distance relays, where separate relays are required to provide the zone 2 and zone 3 characteristics. In several schemes, the zone 3 characteristics are omitted and back-up protection is provided by the time delayed zone 2 tripping or by IDMT overcurrent relays.

Chapter 5: Current Differential Protection

SAQ 5.1 a

In unit protection schemes such as current differential, it is necessary to transmit instantaneous values of the currents from one end of the protected line to the other, for comparison purposes. The use of three separate channels for the transmission of three-phase currents is severely expensive and often not very practicable. Hence, we arrange to condense the three-phase currents into a single quantity for transmission over the communication channel. The two most common methods employed are:

1. summation transformer shown below **SAQ 5.1 figure a,**

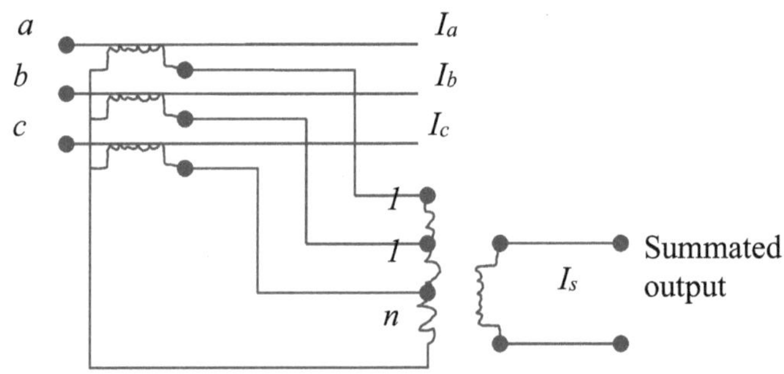

SAQ 5.1 figure a

The winding ratio of the transformer is 1:1:n so that the summated output is given as:

$$I_s \propto (n+2)I_a + (n+1)I_b + nI_c$$

Two-phase sequence segregating network: shown below in **SAQ 5.1 figure b**

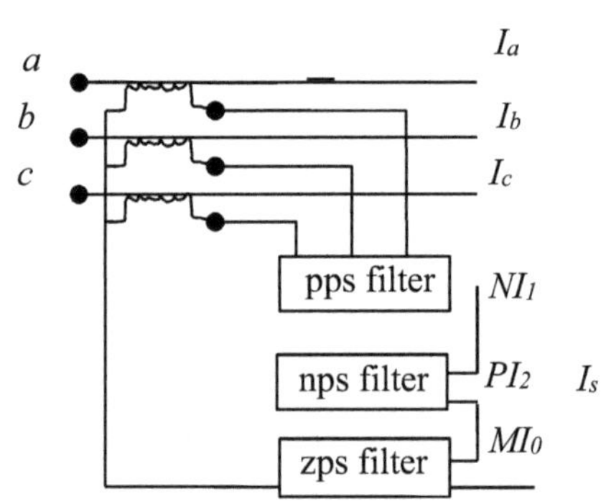

SAQ 5.1 figure b

The ratio of the phase currents is arranged such that the summated output is given as:

$$I_s \propto NI_1 + MI_2 + PI_0$$

In the summation transformer case, it is quite possible that, under certain fault conditions, the faulted phase current can be masked by the currents flowing in the healthy phases giving rise either to a zero or a misleading signal (i.e., a signal of the wrong polarity) through the summated transformer. This occurs because the transformer ratio employed is not conducive to all the fault conditions encountered in practice. A phase sequence segregating network is far less likely to cause blind spot problems.

SAQ 5.1 b

For the "c" phase – earth fault, the phase sequence network connection is as follows in SAQ 5.1 figure c.

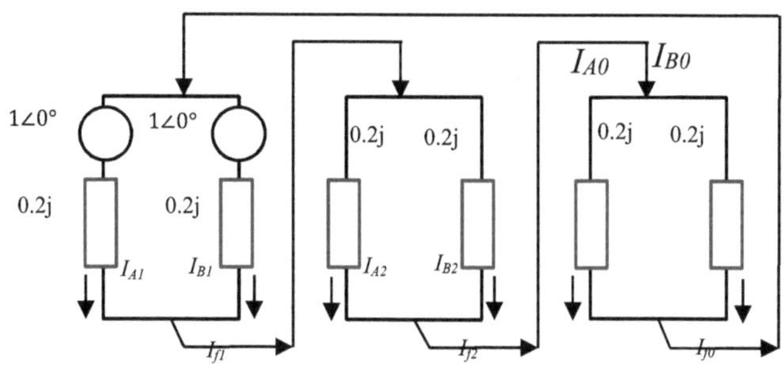

SAQ 5.1 figure c

In a simplified form, this circuit becomes (as shown in SAQ 5.1 figure d):

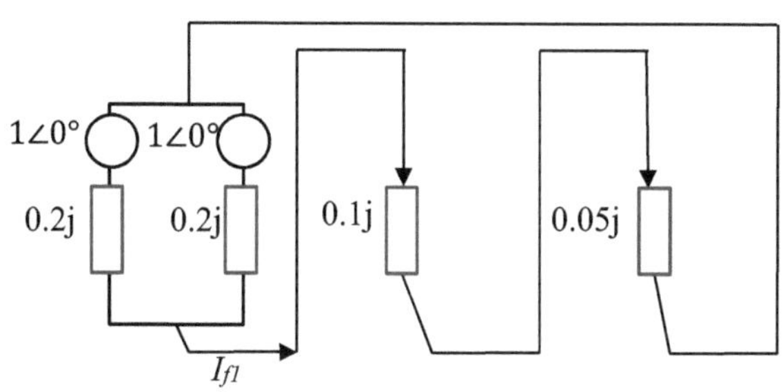

SAQ 5.1 figure d

This gives

$$I_{f1} = I_{f2} = I_{f0} = j4.0\,p.u.$$

Thus for the two ends we have:

$$I_{A1} = I_{B1} = j2.0$$

$$I_{B1} = I_{B2} = j2.0$$

$$I_{A0} = j0.25$$

$$I_{B0} = j3.75$$

Now in the case of the summation transformer:
For end A

$$I_A \propto 5I_a + 4I_b + 3I_c$$

or

$$I_A \propto 5\left[a^2 I_{c1} + aI_{c2} + I_{c0}\right] + 4\left[aI_{c1} + a^2 I_{c2} + I_{c0}\right] + 3\left[I_{c1} + I_{c2} + I_{c0}\right]$$

Taking "c" phase as reference
For end "A" we have

$$I_{c1} = I_{c2} = j2.0$$

and

$$I_{co} = j0.25$$

This then gives

$$I_A \propto (-3j)\,p.u.$$

Likewise, for end B, we obtain

$$I_B \propto (39j)\,p.u.$$

The currents through the two CTs are of opposite polarity and hence the phase comparison device will indicate external faults (see notes), that is, unsatisfactory operation.

For the phase sequence network:
For end A:

$$I_A \propto 6I_{a2} - I_{a1} = 6aI_{c2} - a^2 I_{c1}$$

$$I_A \propto j2.0\left(6a - a^2\right)$$

Likewise for end B

$$I_B \propto j2.0\left(6a - a^2\right)$$

The currents through the two CTs are cophasal. The device will therefore indicate an internal fault correctly.

SAQ 5.2 a

The principal of a balanced voltage differential feeder protection can be explained with reference to SAQ 5.2 figure a below:

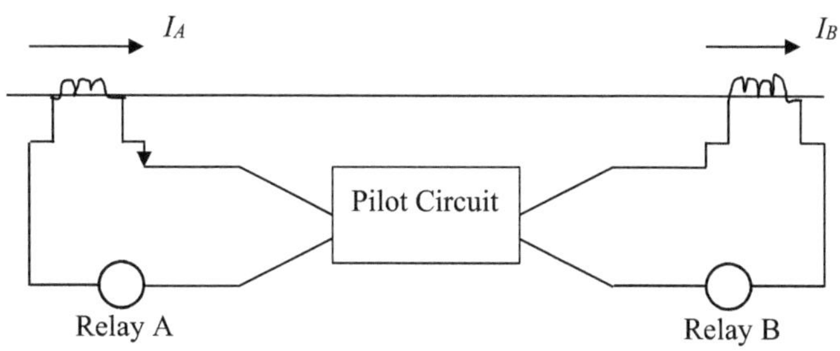

SAQ 5.2 figure a

The CTs are connected in opposition so that under external faults (or normal operations) no current flows in the series connected relays. This is so because the secondary e.m.f.s of CTs are opposed. For an internal fault, the e.m.f.s developed across the secondary of CTs are no longer the same, and a current thus flows through the relays. In a double end-fed system it is essential to have near simultaneous tripping of the two circuit breakers so as to give sufficient time for the fault arc path to deionize and regain the withstand voltage level before a successful reclosure can take place after a prescribed period.

SAQ 5.2 a

The tripping criterion for end A relay is

$$\left|\frac{I_A}{I_B}-1\right| \geq 0.5$$

For end B relay, it is:

$$\left|\frac{I_A}{I_B}-1\right| \geq 0.5$$

or

$$\left|\frac{I_A}{I_B}-\frac{4}{3}\right| \geq \frac{2}{3}$$

The tripping characteristic of the two relays plotted on the same axes are thus:

$$\frac{I_A}{I_B} = 1.5 + j0.5\, is\ at\ X$$

This is outside the stability zone of end A relay but inside the stability zone of end B relay. Hence the fault will not be detected by R_{L2} at the same time as R_{L1}.

An alternative characteristic for R_{L2} is

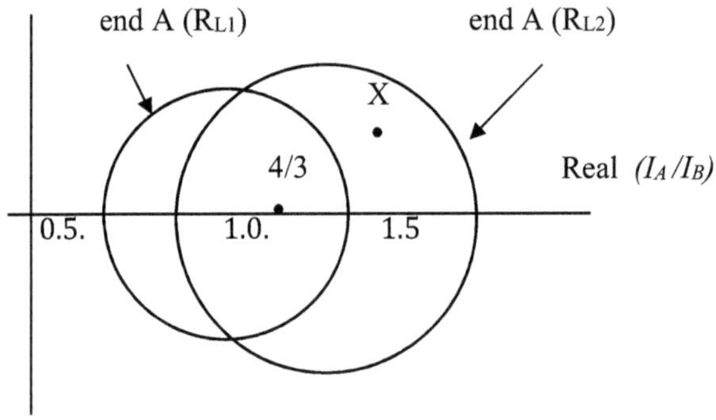

SAQ 5.2 figure b

Clearly, from SAQ 5.2 figure b above, it can be seen that the point corresponding to the fault current ratio.

$$\left|\frac{I_A}{I_B} - \frac{4}{3}\right| \geq \frac{2}{3}$$

This is equivalent to:

$$\left|\frac{I_A}{I_B} - 1\right| \geq 0$$

Chapter 6: Auto-Reclose for Transmission Systems

SAQ 6.1

To discriminate between single-phase and multi-phase faults, the auto-reclose relay needs separate inputs from the a, b, and c phase protection trips or the pole status contacts from the circuit breaker. These three inputs could control repeat relays since multiple contacts are required for the scheme logic.

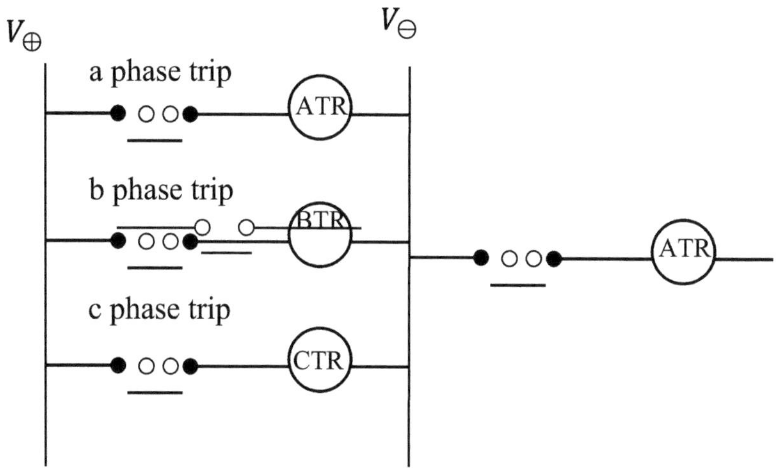

SAQ 6.1 figure a

The first set of contacts from these relays are used to initiate the auto-reclose

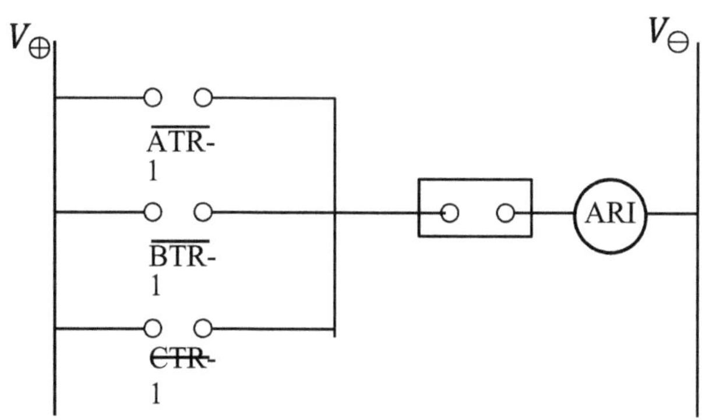

SAQ 6.1 figure b

However, should two phases trip or two poles on the circuit breaker open, the scheme is required to lock-out immediately, therefore:

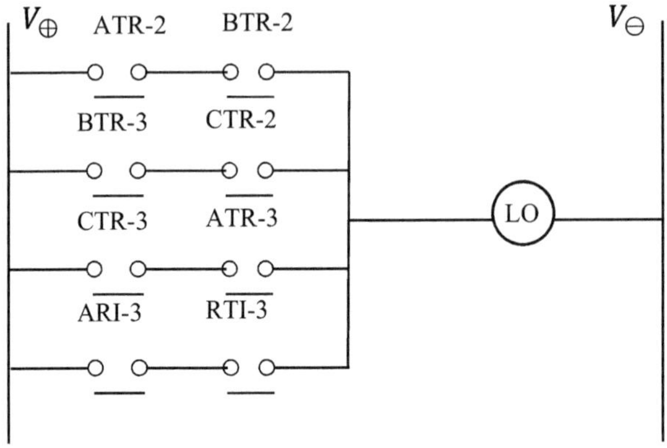

SAQ 6.1 figure c

To include check synchronization and live line/dead bus and deadline/ live bus supervision into the scheme, contacts from these relays needs to be included in the circuit driving the circuit breaker close relay.

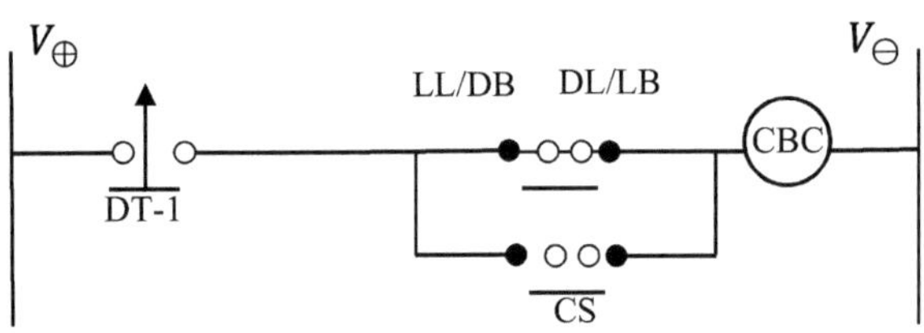

SAQ 6.1 figure d

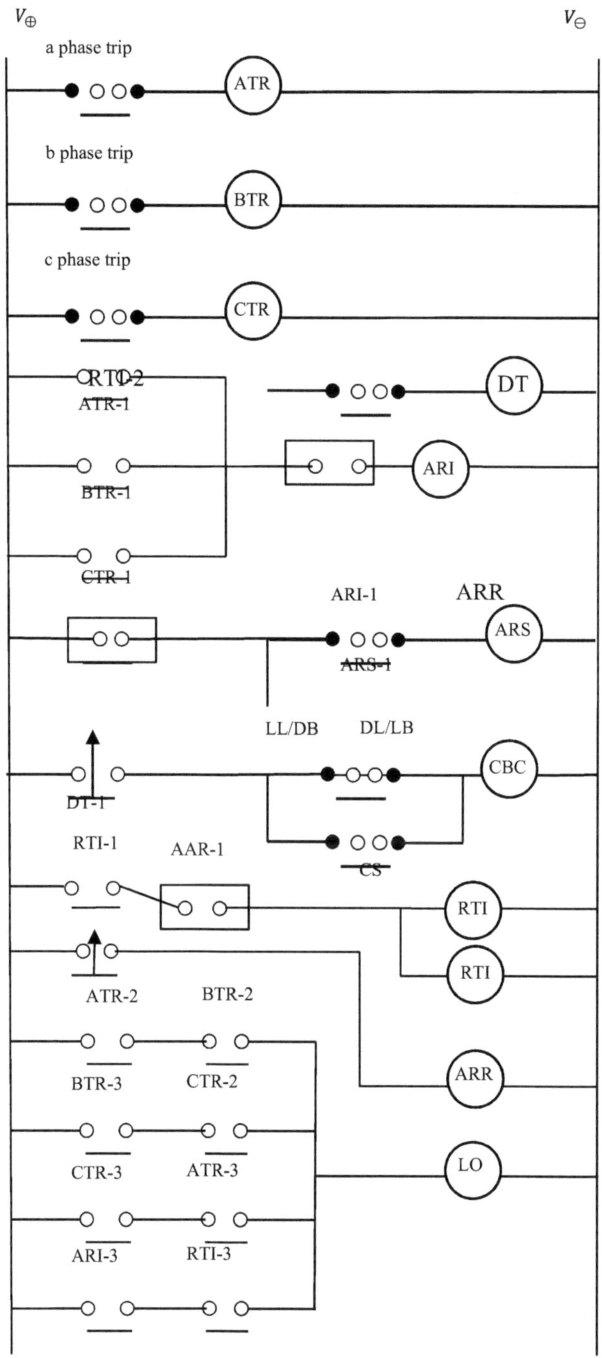

SAQ 6.1 figure e

Chapter 7: Numeric Protection

SAQ 7.1 a

In twos complement from, a binary number is positive if it begins with a zero, and is negative if it begins with a one. Starting from the right-hand side of a binary number and moving to the left the decimal equivalent of the nth bit is given by 2^{n-1}, for example, the first four bits represent:

8 4 2 1

– remember that 2^0 is defined as 1.

The first step is to convert the binary number into decimal from. This is straightforward for positive numbers:

i) decimal equivalent
$$= 2^7 + 2^3 + 2^2 + 2^0 = 128 + 8 + 4 + 1 = 141$$

ii) decimal equivalent
$$= 2^{11} + 2^{10} + 2^6 + 2^5 + 2^1 + 2^0$$
$$= 2048 + 1024 + 64 + 32 + 2 + 1 = 3171$$

For negative twos complement numbers, we have to remember that the leading (i.e., left hand side) bit is negative.

iii) decimal equivalent
iv)
$$= -2^{15} + 2^{14} + 2^{13} + 2^{12} + 2^{11} + 2^{10} + 2^6 + 2^5 + 2^1 + 2^0$$
$$= -32768 + 16384 + 8192 + 4096 + 2048 + 1024 + 64 + 32 + 2 + 1 = -925$$

Now, to find the analog equivalent current, we multiple the decimal equivalent by the current per binary step (0.0375 A)

i) Current = 141 × 0.0375 = 5.288 A
ii) Current = 3171 × 0.0375 = 118.9 A
iii) Current = – 925 × 0.0375 = – 34.69 A

SAQ 7.1 b

i) The programing instructions for the relaying algorithm are in constant use when the relay is monitoring and do not change over time. Furthermore. It would be inconvenient if these instructions were lost every time the relay is powered down. For these reasons, programming instruction are stored within read only memory (ROM). That is memory, the contents of which cannot be changed by the microprocessor. ROM contents are not lost during power down. The contents of a ROM are programed by the manufacturer.
ii) Sampled values of voltage and current have a finite lifetime within the relaying algorithm, that is to say that the samples will not be involved in any calculation

beyond a certain time limit. This limit will vary according to the algorithm, but a typical value is the time period of one cycle (i.e., 20 ms). In order that "old" can be discarded when they are of no further use. It is imported the storage for the sample can be overwritten with "new" samples. Furthermore, it will be apparent that storage of sampled values during a power down will not serve any useful purpose. Random access memory (RAM) used for this storage since it can be both read from and written to by the microprocessor. RAM contents are lost during power down.

iii) In contrast to (i) and (ii) relay setting date requires a memory that can be changed since settings may need occasional operational changes yet need to be retained during power downs since it would be inconvenient to have to reprogram the settings continually. This requirement is not full filled by either ROM or RAM. Instead, a form of nonvolatile RAM—the contents can be changed by the microprocess—but is also capable of power -down storage. Atypical implementation of a NVRAM to be found in a digital protection relay is the electrically alterable read only memory (EAROM)

SAQ 7.2 a

A typical solution to this problem is shown in the diagram below

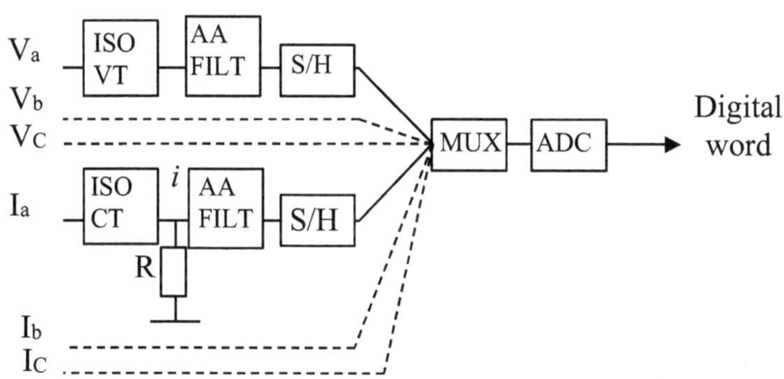

SAQ 7.2 figure

The components blocks are further described

ISO_{VT} or ISO_{CT}	Isolating transformer, either a voltage transformer (VT) type of a current transformer (CT) type. The isolating transformer provides galvanic between the substation environment and the relay electronics
AAFILT	Anti-aliasing filter used to prevent frequency of greater than half of the sampling frequency being converted to discrete form
S/H	Sample and hold amplifier used to "hold" the input analog signal at a constant level for the duration of the analog-to-digital conversion process. Each analog channel has its own S/H thus enabling the signals to be converted at the same instant in time. The S/H is under control of the microprocess that 5 sets all the S/H into "Hold" mode at the required instants.

MUX	Multiplexer is used to direct each of the six analog input signals into the single analog-to-digital converter (ADC) sequentially. The use of one MUX and one ADC is more cost-effective than using six ADCs since these devices are relatively expensive
ADC	Analog-to-digital converter converts the analog signal into discrete time form. ADCs used in digital protection relays are usually of the successive approximately type and typically have a resolution of 14–16 bits

SAq 7.2 b

Since there are six input channels in this relay, the total number of instruction cycles to perform all activities except the house keeping tasks

$$\left[6(153+135)+961+720\right]=35409$$

However, housekeeping tasks will add an extra 10% to this making

$$1.1\times 3409=3750$$

Now, the time for one instruction cycle is equal to the time for one clock cycle. Hence, the time for one instructions cycle is:

$$=\frac{1}{15\times 10^{6}}s$$

Therefore, the total execution time is

$$=\frac{3750}{15\times 10^{6}}s$$

which is the time required to process one sample. Thus, the maximum sample frequency is the reciprocal of the time required to process one sample:

$$=\frac{15\times 10^{6}}{3750}=4\,kHz$$

SAQ 7.3

The equation for finite impulse response filter having six coefficient is:

$$i_{filt}\left[nT\right]=i\left[nT\right]\times H_1+i\left[(n-1)T\right]\times H_2+i\left[(n-2)T\right]$$
$$\times H_3+i\left[(n-3)T\right]\times H_4+i\left[(n-4)T\right]\times H_5+i\left[(n-5)T\right]\times H_6$$

With the first filter a), it is easy to calculate the results since all the coefficients are 1.0, thus the filter output at $n=5$ is given by:

$$i_{filt}\left[5T\right]=1.440+1.141+0.717+0.301+0.029+0.0=3.628$$

Similarly, the output at $n=6$ is given by:

$$i_{filt}\left[6T\right]=1.513+1.440+1.141+0.717+0.301+0.029=5.141$$

and so on

A table of results for filter a) is shown below

Results for filter a)

n	5	6	7	8	9	10
$i_{filt}[nT]$	3.628	5.141	6.438	7.048	6.699	5.387

11	12	13	14	15	16	17	18
3.375	1.126	−0.830	−2.030	−2.209	−1.369	0.220	2.091

With the second filter b) it is slightly more complicated since the coefficients differ. The filtered output at n = 5 is now given by:

$$i_{filt}\left[5T\right] = -0.5\times1.440 - 1.0\times1.141 - 0.5\times0.717 + 0.5\times0.301 + 1\times0.029 + 0.5\times0.0 = -2.040$$

Similarly, the output at $n = 6$ is given by:

$$i_{filt}\left[6T\right] = -0.5\times1.513 - 1.0\times1.440 - 0.5\times1.141 + 0.5\times0.717 + 1.0\times0.301 + 0.5\times0.029 = -2.093$$

and so on

A table of results for filter a) is shown below

Results for filter b)

n	5	6	7	8	9	10
$i_{filt}[nT]$	−2.040	−2.093	−1.458	−0.319	1.009	2.158

11	12	13	14	15	16	17	18
2.811	2.784	2.075	0.870	−0.516	−1.716	−2.415	−2.429

A graph of the original and two filtered versions of the waveform is shown below. For clarity, the original waveform has been drawn on an expanded scale. It will be apparent from the graph that the original waveform contains an exponential offset. Filter a) fails to remove the offset whereas filter b) does. Filter a) introduces an approx. 90° phase shift, filter b) introduces a 180° phase shift. The filters have differing magnitude gains. In terms of protection, filter b) would be preferable to filter a) since it removes the exponential offset.

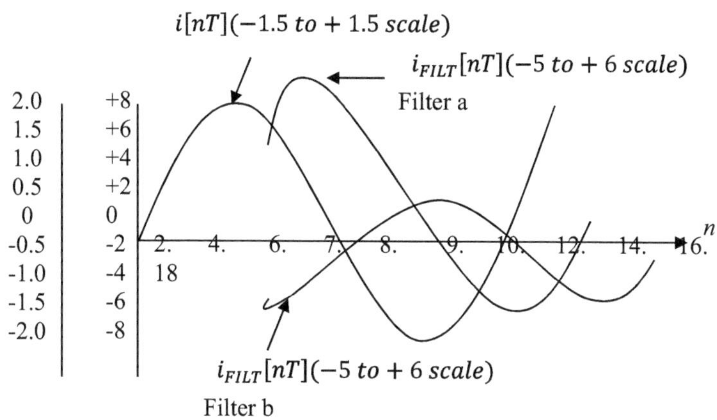

SAQ 7.3 figure

SAQ 7.4 a

The discrete Fourier transform (DFT) converts discrete time sampled values into phasor form.

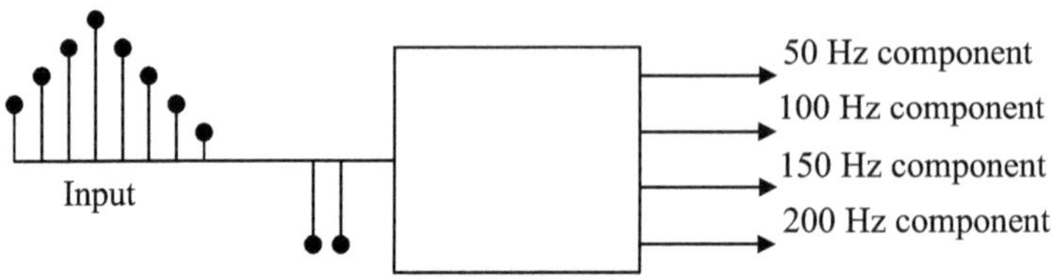

SAQ 7.4 figure a

The basic equation is:

$$I(m) = \int_{n=0}^{N-1} i(k) e^{\left(\frac{j2\pi mn}{N}\right)}$$

where

$i(k)$ is input waveform time series consisting of N points
$I(m)$ is the corresponding phasor value
M is the harmonic order, that is, setting $m = 0$ gives the d.c. content
$M = 1$ gives the fundamental, etc.

The fundamental is made to correspond to the 50-Hz phasor. This is achieved by making the sampling frequency equal to N times 50 Hz

Setting m = 1, and adapting basic equation for convenient use, it can be shown that the real and imaginary components of the phasor I are given by

$$\Re(I) = \frac{2}{N} \int_{n=0}^{N-1} i(k) \sin(n\omega T)$$

$$\Re(I) = \frac{2}{N} \left(\frac{i(0)}{2} + \frac{i(N)}{2} + \int_{n=0}^{N-1} i(k) \sin(n\omega T) \right)$$

SAQ 7.4 b

The sampling frequency for this data is:

$$\frac{1}{T} = \frac{1}{2.5 \times 10^{-3}} = 400\, Hz$$

The variable N in the DFT equation is the sampling frequency divided by the power system frequency,

That is,

$$N = \frac{400}{50} = 8$$

Thus, for the data from end S, the real part of I is given by:

$$\Re(I) = \frac{2}{N}\left(i(0)\sin 0 + i(1)\sin \omega T + i(2)\sin 2\omega T + i(3)\sin 3\omega T\right)$$
$$+\frac{2}{N}i\left[(4)\sin 4\omega T + i(5)\sin 5\omega T + i(6)\sin 6\omega T + i(7)\sin 7\omega T\right]$$

where

$$\omega = 2\pi \times 50.\frac{rad}{s}, T = 2.5\,ms$$

This then gives:

$$\Re(I) = \frac{2}{8}\begin{bmatrix} 1732\times 0 + 1932\sin\frac{\pi}{4} + 1000\sin\frac{\pi}{2} - 518\sin\frac{3\pi}{4} \\ -1732 sin\pi - 1932\sin\frac{5\pi}{4} - 1000\sin\frac{3\pi}{4} + 518\sin\frac{7\pi}{4} \end{bmatrix}$$

Which solve to give:

$$\Re(I) = 1000.0$$

Similarly, the real component solves to give:

$$\Re(I) = 1732$$

$$I_s = 1732 + j1000$$

Application of the Fourier equations to the end R data yields:

$$I_R = -1060 - j1696$$

If we draw these vectors, it will be apparent that they represent current phasors which are almost in anti-phase and thus, this represents and internal fault to the differential relay

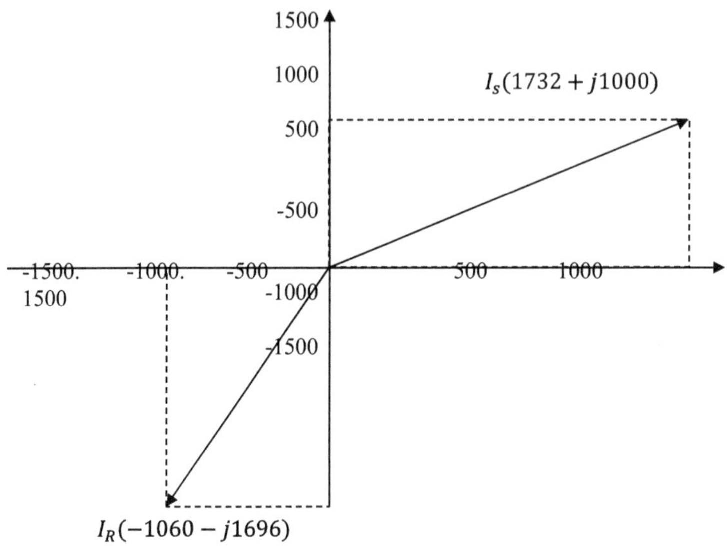

7.4 figure b

SAQ 7.5 a

The fundamental line equation to be solved by a numeric distance relay is:

$$v[nT] = R\,i[nT] + L\frac{di[nT]}{dt}$$

However, to solve this equation for R and L, we need two separate solutions for $v[nT]$ and $i[nT]$. Another solution can be found by using values of voltage and current at an earlier point in time, that is, K samples earlier $v[(n-k)T]$ and $i[(n-k)T]$. Letting $i'[nT]$ represent the time derivative

$$\frac{di[nT]}{dt}$$

We can represent the two solutions of the fundamental line equation in matrix form

$$\begin{bmatrix} v[nT] \\ v[(n-k)T] \end{bmatrix} = \begin{bmatrix} v[nT] & i'[nT] \\ v[(n-k)T] & i'[(n-k)T] \end{bmatrix} \begin{bmatrix} R \\ L \end{bmatrix}$$

By inverting the current matrix and pre-multiplying both sides of the equations, we obtain:

$$\begin{bmatrix} R \\ L \end{bmatrix} = \frac{1}{D} \begin{bmatrix} i'[nT] & i'[nT] \\ -i'[(n-k)T] & i'[(n-k)T] \end{bmatrix} \begin{bmatrix} v[nT] \\ v[(n-k)T] \end{bmatrix}$$

where the determinant of the matrix D, is given by:

$$D = i[nT]i'[(n-k)T] - i'[nT]i[(n-k)T]$$

and time derivative is calculated from

$$i'[nT] = \frac{i[(n-1)T] - i[(n-1)T]}{2T}$$

The reactance may be calculated from the inductance values using

$$X = 2\pi f_0 L$$

where f_0 is the nominal power system frequency?

SAQ 7.5 b

A graph of the behavior of the "a" phase-to-earth trip counter is shown below. The trip counter is assumed to be at zero initially since no other fault is in progress. Note that the counter transient counts to 9 at time $t = 1.5$ ms since both R and X are zone. At the next two sampling instants, the INC is −9 since R is less than the lower resistance line of the characteristic. However, the counter never goes below zero. The counter exceeds the trip threshold of 45 at time $t = 5.5$ ms and thus the relay trips

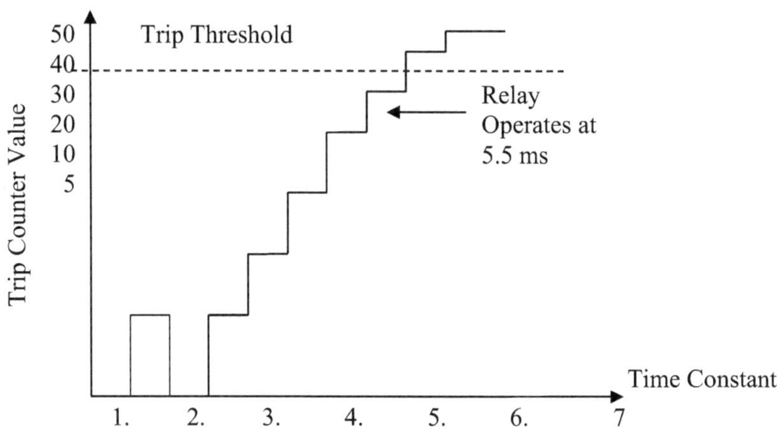

SAQ 7.5 figure

Chapter 8: Numeric Protection

? Self-assessment question 8.1

Figure 9.12 shows a feeder arrangement on which overcurrent protection relays R1 and R2 are employed with CT ratio.

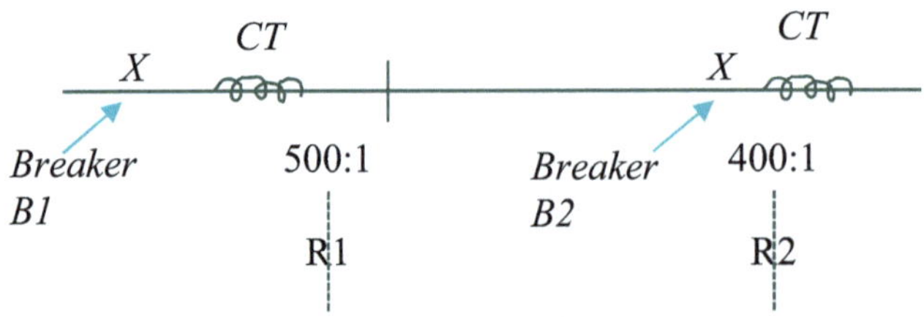

Figure 8.1

Each breaker has an operating time 0.2 s and the plug setting on both relays is 125% for normal overcurrent. If each relay has the inverse characteristics as shown in Figure 9.3 determine the following:

(a) The time multiplier setting for relay R2 if the three-phase fault current is 2 kV and the corresponding fault clearance time is to be 1 s
(b) The time multiplier setting for relay R1 necessary for discrimination
(c) The fault clearance time for a 4-kV three-phase occurring beyond breaker B2 and also for a similar fault occurring between the two breakers.

Will the same time multiplier§ settings be suitable when dealing with earth fault currents of 1 kA if the plug setting for the earth fault element is 40%

SAQ 8.1

(a) The current in relay R_2 as a multiple of plug setting is given by:

$$\frac{100 \times 2000}{400 \times 125} = 4$$

Using the equation for the universal relay characteristic of Figure 9.3, we obtain:

$$t = \frac{0.8}{4^{0.02} - 1} = 4.98\,s$$

The relay must operate in (1 – 0.2) = 0.8 s, therefore the time multiplier setting for the relay must be:

$$\frac{0.8}{4.98} = 0.16$$

(b) The current in relay R1, as a multiple of plug settings is given by:

$$\frac{100 \times 2000}{500 \times 125} = 3.2$$

Using the equation for the universal relay characteristic of Figure 9.3, we obtain:

$$t = \frac{0.14}{3.2^{0.02} - 1} = 5.95\,s$$

R1 must have minimum allowable operating time of 1.0 s. to ensure stability let us assume a safety margin of 0.1 s so that the minimum allowable time for operation becomes 1.1 s. then the time multiplier setting for relay R1 is:

$$\frac{1.1}{5.95} = 0.185$$

(c) The current in relay R2 as a multiplier of plug setting is given by:

$$\frac{100 \times 4000}{400 \times 125} = 8$$

The operating time is 3.3 s

Thus, the fault clearing time associated with R2, for fault beyond B2, is given as:

$$(3.3 \times 0.163 + 0.2) = 0.728\,s$$

Likewise, current in R1 as multiple of plug setting is:

$$\frac{100 \times 4000}{500 \times 125} = 6.4$$

The operating is calculated 3.7 s.

Therefore, the fault clearing time associated with relay r1 for a fault between the breaker, is given as:

$$(3.7 \times 0.185 + 0.2) = 0.88\,\text{s}$$

(d) For an earth fault, the current in R1 as a multiple of plug setting is:

$$\frac{100 \times 1000}{500 \times 40} = 5$$

The operating time, from the characteristic equation is 4.28 s.

Actual operating time of the relay is:

$$4.28 \times 0.185 = 0.79\,\text{s}$$

For an earth fault, the current in R2 as a multiple of plug setting is:

$$\frac{100 \times 1000}{400 \times 40} = 6.25$$

The operating time, from the characteristic equation is 3.75 s.

The actual operating time of the relay is 3.75 × 0.16 = 0.6 s

The clearance time at end B is therefore 0.6 + 0.2 = 0.8 s so that the relay at end A will have operated and initiated breaker action just before the breaker at end B has cleared the fault.

? Self-assessment question 8.2

(a) Outline the principles of distance protection and explain why it is useful for performance remote back-up protection functions
(b) A distance relay with MHO characteristics is situated at end A of a 400 kV single circuit transmission line AB and has first zone setting of 80%. The positive phase sequence impedance of the line Ab is (2.0 + j36.0) and the relay is set to protect 80% of this line. If a single-phase earth fault of resistance 20 Ω occurs, three quarter of the way along the line from end A show that the relay at end A under reaches to such an extent that it will not "see" this as a fault in zone 1 (the remote end infeed from end B can be neglected)

SAQ 8.2

(a) This question is answered in full in Chapter 9.4 of the main text.
(b) The relay setting characteristic is obtained by constructing the line $AB = (1.6 + j28.8)\Omega$ and constructing a circle on AB as diameter. This is shown in Figure 45.

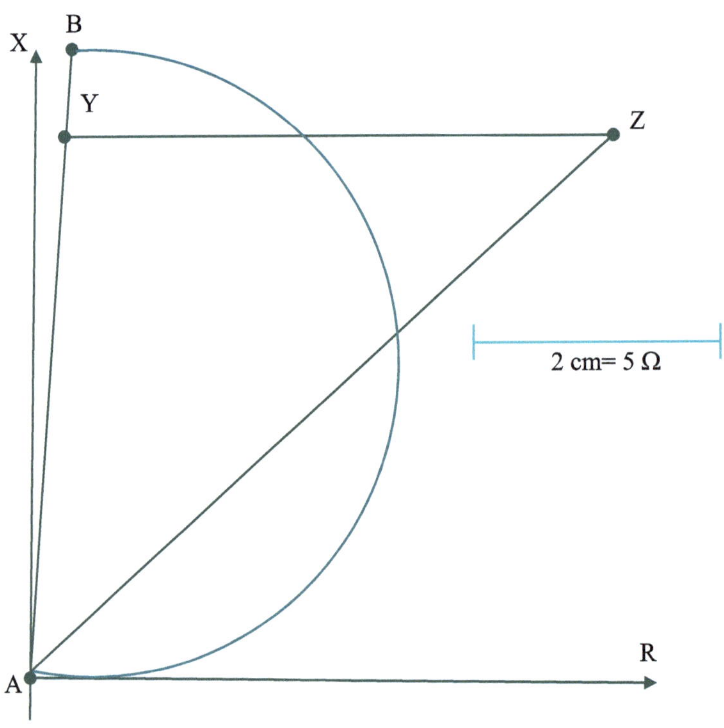

Figure 8.2

At a position three quarter the way along the line, the impedance, measured from end A, is $Z = (1.5 + j2.7)\Omega$ this is shown as the line AY on the solid locus of Figure 45. Clearly, a resistance-free fault lies well within the relay characteristic.

However, if we add a fault resistance of 20 Ω, then the impedance measured from end A is $(21.5 + j27.0)\Omega$ and this shown the line AZ in Figure 45.

The point Z is well outside the relay characteristic and the relay will fail to see this is as an internal fault.

? Self-assessment question 8.3

(a) Outline the principle of a simple differential protection scheme and explain why it is necessary to use a bias type characteristics.
(b) A low voltage distribution line is protected by a bias current differential protection as shown in Figure 9.13

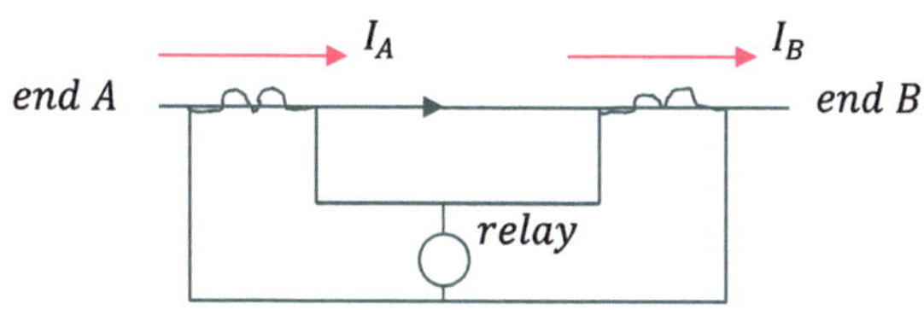

Figure 8.13

Inter fault occurs on the feeder such that

$$I_A = (40 - j30)A$$

$$I_B = (20 - j15)A$$

If the tripping criterion for the relay is such that;

$$|I_{diff}| = |I_A + I_B| \geq |I_{bias}| = k\{|I_A| + |I_B|\} \tag{8.1}$$

Deduce a value of the constant k for which the relay would "just" operate

SAQ 8.3

(a) A simple current differential scheme derives from the **Merz and Price** Calculation system. In its simplest form, the CTs at each end of the line are connected as shown in figure 46 so that, in theory, the end A current I_A is the same as the end B current I_B for no fault or for fault outside the protected zone; hence there is no operation.

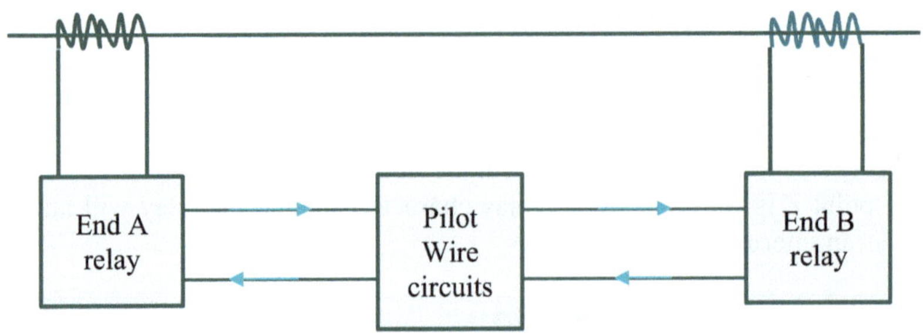

Figure 8.3

A bias type of characteristic is, however, required for two main reasons:

i. Under heavy through fault conditions, the secondary currents are not exactly equal due to small difference in the CT ratios
ii. The pilot wire circuit has shunt capacitance so that the currents transmitted to either end for comparison purposes will be slightly different under no fault or through fault conditions.

(b) $|I_{diff}| = |I_A + I_B| = |40 - j30 + 20 + j15| = |60 - j15| = 61.8\ A$

$$|I_{bias}| = k\left(|I_A| + |I_B|\right) = k(50 - 25) = k75\ A$$

Hence the value of K for which the relay would just operate is:

$$k = \frac{61.8}{75} = 0.824$$

? Self-assessment question 8.4

A 30 MVA transformer feeds several feeders through an 11 kV busbar. The transformer circuit breaker is provided with a 250:1 current transformer and each feeder is provided with 100:1 current transformer. Each set of current transformers feeds and induction type overcurrent relay and each relay has a time–current characteristic given by:

$$t = \frac{0.14}{\left(psm^{0.02} - 1\right)} s \tag{8.2}$$

where psm is the operating current in multiples of plug setting.
All relays have plug settings of 125%
A fault on one feeder results in a phase current of 1000 A
Determine:

(i) Operating time of the feeder relay if its time multiplier setting is 0.5

The time setting required on the transformer relay to provide o.5 s discrimination time margin

SAQ 9.4

(a) The plug setting defines the sensitivity of the relay. The time multiplier setting controls the of operations of the relay.
(b) The secondary current in the feeder CT is $1000/100 = 10\,A$

The plug setting multiplier is therefore $10/1.25 = 8.0$

The time of operations from the relay characteristic is:

$$t = \frac{0.14}{8^{0.02} - 1} = 3.30\,s$$

The operating time is therefore $3.30 \times 0.5 = 1.65\,s$.

The secondary current in the transformer CT is $\frac{100}{250} = 4\,A$

The Plug setting multiplier is therefore $\frac{4}{1.25} = 3.2$

The time of operations from the relay characteristic is:

$$t = \frac{0.14}{3.2^{0.02} - 1} = 5.95\,s$$

If we allow 0.5 s margin, then the transformer relay should operate after $1.65 + 0.5 = 2.15\,s$, so that setting must be:

$$\frac{2.15}{5.95} = 0.36$$

? Self-assessment question 8.5

A distance protection relay with a MHO characteristic is positioned at the sending end of a 400-kV single-circuit transmission line and set to protect 80% of the line. The line has a positive phase sequence impedance of $(2 + j36)\Omega$

A single phase to earth fault of impedance $(10 + j3.75)\Omega$ occurs at a point 60% of the way along the line from the sending end.

(a) Construct the MHO characteristic for the relay and use it to show that the relay will not operate for this fault.
(b) Determine the maximum value for a resistive fault at the point concerned for the relay just to detect the fault

SAQ 8.5

The impedance for zone 1 setting is $0.8(2+j36)=1.6+j28.8\Omega$

The impedance line is drawn as the line *OA* in Figure 47 and the MHO characteristic circle is drawn on *OA* as diameter.

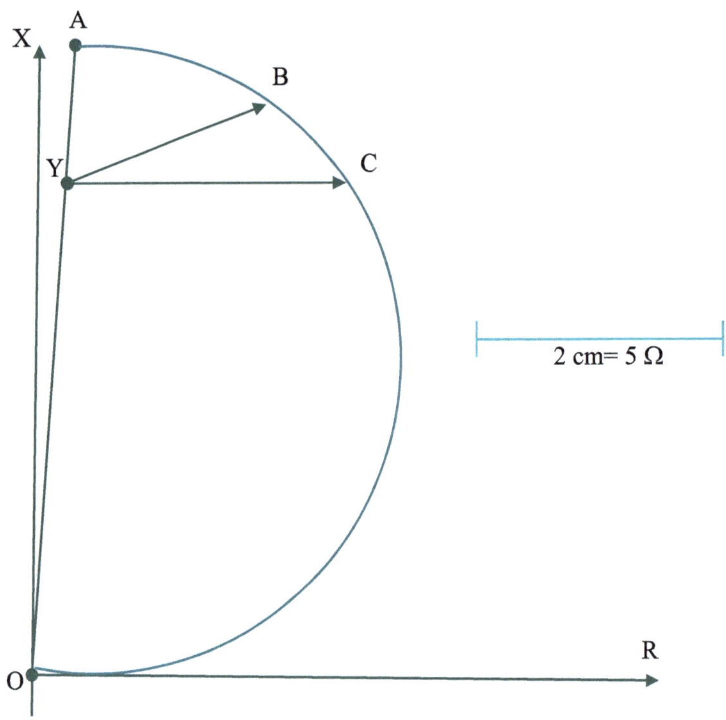

Figure 8.4

The impedance along the line to the point of fault is $1.2+j21.6\Omega$, this is shown as *OY* as the vector *YB*. The point B lies outside the circle and the relay will not operate.

The maximum resistance value of a resistive fault at the point concerned is defined by the line *YC*. Measurement gives $YC=12.0\Omega$

Chapter 9: Transmission Line Characteristic

? Self-assessment question 10.1

An overhead line has a total series impedance *Z of* $(5+j13)\Omega$ and a shunt admittance of $(j0.75\times10^{-2})s$

(a) The approximate **A, B, C, D** constant of the line based upon **Π** and **T** representation; and
(b) The **A, B, C, D** constants of the line when its capacitance is neglected.

Determine also the magnitude and phase of the sending-end voltage relative to the receiving end voltage when a load of 30 MW at 0.85 power factor lagging is supplied. The receiving end line voltage 132 kV. Compare the results obtained by using the approximate equivalent circuits of the line and comment on the comparison. The given parameters are consistent with a line length of approximately 30 km

SAQ 10.1

(a) For the π network, the **A, B, C, D** constants are given as:

$$\begin{bmatrix} A & B \\ C & D \end{bmatrix} = \begin{bmatrix} 1+\frac{ZY}{2} & Z \\ Y+\frac{ZY^2}{4} & 1+\frac{ZY}{2} \end{bmatrix}$$

In which $Z = (5 + j13) and\ Y = (j0.75 \times 10^{-2})\ S$

We thus obtain:

$$\begin{bmatrix} A & B \\ C & D \end{bmatrix} = \begin{bmatrix} (0.9513 + jo.o188) & 5 + j13 \\ (0.0002 + j0.0073) & 1+\frac{ZY}{2}0.9513 + j0.0188 \end{bmatrix}$$

For the T network, the **A, B, C, D** constant are given by:

$$\begin{bmatrix} A & B \\ C & D \end{bmatrix} = \begin{bmatrix} 1+\frac{ZY}{2} & Z+\frac{Z^2Y}{4} \\ Y & 1+\frac{ZY}{2} \end{bmatrix}$$

We thus obtain:

$$\begin{bmatrix} A & B \\ C & D \end{bmatrix} = \begin{bmatrix} 0.9513 + j0.0188 & 4.7562 + j12.73 \\ 0.0075 & 0.9523 + j0.0188 \end{bmatrix}$$

(b) If the line shunt capacitance is neglected, then the **A, B, C, D** constants are given as:

$$\begin{bmatrix} A & B \\ C & D \end{bmatrix} = \begin{bmatrix} 1 & (5 + j13) \\ 0 & 1 \end{bmatrix}$$

(c) At the receiving end we have:

$$\sqrt{3}V_{line}I_R cos\phi = 30 \times 10^6\ W$$

which gives:

$$I_R = \frac{30\times10^6}{\sqrt{3}\times0.85\times132\times10^3} = 155\,A$$

The power factor angle is 31.8°. Thus, for a lagging power factor, we obtain:

$$I_R = 131.75 - j81.68\,A$$

$$V_R = \frac{132}{\sqrt{3}} = 76.2\,kV$$

and the relationship between the sending-end voltage and current and the receiving-end voltage and current is given by:

$$\begin{bmatrix} V_s \\ I_s \end{bmatrix} = \begin{bmatrix} A & B \\ C & D \end{bmatrix} \begin{bmatrix} V_R \\ I_R \end{bmatrix}$$

This gives:

$$V_s = AV_R + BI_R$$

For the **π network** $A = (0.9513 + j0.0188)$ and $B = (5 + j13)$,
Thus

$$V_s = \left[76.2(0.9513 + j0.0188) + 0.155(0.85 - j0.53)(5 + j13)\right] kV$$

Or

$$V_s = (74.216 + j2.734)\,kV$$

Or

$$V_s = 74.27\angle 2.1°\,kV \text{ per Phase.}$$

Or

$$V_s = 128.6\angle 2.1°\,kV\,(\text{line value})$$

If the shunt admittance is neglected, then $A = 1.0$ and $B = (5 + j13)$, and

$$V_s = \left[76.2(1.0) + 0.155(0.85 - j0.53)(5 + j13)\right] kV$$

Or

$$V_s = (77.93 + j1.30)\,kV$$

Or

$$V_s = 77.94\angle 1.1°\,kV \text{ per Phase.}$$

Or

$$V_s = 135.0\angle 1.1°kV\left(\text{line value}\right)$$

The results show that, as expected, there is little difference in the sending-end voltage using either the **π or T network** equivalent circuits. However, the result obtained when shunt admittance is neglected, this is significantly different. This follows because the latter line model representation is only very approximate.

? Self-assessment question 10.2

Outline the basis of a line performance chart for an overhead line.

A balanced three-phase overhead line has constants $A = 0.95 \angle 10°$and $B = 110 \angle 65°$ if the receiving end phase voltage is held constant at 325 kV, determine graphically or otherwise:

(a) The sending-end phase voltage necessary to deliver a load of 1500 MW at 0.9 power factor leading at the receiving-end; and
(b) The maximum power that can be transmitted if the sending-end voltage is held constant at the value determined in part (a)

SAQ 8.2

(a) The sending quantities are related to the receiving end quantities by the following:

$$\begin{bmatrix} V_s \\ I_s \end{bmatrix} = \begin{bmatrix} A & B \\ C & D \end{bmatrix} \begin{bmatrix} V_R \\ I_R \end{bmatrix}$$

Which gives

$$V_s = AV_R + BI_R$$

We can now draw a phasor diagram based on the above equation. For this, it is normal to take V_R as the reference phasor. For a leading power factor, the receiving-end current can be written as:

$$I_R = I_{RP} + jI_{RQ}$$

and we have:

$$B = |B|\angle\beta \text{ and } = |B|\angle\alpha$$

The phasor representation is shown in Figure 42.

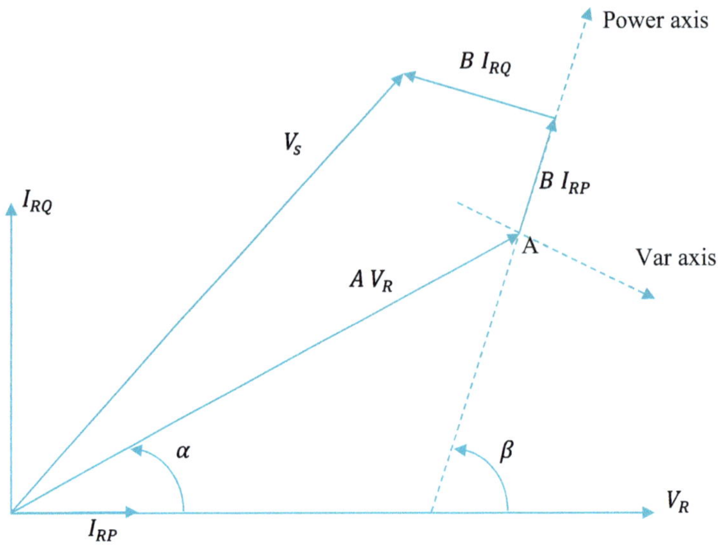

Figure 42

The real power at the receiving end is proportional to $|B|\ I_{RP}$

The reactive power at the receiving end is proportional to $|B|\ I_{RQ}$

Since 'B" is a constant, it is apparent that the axis along the $|B|\ I_{RP}$ phasor represents real power and the axis along the $|B|\ I_{RQ}$ phasor represents the reactive power.

(b) We will follow an analytical solution for this part. You may wish to try a graphical solution to see if you can make it agree with following result:

For a phase voltage of 325 kV at the receiving end,

$$I_R = \frac{1500 \times 10^6}{3 \times 0.9 \times 325 \times 10^6} = 1.71\,kA \text{ at a leading power factor angle of } 25.84^\circ$$

Hence

$$V_s = 0.95\angle 10^\circ \times 325\angle 0 + 110\angle 65^\circ \times 1.71\angle 25.84^\circ$$

$$V_s = 308.75\angle 10^\circ + 188.1\angle 90.84^\circ$$

$$V_s = (304.06 + j53.614) - (2.758 + j188.08)$$

$$V_s = (301.30 + j241.69) = 386.3\angle 38.7^\circ kV$$

Now,

$$V_s = AV_R + BI_R$$

So,

$$I_R = \frac{V_s}{B} - \frac{AV_R}{B}$$

and

$$I_R = \frac{V_s V_R}{B} - \frac{AV_R^2}{B}$$

or

$$V_R I_R \angle \varnothing = \left|\frac{V_s V_R}{B}\right| \cos\angle(\delta - \beta) - \left|\frac{AV_R^2}{B}\right| \cos\angle(\alpha - \beta)$$

From this, we obtain the equation for the receiving-end power:

$$P_R = \left|\frac{V_s V_R}{B}\right| \cos\angle(\delta - \beta) - \left|\frac{AV_R^2}{B}\right| \cos\angle(\alpha - \beta)$$

Once we have fixed the voltage, the only variable in this equation is the load angle δ, so for maximum power at the receiving end we need the condition $\delta = \beta$
Hence:

$$P_R(\max) = \left|\frac{V_s V_R}{B}\right| - \left|\frac{AV_R^2}{B}\right| \cos\angle(\alpha - \beta)$$

or

$$P_R(\max) = \frac{386.3 \times 325}{110} - \frac{386.3 \times 325^2}{110} \times 0.5736$$

$$P_R(\max) = \frac{386.3 \times 325}{110} - \frac{386.3 \times 325^2}{110} \times 0.5736$$

$$= 1141.34 - 523.247 = 618.1\,MW\ per\ Phase$$

The maximum power transfer is 3 × 618.1 = 1854 MW

? Self-assessment question 10.3

An overhead transmission line consists of two long, parallel conductors each of radius r separated by a distance D. The conductors from a g0-and return circuit.

Starting with the magnetic circuital law, show that the self-inductance of the line is given by:

$$L = \frac{\mu_0}{\pi} In\left(\frac{D - r}{r}\right) Hm^{-1}$$

where $\mu_0 = 4\pi \times 10^{-7} Hm^{-1}$ is the permittivity of the free space.

(c) A Four wire three-phase, 50 Hz overhead line ABCN is represented in Figure 8.14 with separation as shown. A communication line DE runs parallel to the three-phase conductor A,B,and C.

A third harmonic current of 100 A **r.m.s** flows in the transmission line. Calculate the voltage induced in the communication line per kilometer length.

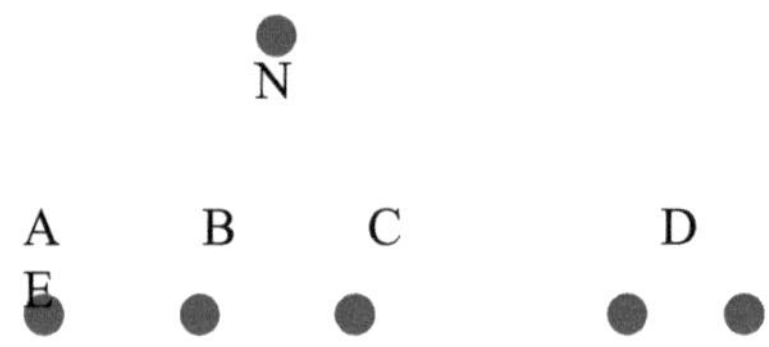

Figure 8.1

$$D_{AB} = D_{BC} = D_{BN} = 1.2$$

$$D_{CD} = 1m$$

$$D_{DE} = 0.5$$

SAQ 10.3

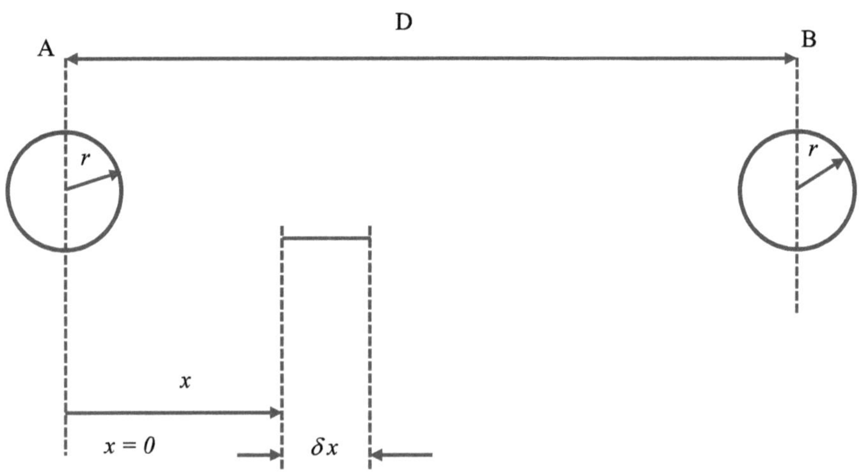

Figure 43

(a) Using the circuital law, the flux density at a radius **X** from the axis of conductor **A** is:

$$B_x = \frac{\mu_0 I}{2\pi X} + \frac{\mu_0 I}{2\pi (D - X)}$$

The flux in the incremental area of width δx for a unit

$$\delta\varnothing = \frac{\mu_0 I}{2\pi} + \left[\frac{1}{x} + \frac{1}{D - X}\right]$$

and the total flux linking the two conductors is:

$$\Phi = \frac{\mu_0 I}{2\pi} \int_{x=r}^{x=D-r} \left[\frac{1}{x} + \frac{1}{D - X}\right] dx$$

$$\Phi = \frac{\mu_0 I}{\pi} \ln\left(\frac{D - r)}{r}\right)$$

The self-inductance is:

$$L = \frac{\Phi}{j} = \frac{\mu_0 I}{\pi} \ln\left(\frac{D - r)}{r}\right)$$

(b) The flux linking loo DE from ant parallel current carrying conductor **X** is:

$$\Phi = \frac{\mu_0 I}{2\pi} \ln\left(\frac{XE}{XD}\right)$$

Applying this to the given arrangement of conductors, we obtain

$$\Phi = \frac{\mu_0 I sin 3\omega t}{2\pi} \ln\left(\frac{d_{AE} \times d_{BE} \times d_{CE} \times (d_{ND})^3}{d_{AD} \times d_{BD} \times d_{CD} \times (d_{NE})^3}\right)$$

In which $\omega = 2\pi f$ and $f = 50\ Hz$

Note that the third harmonic currents in the lines are cophasal and all current retune via the neutral wire is $-3Isin3\omega t$

$$\Phi = 52.6 \times 10^{-7} \times \sin 3\omega t \text{ Wb per meter length}$$

The e.m.f. induced per meter length is therefore:

$$e = \frac{d\Phi}{dt} = 3 \times 52.6 \times 10^{-7} = 4.96\, mV / m$$

? Self-assessment question 10.4

Conditions on a long balanced three-phase overhead transmission line are described by the following equation:

$$V_1 = AV_2 + BI_2$$

$$A = 0.95\angle 10° \text{and } B = 110\angle 65°\Omega$$

The phase voltage at end 3 is held constant at 232 kV and the following operating conditions are applied:

(a) The phase voltage at the sending-end V1, is constrained so that

$$0.95 < \left|\frac{V_1}{V_2}\right| < 1.05$$

and

(b) the phase angel, δ, between V1 and V2 constrained so that

$$-30° < \delta < 30°$$

Determine graphically or otherwise the limiting values of reactive power that can be transferred by the line to the receiving end under all allowed operating conditions

SAQ 10.4

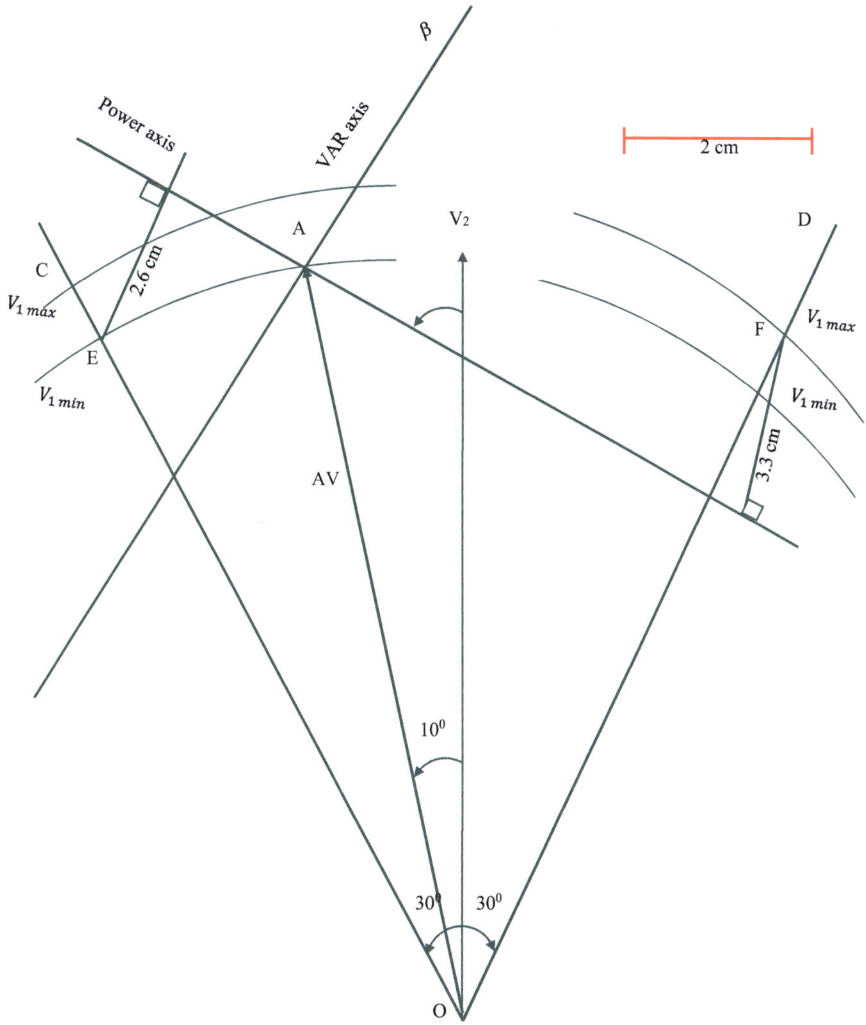

Fig. 9.5

$$A = 0.95\angle 10°, B = 110\angle 65°\Omega, V_2 = 232\,kV \text{ per phase}, so\, AV_2 = 220\angle 10°kV$$

The diagram of Figure 44 is constructed as follows to a scale of 1 cm ≡ 20 kV.

Draw the reference phasor $V_2 = 232 \angle 10\,°\,kV$.

Construct the phasor $AV_2 = 220 \angle 10\,°\,kV$. This is the line OA.

Construct the line through A at angle $\beta = 65\,°$ *to* V_2 this is the axis for power measurement with origin at A.

Construct a line through A at a right angle to the power axis. This is the var axis with origin at A.

Draw in lines OC and OD at±30° to present the limits of the load angle δ.

We also know that the limits on the sending-end voltage are:

$$V_{1(\max)} = 1.05 \times 232 = 243.6\,kV$$

$$V_{1(\min)} = 0.95 \times 232 = 220.4\,kV$$

We now construct circle of radius 243.6 kV and 220.4 kV (to scale) to represent these limiting conditions.

Intercepts at point E and F then identify the limiting points of operation. The var at each point are represented by the perpendicular distance to the power axis. The measured distances are 2.6 cm and 3.3 cm for the case of leading and lagging var respectively. These correspond to the voltage BI_{RQ}.

The corresponding VAR are:

$$\frac{3 \times \left(2.6 \times 20 \times 10^3\right) \times 232 \times 10^3}{110} = 329\,\text{Mvar leading}$$

and

$$\frac{3 \times \left(3.3 \times 20 \times 10^3\right) \times 232 \times 10^3}{110} = 418\,\text{Mvar lagging}$$

Assignment A

TMA A.1

A CT with a ratio 2000:5 is to transduce a fault current of 40 kA. If the secondary winding resistance is 0.31 W and the relay burden including the secondary leads os 2 W, what will be the secondary voltage produced by the CT?

A) 46.2 V
B) 231 V
C) 9.24 V

TMA A.2

The flux developed across a CT core many cycles after a fault is given by:

$$\Phi_A\left(1+\frac{X}{R}\right)$$

where
Φ_A = steady-state peak flux
X,R are the power frequency reactance and resistance, respectively.

The saturation flux of a CT is 100 times that associated with full load current I_{fl}. what is the maximum primary system X:R ratio for which satisfactory transformation for fault current of 10 I_{fl} will occur?

A) 9
B) 99
C) 10

T. Masood et al., *Protection of Electrical Power Transmission Systems*, Power Systems, https://doi.org/10.1007/978-3-031-94312-6

TMA A.3

A CVT has a high-voltage capacitance of 2000 pF and is connected to a 400-kV 50-Hz system, the interstage voltage being 10 kV. What is the low-voltage capacitance?

A) 78 pF
B) 94.7 pF
C) 44.2 nF

TMA A.4

In question 3 above, what is the tuning inductance?

A) 219 H
B) 229 H
C) 209 H

TMA A.5

Which of the principal IDMT relay characteristics offers?

TMA A.5.1

The greatest tolerance to current transformer saturation

A) Standard inverse
B) Very inverse
C) Extremely inverse

TMA A.5.2

The closed match to the characteristic of a fuse

A) Standard inverse
B) Very inverse
C) Extremely inverse

TMA A.6

When used to protect a network of feeder circuits, IDMT overcurrent relays provide:

A) a unit protection scheme
B) a non-unit protection scheme

TMA A.7

As a general rule, the current setting of an IDMT overcurrent relay I_R is chosen such that:

$$I_R > x\left(I_{load}\right)_{MAX}$$

$$I_R < y\left(I_{fault}\right)_{MIN}$$

where x=

A) 0.5
B) 1.0
C) 2.0

and y=

A) 0.5
B) 1.0
C) 2.0

Which are the correct values for x and y

TMA A.8

The traditional time used for the grading margin when choosing the settings of IDMT overcurrent relays

A) 40 ms
B) 400 ms
C) 4 s?

TMA A.9

Distance protection is used at which of the following voltages levels

A) 11 kV

B) 33 kV
C) 66 kV
D) 132 kV
E) 275 kV
F) 400 kV
G) 765 kV
H) All of these

TMA A.10

Using an amplitude comparator, an impedance characteristic is obtained using which of the following measurands?

A) $S_0 = \frac{IZ_R}{2}$ $S_R = V - \frac{IZ_R}{2}$
B) $S_0 = IZ_R$ $S_R = V$
C) $S_0 = IZ_R - V$ $S_R = V$
D) $S_0 = IZ_R - V$ $S_R = IZ_R + V$

TMA A.11

Using the phase angle comparator with trip limits of +/– 90, a MHO circle characteristic can be obtained using which of the following measurands?

A) $S_1 = \frac{IZ_R}{2}$ $S_R = V - \frac{IZ_R}{2}$
B) $S_1 = IZ_R$ $S_2 = V$
C) $S_1 = IZ_R - V$ $S_2 = V$
D) $S_1 = IZ_R - V$ $S_R = IZ_R + V$

TMA A.12

The induction cup comparator is:

A) an amplitude comparator
B) a cosine phase angle comparator
C) a sine phase angle comparator?

TMA A.13

A full scheme, three zone, three-phase distance relay typically uses how many comparators?

A) 1
B) 3
C) 6
D) 12
E) 18
F) 24

TMA A.14

The zone one reach of distance relay is typically set to which percentage of the protected feeder's impedance?

A) 60%
B) 80%
C) 100%
D) 120%

TMA A.15

A zero impedance short-circuit fault in the end-zone of protected transmission line will cause which zone of a three-zone distance protection to trip?

A) Zone 1
B) Zone 2
C) Zone 3
D) All of these

TMA A.16

A directional earth fault protection must be inhibited in the event of communications channel failure

A) True
B) False

TMA A.17

For an "a" phase – earth fault on a three-phase transmission system, the fault current at one end of the line is terms of its sequence components is given as:

$$I_{a1} = I_{a2} = 2.0\, p.u.$$

$$I_{a0} = 0.25\, p.u.$$

If a 1:1:3 summation transformer is used for summating the three-phase quantities into single quantity, what is the summated current proportional to?

A) 37 p.u.
B) 9 p.u.
C) 2.75 p.u.

TMA A.18

A plain feeder AB is protected by a balanced voltage differential protection scheme with a tripping criterion given by:

$$\left| \frac{I_A}{I_B} - 1 \right| \geq 0.5$$

If the fault current at the two ends of the line are:

$$I_A = (600 + j100)\, A$$

$$I_B = (400 - j50)\, A$$

Deduce whether the fault current will be detected by the relay

A) Yes
B) no

TMA A.19

If the tripping criterion is question 18 above were to be

$$\left| \frac{I_A}{I_B} - 1 \right| \geq 0.5$$

Show that the relay would fail to detect this fault

TMA A.20

An auto reclose scheme can restore a power system to its pre-fault condition when the fault is of which type.

A) a transient fault
B) a permanent fault

TMA A.21

The time delay between the auto-reclose being initiated and the instruction from the circuit breaker to close is referred to as which time.

A) the dead time?
B) the reclaim time.
C) the reclose time?

TMA A.22

An auto-reclose scheme with a dead time set to 900 ms is defined as:

A) a delayed auto-reclose scheme
B) a high-speed auto-reclose scheme

multiple choice questions (more than one answer may be correct for questions 23 to 29):

TMA A.23

The relay algorithm of a numeric relay would be stored in:

A) the program memory
B) the data memory
C) the random access memory
D) the nonvolatile memory

TMA A.24

The binary number 10110010 is equivalent to:

A) 178 in unsigned integer format
B) -78 in signed (twos complement integer format)
C) -50 in signed (twos complement) integer format
D) -9/32 in signed (twos complement) fractional format

TMA A.25

The use of isolated transformers for busbar signals entering a numeric relay is needed.

A) To prevent the high-frequency clock signal of the relay microprocessor reaching the substation environment
B) To transform signals at busbar voltage levels to acceptable, safe levels
C) To protect the 3 relay electronics against potentially damaging high-voltage spikes from the substation
D) To reduce the burden of relay

TMA A.26

A 14 bit analog-to-digital converter has an analog input signal range of 0–10 V and output unsigned values linearly proportional to the analog input (i.e., an input of) V will result in a converted value of 0) what input voltage will correspond to a convert value of 64 (000 000 0100,000)?

A) 39 mV
B) 78 mV
C) 0.39 V
D) 0.78 V

TMA A.27

An analog signal at a frequency of 6.1 kHz is sampled at a frequency of 5.4 kHz this causing aliasing. What frequency will be apparent in the sampled signal?

A) 4.7 kHz
B) 3.4 kHz
C) 2.0 kHz
D) 0.7 kHz

TMA A.28

Equation 1 represents which of the following digital signal processes?

A) a finite impulse response (FIR) filter
B) an infinite response filter (IIR) filter
C) a discrete Fourier transform'
D) a fast Fourier transforms

TMA A.29

Which of the following digital signal processes can be used to convert discrete time values into phasor form?

A) a finite impulse response (FIR) filter
B) an infinite response filter (IIR) filter
C) a discrete Fourier transform'
D) a fast Fourier transforms

Assignment B

TMA B.1a

Using rough sketches, explain the principle of a capacitor voltage transformer (CVT) used in deriving signals for protection schemes State the factors that determine the transient response of CVTs.

TMA B.1b

The voltage on one phase of a 50 Hz 4000/0.11 kV CVT is suddenly reduced to zero by the occurrence of a fault at voltage maximum. If the interstage is 12 kV and the high voltage capacitance is 2000 pF, determine:

i. the low-voltage capacitance
ii. the tuning inductance
iii. given that the magnetizing inductance of the interstage transformer is 2.4 kH, the frequency at which this might what circumstances is this likely to occur and how can such a condition be avoided.

TMA B.2

Derive the settings for the phase fault IDMT overcurrent relays at location A and B used to protect the 11 kV, three-phase system shown in TMA 3 figure a given that:

- the short-circuit capacity of the incoming feeder can vary from 50 MVA to 100 MAV

T. Masood et al., *Protection of Electrical Power Transmission Systems*, Power Systems, https://doi.org/10.1007/978-3-031-94312-6

- the maximum local current in feeder AB is 300 A and in feeder BC is 150 MVA the IDMT overcurrent relay characteristics is defined by

$$t = \frac{13.5K}{\frac{I_f}{PI_N} - 1} s$$

- where:

I_f = *the relay input current in amperes*
I_N = *the relay nominal rated current, in this case* 5 ° *A*
K = *time multiplier setting* (*TMS*)
P = *plug scale multipier* (*PSM*)

- the available TMS settings are: 0.1 to 1.0 in steps of 0.1
- the available PSM settings are: 0.5, 0.75, 1.0, 1.25, 1.5, 1.75 and 2.0
- the relay at C has settings of TMS = 0.2 PSM = 1.0
- 1 the acceptable grading margin is o.4 s.

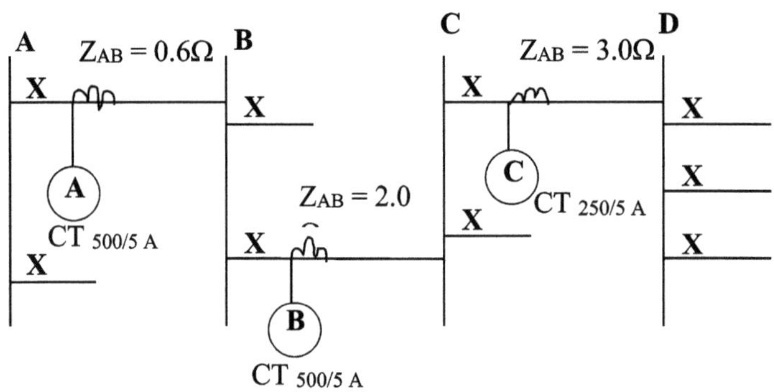

TMA B.3 Figure a

TMA B.3

For the 132 kV network shown in TMA 5 figure a suggest suitable settings for the zone 1, zone 2, and zone 3 relays and residual compensation for MHO characteristics three-zone distance relay connected at point R, given that:

$$Z_1 = 0.177 + j0.402 \frac{\Omega}{km} (primary)$$

$$CT\ ratio = 600 : 1A$$

$$Z_0 = 0.354 + j1.022\frac{\Omega}{km}(primary)$$

$$CT\,ratio = 132\,000 : 110\,A$$

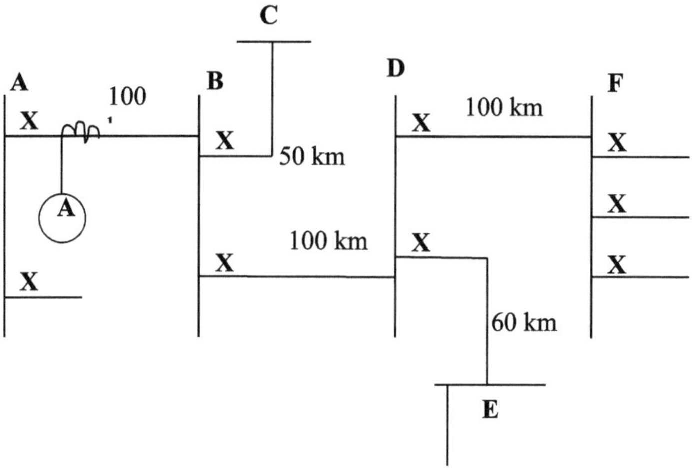

TMA B.5 Figure a

TMA B.4

A teed circuit is protected by current differential protection which has a tripping criterion.

$$I_{diff} \geq KI_{bias}$$

where

$$I_{diff} = |I_A + I_B + I_C|$$

$$I_{diff} = \left||I_A| + |I_B| + |I_C|\right|$$

If the fault current at the three ends given by:

$$I_A = (600 + j300)\,A$$

$$I_B = (400 + j150)\,A$$

$$I_C = (200 + j150)\,A$$

What is the value of K for which the relay would just detect the fault?

TMA B.5a

Outlines some of the disadvantages of using rented telephone company pilot links in differential feeder protection and briefly discuss how they are overcome by the use of fiber optical links

TMA B.5b

Give reasons why it is sometimes necessary to summate the three-phase currents into a single quantity when employing current differential protection, briefly outline the two methods of summation used in practice

TMA B.5c

A summation transformer used for deriving relaying quantities for differential feeder protection scheme is supplied with an unbalanced three-phase input of phase sequence a, b, c in the normal manner. If system conditions are such that the **nps and pps** components of current in the "c" phase are identical, find the minimum value of n to be used with a 1:1:n primary turns ratio transformer if positive output is to be maintained when

$$\frac{I_{c1}}{I_{c0}} = 4$$

where I_{C1} and I_{C0} are the **pps** and **zps** components of "c" phase current

TMA B.6

Describe the sequence of events associated with a single shot auto-reclose relays following

a) a transient fault
b) a permanent fault

TMA B.7

Table provides the sampled values of voltage and current recorded form a busbar. The sampling intervals is 2.5 ms

Table 1

Time (ms)	Voltage (kV)	Current (A)
0	546.4	1732
2.5	282.8	1932
5.0	−146.4	1000
7.5	−489.9	−518
10.0	−546.4	−1732
12.5	−282.8	−1932
15.0	146.4	−1000
17.5	489.9	518
20.0	546.4	1732

a) describe the discrete Fourier transform equation and apply it to the values of Table 1, clearly explaining each step, in order to calculate the 50-Hz voltage and current phasors.
b) Assuming that the voltage, v, and current I, samples of table 1 are related by an expression of the form:

$$v = Ri + L\frac{di}{dt}$$

Drive an expressing for R, the apparent resistance, and L, the apparent inductance. Thus, calculate R and L for the values in Table 1.

MIX
Papier aus verantwortungsvollen Quellen
Paper from responsible sources
FSC® C105338

If you have any concerns about our products,
you can contact us on
ProductSafety@springernature.com

In case Publisher is established outside the EU,
the EU authorized representative is:
Springer Nature Customer Service Center GmbH
Europaplatz 3, 69115 Heidelberg, Germany

Printed by Libri Plureos GmbH
in Hamburg, Germany